QMW Library
23 1096148 X

D1348611

MULTILINGUAL ILLUSTRATED GUIDE TO THE WORLD'S COMMERCIAL COLDWATER FISH

Also available

MULTILINGUAL ILLUSTRATED GUIDE TO THE WORLD'S COMMERCIAL WARMWATER FISH

MULTILINGUAL ILLUSTRATED GUIDE TO THE WORLD'S COMMERCIAL COLDWATER FISH

Claus Frimodt

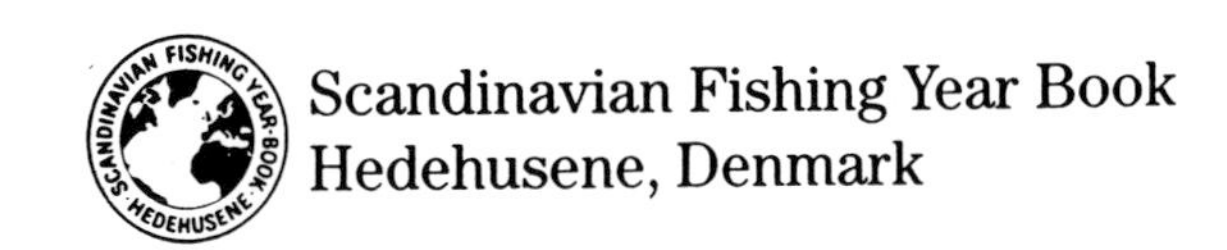
Scandinavian Fishing Year Book
Hedehusene, Denmark

Fishing News Books

A division of Blackwell Science Ltd
Editorial Offices:
Osney Mead, Oxford OX2 0EL
25 John Street, London WC1N 2BL
23 Ainslie Place, Edinburgh EH3 6AJ
238 Main Street, Cambridge,
Massachusetts 02142, USA
54 University Street, Carlton,
Victoria 3053, Australia

Other Editorial Offices:
Arnette Blackwell SA
1, rue de Lille, 75007 Paris
France

Blackwell Wissenschafts-Verlag GmbH
Kurfürstendamm 57
10707 Berlin, Germany

Feldgasse 13, A-1238 Wien
Austria

First published 1995

Printed and bound by
Vincenzo Bona sri, Turin, Italy

DISTRIBUTORS
Marston Book Services Ltd
PO Box 87
Oxford OX2 0DT
(*Orders:* Tel: 01865 206206
Fax: 01865 721205
Telex: 83355 MEDBOK G)

USA
Blackwell Science, Inc.
238 Main Street
Cambridge, MA 02142
(*Orders*: Tel: 800 215-1000
617 876-7000
Fax: 617 492-5263)

Canada
Oxford University Press
70 Wynford Drive
Don Mills
Ontario M3C 1J9
(*Orders*: Tel: 416 441 2941)

Australia
Blackwell Science Pty Ltd
54 University Street
Carlton, Victoria 3053
(*Orders*: Tel: 03 347-0300
Fax: 03 349-3016)

A Catalogue record for this book is available from the British Library

ISBN 0-85238-213-8

Library of Corgress
Cataloging in Publication Data
Frimodt, Claus,
Multilingual illustrated guide to the world's commercial coldwater fish / Claus Frimodt.
p. cm.
Includes bibliographical references and index.
ISBN 0-85238-213-8
1. Fisheries—Dictionaries—Polygolt. 2. Fishes—Dictionaries—Polyglot. 3. Dictionaries. Polyglot.
I. Title.
SH201.F73 1995
641.3'92'03—dc20 95-35316
CIP

CONTENTS

Species descriptions:

Scientific Names

How to use this book

This is one of two multilingual volumes (*An Illustrated Guide to the World's Commercial Coldwater Fish* and *An Illustrated Guide to the World's Commercial Warmwater Fish*) which together cover over 300 species of finfish around the world. The species were selected in consultation with experts from many countries and are intended to be as representative as possible of the thousands of species that are used and traded in and between countries. Highlights of every species description are given in Danish, German, Spanish, French and Italian.

Please read the following notes; they will help you derive the greatest possible value from these volumes.

Names: Common Names in up to 18 languages have been chosen on the basis that they are the most widely used names commercially. In addition, **Local Names** are given in languages appropriate to the range and usage of the fish. All languages have been transliterated into the Roman alphabet.

Symbols: Symbols are used to indicate fishing methods, forms of preparation and usage. Please see the lists of symbols and language codes immediately following this section for definitions in six languages.

Maps: Maps give approximate overall current range of each species, including, for farmed fish, the areas where it is grown even if these are far from the original native range.

Indexes: These are the most important part of the book as they will enable you to find information quickly and accurately. There is a scientific name index which lists every WORD in each scientific name; and there is a single integrated index covering all the Common Names used in the entire volume.

Nutrition information: The nutrition information is not suitable for commercial use on labels. It is presented only as a guide to comparative levels of nutrients and is not intended to be scientifically accurate source material. For this reason, we have not indicated individual sources of the numbers. Treat all the nutrient information as estimates only. So far as possible, the numbers are based on edible meat or fillets portions, not on whole fish. In some cases, however, the only numbers available are for whole fish. Fat content of many fishes changes greatly according to breeding, seasonal and feeding cycles. Bykov, for example, found Pacific herring varying between 3 and 33 percent fat. This is one reason why such wide variations and discrepancies are found in published work on fish nutrient levels. Omega-3 numbers are frequently unavailable. Where they are missing, we have assumed that omega-3 fatty acids are 20 parts per thousand of the fat.

The authors are indebted to the Royal Society of Chemistry in the United Kingdom for providing access prior to publication of their most valuable *Fish and Fish Products* and for giving permission to use some of their data. Please note that their information has been merged with that from many other sources: the authors of these volumes bear sole responsibility for any errors.

Brugervejledning

De to flersprogede bøger (*En Illustreret, Flersproget Guide til Verdens Kommercielle Koldtvandsfisk* og *En Illustreret, Flersproget Guide til Verdens Kommercielle Varmtvandsfisk*) dækker tilsammen mere end 300 af verdens vigtigste fisk. Arterne er fundet i samarbejde med eksperter fra mange lande og repræsenterer et udvalg baseret på tusindvis af arter af stor kommerciel betydning såvel nationalt som internationalt. For hver enkel art finder De et kort resumé på dansk, tysk, spansk, fransk og italiensk.

Vi anbefaler, at De læser følgende noter for at opnå størst muligt udbytte af bøgerne.

Common Names: De mest almindelige kommercielle navne på indtil 18 sprog. **Local Names:** Lokale navne relateret til fiskens udbredelse og anvendelse. Samtlige sprog er omskrevet til det latinske alfabet.

Symboler: Symbolerne viser fangstmetoder, tilberedning og anvendelsesmuligheder. Umiddelbart efter denne vejledning følger symbolforklaring og sprogkoder på 6 sprog.

Landkort: Viser den generelle udbredelse og områderne, hvor arterne evt. opdrættes.

Index: Dette er bogens vigtigste afsnit - nøglen til hurtige og præcise opslag. I det latinske index kan De søge på hvert enkelt ord i navnet, og i det generelle index på de mest almindelige kommercielle navne på indtil 18 sprog.

Næringsværdi: Oplysningerne om næringsværdier er kun vejledende og bør ikke betragtes som videnskabelige fakta. I nogle tilfælde er tallene et anslået gennemsnit baseret på flere kilder. Værdien angives undtagelsesvis for den hele fisk og ikke, som for flertallets vedkommende, for det anvendelige kød eller filetterne. Fedtprocenten varierer alt efter yngleperiode, sæson og føde. Faglitteraturen indeholder divergerende oplysninger - Bykov, for eksempel, har i sild konstateret en fedtprocent, der svinger fra 3 til 33. Omega-3 tal er ofte utilgængelige. Hvor tallet mangler, estimerer vi, at omega-3 tallet er 20 promille af fedtindholdet.

Forfatterne takker Royal Society of Chemistry, England, for tilladelse til at benytte oplysninger fra det endnu ikke offentliggjorte værk *Fish and Fish Products*. Disse oplysninger er vurderet uafhængigt af andet kildemateriale. Forfatterne til begge bind bærer ansvaret for enhver fejl.

Hinweise zur Benutzung

Die beiden mehrsprachigen Bücher (*Ein illustriertes, mehrsprachiges Nachschlagewerk über die kommerziellen Kaltwasserfische der Welt* und *Ein illustriertes, mehrsprachiges Nachschlagewerk über die kommerziellen Warmwasserfische der Welt*) erfassen insgesamt mehr als 300 der wichtigsten Fische der Welt. Die Arten wurden in Zusammenarbeit mit Experten aus vielen Ländern ausgewählt und repräsentieren einen Ausschnitt der tausenden von Arten mit grosser wirtschaftlicher Bedeutung im nationalen und internationalen Massstab. Für jede einzelne Art finden Sie eine kurze Zusammenfassung in Dänisch, Deutsch, Spanisch, Französisch und Italienisch.

Wir empfehlen Ihnen folgende Hinweise zu berücksichtigen, damit die Bücher sich Ihnen voll erschliessen.

Common Names: Die gebräuchlichsten kommerziellen Namen in bis zu 18 Sprachen. **Local Names:** Ortsbedingte Namen abhängig von der Verbreitung und Anwendung des Fisches. Alle Sprachen gehen vom lateinischen Alphabet aus.

Symbole: Die Symbole geben die Fangmethoden, Zubereitung und Verwendungsmöglichkeiten an. Im Anschluss an die Hinweise zur Benutzung folgen die Symbolerläuterungen sowie die Sprachabkürzungen in sechs Sprachen.

Landkarte: Zeigt die allgemeine Verbreitung sowie die Gebiete, wo die Arten eventuell gezüchtet werden.

Register: Das ist der wichtigste Abschnitt des Buches - der Schlüssel zum schnellen und sicheren Nachschlagen. Im lateinischen Register können Sie die einzelnen Wörter des Namens nachschlagen und im allgemeinen Register die gebräuchlichsten kommerziellen Namen in bis zu 18 Sprachen.

Nährwert: Die Angaben zum Nährwert sind nur Richtwerte und dürfen nicht als wissenschaftlich begründet aufgefasst werden. Bei einigen handelt es sich um geschätzte Durchschnittswerte, die aus mehreren Quellen stammen. In Ausnahmefällen erfolgt die Angabe für den gesamten Fisch und nicht wie bei der Mehrzahl für das verwendbare Fleisch oder die Filets. Der Fettgehalt richtet sich nach Laichzeit, Saison und Nahrung. Die Fachliteratur macht unterschiedliche Angaben - Bykov, um nur ein Beispiel zu nennen - hat bei Hering einen Fettgehalt festgestellt, der zwischen 3 und 33 Prozent schwankt. Omega-3-Zahlen sind häufig nicht zugänglich. Wo die Angabe fehlt, schätzen wir die Omega-3-Zahl auf 20 Tausendstel des Fettgehalts.

Die Herausgeber bedanken sich bei der Royal Society of Chemistry, Grossbritannien, für die Genehmigung, Angaben aus deren zu dem Zeitpunkt noch nicht veröffentlichten Werk *Fish and Fish Products* verwenden zu dürfen. Wir machen darauf aufmerksam, dass diese Angaben zusammen mit anderem Quellenmaterial verarbeitet worden sind; die Herausgeber der vorliegenden Nachschlagewerke tragen die alleinige Verantwortung für eventuelle Fehler.

Instrucciones para el uso

Los dos volúmenes plurilingües (*Una guía ilustrada plurilingüe sobre los pescados comerciales de agua fría de todo el mundo* y *Una guía ilustrada plurilingüe sobre los pescados comerciales de agua caliente de todo el mundo*) abarcan en total más de 300 de las especies más importantes del mundo. Las especies han sido elegidas en colaboración con expertos de muchos países y representan una selección de los miles de especies de gran importancia comercial que existen a nivel nacional y a nivel internacional. Para cada especie encontrará un breve resumen en danés, alemán, español, francés e italiano.

Le recomendamos leer las siguientes instrucciones para sacar el máximo provecho posible de estos volúmenes.

Common Names: Los nombres comerciales más corrientes en hasta 18 lenguas. **Local Names:** Nombres locales relacionados con la extensión y el uso del pescado. Todas las lenguas han sido transcritas en el alfabeto latino.

Símbolos: Los símbolos indican los métodos de pesca y las posibilidades de preparación y de uso. Después de estas instrucciones sigue una explicación de los símbolos y de los códigos lingüísticos en 6 lenguas.

Mapas: Muestran la extensión general y las eventuales zonas de cría de las varias especies.

Indices: Esta es la parte más importante de la obra - la clave para una rápida y precisa consulta. En el índice latino podrá buscar cada una de las palabras del nombre, y en el índice general están indicados los nombres comerciales más corrientes en hasta 18 lenguas.

Valor nutritivo: Las informaciones sobre el valor nutritivo sólo son indicativas, y no deberán ser consideradas como datos científicos. En algunos casos las cifras representan un promedio estimado, basado en diversas fuentes. Para algunas especies el valor indicado se refiere al pescado entero, pero en la mayoría de los casos el valor está basado en la carne comestible o sus filetes. El contenido de grasa varía según la época de reproducción, la estación y la alimentación. La literatura científica contiene informaciones divergentes. Bykov, por ejemplo, ha comprobado un porcentaje de grasa en el arenque que varía del 3 al 33 por ciento. Las cifras omega-3 a menudo son inaccesibles. Donde faltan, se puede suponer que la cifra omega-3 sea del 20 por mil del contenido de grasa.

Los autores agradecen a la Royal Society of Chemistry, Reino Unido, su permiso de usar las informaciones contenidas en la obra *Fish and Fish Products* recién publicada, pero todavía no publicada en el momento de la elaboración de los volúmenes. Dichas informaciones han sido objeto de una evaluación independiente junto con otras fuentes; los autores de los volúmenes asumen la responsabilidad absoluta de posibles errores.

Avis aux usagers

Les deux ouvrages multilingues (*Guide illustré multilingue des poissons commerciaux des eaux froides du monde* et *Guide illustré multilingue des poissons commerciaux des eaux chaudes du monde*) couvrent ensemble plus de 300 des plus importants poissons du monde. Retenues avec la collaboration d'experts venant d'un grand nombre de pays, les espèces représentent une sélection parmi des milliers d'espèces d'un grand intérêt commercial sur le plan aussi bien national qu'international. Pour chaque espèce, vous trouverez un bref résumé en danois, allemand, espagnol, français et italien.

Nous vous conseillons de lire les notes explicatives qui suivent afin de tirer plein profit des ouvrages.

Common Names: Les noms commerciaux les plus usuels sont indiqués dans jusqu'à 18 langues, avec une translittération en caractères latins le cas échéant. **Local Names:** Des termes se rapportant à l'aire d'habitation du poisson et à son emploi.

Symboles: Les symboles, reproduits à la suite de cet avis, représentent les méthodes de pêche et les modes d'utilisation et de préparation; ils sont suivis des codes-langue (6).

Cartes géographiques: Montrent l'habitat général et, éventuellement, les zones d'élevage des espèces.

Index: La section la plus importante de l'ouvrage - la clef des recherches rapides et précises. Dans l'index en latin, vous pourrez chercher au moyen de chacun des éléments du nom, et dans l'index général au moyen des noms commerciaux les plus usuels dans jusqu'à 18 langues.

Valeur nutritive: Les informations sur les valeurs nutritives ne sont qu'indicatives et ne doivent pas être considérées comme des données scientifiques. Dans certains cas, les chiffres constituent une moyenne supputée basée sur plusieurs sources. Dans ces cas exceptionnels, la valeur correspond au poisson tout entier, et non pas, comme c'est la règle, à la fraction de chair utilisable ou aux filets. Le taux de lipides varie en fonction de l'époque du frai, de la saison et de la nourriture. Les manuels apportent des chiffres divergeants; à titre d'exemple, Bykov a constaté un taux de lipides dans le hareng qui varie de 3 à 33. Les chiffres d'oméga 3 sont souvent inaccessibles. Dans les cas où ce renseignement manque, nous jugeons qu'il représente 20 pour mille du taux de matière grasse.

Les auteurs sont redevables à la Royal Society of Chemistry du Royaume-Uni d'avoir bien voulu leur donner accès, avant même sa publication, au précieux ouvrage *Fish and Fish Products* et de leur avoir autorisé à utiliser certaines données de cet ouvrage qui vient de sortir. Il convient de préciser, que les informations tirées de cet ouvrage ont été confondues avec celles d'un grand nombre d'autres sources, et que les auteurs de ces volumes sont les seuls responsables de toute erreur éventuelle.

Istruzioni per l'uso

I due volumi multilingue (*Guida multilingue illustrata ai pesci commerciali d'acqua fredda di tutto il mondo* e *Guida multilingue illustrata ai pesci commerciali d'acqua calda di tutto il mondo*) coprono complessivamente più di 300 delle specie più importanti del mondo. La selezione delle specie è stata fatta in collaborazione con esperti di molti paesi e rappresenta quindi una scelta basata su migliaia di specie di grande importanza commerciale sia a livello nazionale che a livello internazionale. Per ogni specie troverete un breve riassunto in danese, tedesco, spagnolo, francese e italiano.

Vi raccomandiamo di leggere le seguenti istruzioni per poter trarre il maggior vantaggio possibile da questi volumi.

Common Names: I nomi commerciali più comuni in ben 18 lingue. **Local Names:** I nomi locali usati nelle zone di diffusione e di impiego del pesce. Tutte le lingue sono trascritte nell'alfabeto latino.

Simboli: I simboli mostrano i metodi di pesca, le possibilità di preparazione e di impiego. Queste istruzioni sono seguite da un elenco dei simboli e dalle sigle linguistiche in 6 lingue.

Carte geografiche: Mostrano la diffusione generale e le eventuali zone di allevamento delle varie specie.

Indici: Questa è la parte più importante del volume - la chiave per una rapida e precisa consultazione. Nell'indice scientifico si potrà ricercare ogni singola componente del nome e nell'indice generale sono riportati i nomi commerciali più comuni in ben 18 lingue.

Valore nutritivo: Le informazioni sul valore nutritivo sono solo indicative e vanno intese come una guida per la comparazione del valore nutritivo e non come dati scientifici. In alcuni casi le cifre rappresentano una media stimata, basata su diverse fonti. Per alcune specie il valore riportato si riferisce al pesce intero, ma per la maggior parte il valore è basato sulle carni commestibili o sui filetti. Il contenuto di grasso varia a seconda dell'epoca della riproduzione, della stagione e del cibo. La letteratura scientifica contiene informazioni divergenti - Bykov, per esempio, ha constatato una percentuale di grasso nell'aringa che varia dal 3 e al 33 per cento. Le cifre omega-3 sono spesso inaccessibili. Dove mancano, si potrà presumere che la cifra omega-3 sia del 20 per mille del contenuto di grasso.

Gli autori ringraziano la Royal Society of Chemistry, Regno Unito, per aver messo a disposizione i loro dati contenuti nell'opera non ancora pubblicata *Fish and Fish Products* uscita di recente, ma non ancora pubblicata al momento di elaborazione dei volumi. Dette informazioni sono state oggetto, insieme alle altre fonti, dello studio su cui si basano i due volumi e gli autori ne assumono l'assoluta responsabilità.

Abbreviations used

	GB — ENGLISH	DK — DANSK	D — DEUTSCH
Local names			
AN:	Angola	Angola	Angola
AR:	Argentina	Argentina	Argentinien
AU:	Australia	Australien	Australien
BR:	Brazil	Brasilien	Brasilien
BU:	Burma	Burma	Burma
CA:	Canada	Canada	Kanada
CH:	China	Kina	China
CL:	Chile	Chile	Chile
CM:	Cambodia	Cambodja	Kamputschea
EC:	Ecuador	Ecuador	Ecuador
EG:	Egypt	Egypten	Ägypten
GU:	Guam	Guam	Guam
HK:	Hong Kong	Hong Kong	Hongkong
HW:	Hawaii	Hawaii	Hawaii
IA:	India	Indien	Indien
IL:	Israel	Israel	Israel
IN:	Indonesia	Indonesien	Indonesien
IV:	Ivory Coast	Elfenbenskysten	Elfenbeinküste
KE:	Kenya	Kenya	Kenia
KO:	Korea	Korea	Korea
KU:	Kuwait	Kuwait	Kuwait
MA:	Malaysia	Malaysia	Malaysia
ME:	Mexico	Mexico	Mexiko
MO:	Morocco	Marokko	Marokko
NC:	New Caledonia	Ny Caledonien	Neukaledonien
NZ:	New Zealand	New Zealand	Neuseeland
PA:	Pakistan	Pakistan	Pakistan
PE:	Peru	Peru	Peru
PH:	Philippines	Filippinerne	Philippinen
PO:	Poland	Polen	Polen
SA:	Saudi Arabia	Saudi Arabien	Saudiarabien
SM:	Samoa	Samoa	Samoa
SO:	South Africa	Sydafrika	Südafrika
TA:	Tanzania	Tanzania	Tansania
TH:	Thailand	Thailand	Thailand
TU:	Tunisia	Tunesien	Tunesien
VE:	Venezuela	Venezuela	Venezuela
VI:	Viet Nam	Vietnam	Vietnam
Common names			
D:	Germay	Tyskland	Deutschland
DK:	Denmark	Danmark	Dänemark
E:	Spain	Spanien	Spanien
F:	France	Frankrig	Frankreich
GR:	Greece	Grækenland	Griechenland
I:	Italy	Italien	Italien
IS:	Iceland	Island	Island
J:	Japan	Japan	Japan
N:	Norway	Norge	Norwegen
NL:	Holland	Holland	Niederlande
P:	Portugal	Portugal	Portugal
RU:	Russia	Rusland	Russland
S:	Sweden	Sverige	Schweden
SF:	Finland	Finland	Finnland
TR:	Turkey	Tyrkiet	Türkei
US:	USA	USA	USA

Abbreviations used

	E — ESPAÑOL	F — FRANÇAIS	I — ITALIANO
Local names			
AN:	Angola	Angola	Angola
AR:	Argentina	Argentine	Argentina
AU:	Australia	Australie	Australia
BR:	Brasil	Brésil	Brasile
BU:	Birmania	Birmanie	Birmania
CA:	Canadá	Canada	Canada
CH:	China	Chine	Cina
CL:	Chile	Chili	Cile
CM:	Camboya	Cambodge	Cambogia
EC:	Ecuador	Equateur	Ecuador
EG:	Egipto	Egypte	Egitto
GU:	Guam	Guam	Guam
HK:	Hong Kong	Hong Kong	Hong Kong
HW:	Hawaii	Hawai	Hawaii
IA:	India	Indes	India
IL:	Israel	Israël	Israele
IN:	Indonesia	Indonésie	Indonesia
IV:	Costa de Marfil	Côte d'Ivoire	Costa d'Avorio
KE:	Kenia	Kenya	Kenya
KO:	Korea	Corée	Corea
KU:	Kuwait	Koweit	Kuwait
MA:	Malaysia	Malaisie	Malaysia
ME:	México	Mexique	Messico
MO:	Marruecos	Maroc	Marocco
NC:	Nueva Caledonia	Nouvelle-Calédonie	Nuova Caledonia
NZ:	Nueva Zelanda	Nouvelle-Zélande	Nuova Zelanda
PA:	Paquistán	Pakistan	Pakistan
PE:	Perú	Pérou	Perù
PH:	Filipinas	Philippines	Filippine
PO:	Polonia	Pologne	Polonia
SA:	Arabia Saudita	Arabie Saoudite	Arabia Saudita
SM:	Samoa	Samoa	Samoa
SO:	Sudáfrica	Afrique du Sud	Sudafrica
TA:	Tanzania	Tanzanie	Tanziania
TH:	Tailandia	Thailande	Thailandia
TU:	Tunicia	Tunésie	Tunesia
VE:	Venezuela	Venezuela	Venezuela
VI:	Vietnam	Viet-Nam	Vietnam
Common Names			
D:	Alemania	Allemagne	Germania
DK:	Dinamarca	Danemark	Danimarca
E:	España	Espagne	Spagna
F:	Francia	France	Francia
GR:	Grecia	Grèce	Grecia
I:	Italia	Italie	Italia
IS:	Islandia	Islande	Islanda
J:	Japón	Japon	Giappone
N:	Noruega	Norvège	Norvegia
NL:	Holanda	Pays-Bas	Olanda
P:	Portugal	Portugal	Portogallo
RU:	Rusia	Russie	Russia
S:	Suecia	Suède	Svezia
SF:	Finlandia	Finlande	Finlandia
TR:	Turquía	Turquie	Turchia
US:	EE.UU	Etats-Unis	USA

Explanation of Symbols

GB — ENGLISH	DK — DANSK	D — DEUTSCH
FISHING METHODS	**FANGSTMETODER**	**FANGMETHODEN**
Aquaculture	Akvakultur	Aquakultur
Trap	Ruse	Reuse
Trawl	Trawl	Schleppnetz
Gill-net	Garn	Setznetz
Seine	Not	Wade
Hook	Krog	Haken
Harpoon	Harpun	Harpune
PREPARATION	**ANVENDELSE**	**VERWENDUNG**
Fresh	Fersk	Frisch
Dried/salted	Tørret/saltet	Getrocknet/gesalzen
Smoked	Røget	Geräuchert
Canned	Konserveret	Konserven
Frozen	Dybfrosset	Gefroren
USED FOR	**TILBEREDNING**	**ZUBEREITUNG**
Steam	Dampning	Dämpfen
Sauté/pan fry	Stegning	Braten
Broil/grill	Grill	Grillen
Boil	Kogning	Kochen
Fry	Friture	Fritieren
Microwave	Mikroovn	Mikrowelle
Bake	Ovnstegning	Backen

E — ESPAÑOL	F — FRANÇAIS	I — ITALIANO
METODOS DE PESCA	**METHODES DE PECHE**	**METODI DI PESCA**
Acuicultura	Aquiculture	Acquicoltura
Nasa	Nasse	Bertuello
Red de arrastre	Chalut	Rete a strascico
Red de enmalle	Filet	Rete a imbrocco
Red de jábega	Seine	Sciabica
Anzuelo	Hameçon	Amo
Arpón	Harpon	Arpone
USOS	**MODES D'UTILISATION**	**IMPIEGO**
Fresco	Frais	Fresco
Secado/salado	Séché/salé	Seccato/salato
Ahumado	Fumé	Affumicato
Enlatado	En conserve	Inscatolato
Congelado	Congelé	Congelato
PREPARACION	**PREPARATION**	**PREPARAZIONE**
Hervir a vapor	Cuisson à l'étuvée	Cottura a vapore
Freír en sartén	Cuisson à poêle	A la sauté
A la parrilla	Grillage	Alla griglia
Hervir	Cuisson en marmite	Lessatura
Freír en freidora	Friture	Frittura
Microondas	Cuisson au four à micro-ondes	Al forno a microonde
Horno	Cuisson au four	Al forno

ALASKA POLLOCK

Scientific name:
Theragra chalcogramma

Family: Gadidae — cods
Typical size: 80 cm, 1.4 kg

Alaska pollock is found in the temperate and colder waters of the North Pacific where it is the basis of one of the largest commercial fisheries in the world.

It is processed into surimi of excellent and consistent quality; surimi made from this species is generally considered the standard against which surimi paste made from other fish is judged.

DESCRIPTION
Alaska pollock is a small relative of the cod. Its large eye is a distinguishing feature and explains the name walleye pollock, which authorities in North America prefer but which fishing and food industries generally disdain in favour of the more commonly accepted name, Alaska pollock (which is also used internationally by FAO).

The species grows rapidly and may live up to 15 years. Dense spawning aggregations may be targeted for their roe. The most productive pollock fisheries are on the outer shelf and slope of the Bering Sea between the eastern Aleutians and Pribiloff Islands, where they are trawled from depths of 50 to 200 m.

COMMON NAMES
D: Pazifischer Pollack
DK: Alaskasej
E: Abadejo de Alaska
F: Lieu de l'Alaska
GR: Bakaliáros tis Aláskas
I: Merluzzo dell'Alaska
IS: Alaskaufsi
J: Suketôdara, suketoudara
NL: Alaska koolvis
P: Escamudo do Alasca
RU: Mintai
S: Alaskasej
SF: Mintai
US: Walleye pollock

HIGHLIGHTS
Alaska pollock, also called walleye pollock, is the most important groundfish species in world fisheries, producing catches of around 6 million tons a year. Once considered too soft for commercial use, the species is now processed into fillets, blocks and surimi.

Pollock roe is an important commodity. It is sold mainly to Japan, where it is salted for consumption. Spawning pollock are often targeted solely for the roe, the carcasses being discarded or used for fishmeal.

Nutrition data:
(100 g edible weight)

Water	81.6 g
Calories	76 kcal
Protein	17.2 g
Total lipid (fat)	0.8 g
Omega-3	0.4 mg

ALASKA POLLOCK *Theragra chalcogramma*

FISHING METHODS

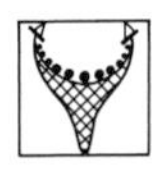 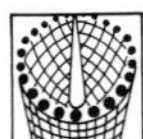 

MOST IMPORTANT FISHING NATIONS

United States, Russia, Japan

PREPARATION

Surimi, roe

USED FOR

EATING QUALITIES

Alaska pollock has soft, white meat, suitable for processed, especially coated, products. Much of the catch is first processed on board factory trawlers, or landed soon after capture at nearby coastal plants. Fast handling is essential if the fish is not to become too soft. The greater part of the catch is filleted, skinned and made into surimi or laminated blocks. Both of these products are standard commodities which are sold for further processing.

The meat is bland and the flake quite small, but pollock is fine for fish-and-chips and similar dishes. Fillets, invariably skinless and boneless, are also sold. They are best baked with sauces. The roe is highly regarded in Japan: roe fisheries are vital to the economic health of many pollock harvesters.

IMPORTANCE

The world's largest and most valuable groundfish fishery is supported by this species. As much as 6 million tons are taken each year, now mostly by the USA and Russia: the greatest part of the resource lives within the 200-mile fishery zones of these two nations.

In the past, Japanese, Korean, Polish and other vessels caught huge quantities of pollock, but these countries have been phased out of American waters in favour of domestic fishermen. In Russia, foreign fishermen are again receiving licenses to harvest pollock.

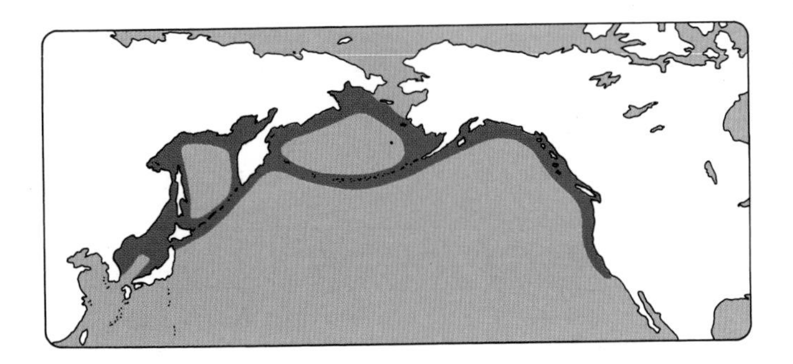

DK: Med årlige fangster på ca. 6 mill. tons er dette den vigtigste bundfisk. Hvor det bløde kød tidligere blev anset for uanvendeligt, forarbejdes det nu til filetter, blokke og 'surimi'

D: Mit jährlichen Fangmengen von rund 6 Mio. Tonnen ist dies der wichtigste Bodenfisch. Das früher als wirtschaftlich nicht verwendbar geltende weiche Fleisch wird heute zu Filets, Blöcken und zu 'Surimi' verarbeitet

E: Con capturas anuales de 6 millones de toneladas es el pescado de fondo más importante. Antes la carne era considerada demasiado blanda para el uso comercial, pero hoy es procesada en filetes, bloques y 'surimi'

F: Avec des pêches annuelles de l'ordre de 6 millions de tonnes, il est le poisson des fonds le plus important. La chair molle, autrefois considérée inutilisable, est maintenant transformée en filets, en blocs et en 'surimi'

I: Con le catture annuali ammontanti a circa 6 milioni di tonnellate, è il pesce di fondo più importante. Le sue carni molli, prima considerate inutilizzabili, vengono ora usate per filetti, blocchi e 'surimi'

ALLIS SHAD

Scientific name:
Alosa alosa

Synonyms: Alewife, rock herring
Family: Clupeidae — herrings, sardines
Typical size: 60 cm, 2.7 kg

France and Portugal produce most of the small catches of this anadromous herring, which swims into rivers in May to spawn — accounting for the name mayfish in numerous European languages.

FISHING METHODS

MOST IMPORTANT FISHING NATIONS
France, Portugal

PREPARATION

USED FOR

 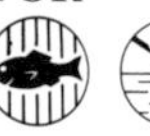 

EATING QUALITIES
Shad are extremely bony fish. It requires considerable expertise to fillet them cleanly. It is almost impossible to eat shad without first filleting it. The meat is delicate, with a medium flake and excellent flavour, but many people do not consider that the culinary rewards merit the effort involved in removing the bones.

COMMON NAMES
D : Maifisch, Alse
DK: Majsild, stamsild
E : Sábalo común, alosa, trisa
F : Alose vraie
GR: Fríssa, sardelomána
I : Alaccia, alosa
IS: Maísíld
N : Maisild
NL: Elft, meivis
P : Sável
RU: Aloza
S : Stamsill, majfisk
SF: Pilkkusilli
TR: Tirsi
US: Allis shad

Nutrition data:
(100 g edible weight)

Water	68.8 g	Total lipid	
Calories	183 kcal	(fat)	12.8 g
Protein	16.9 g	Omega-3	2.6 mg

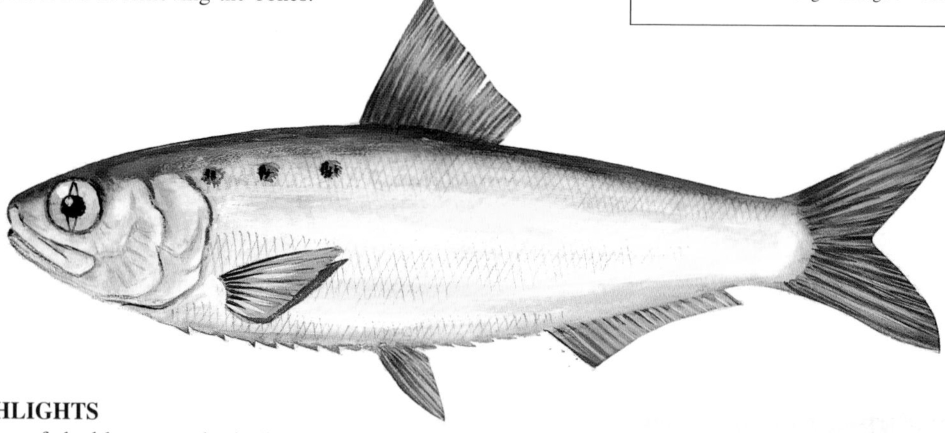

HIGHLIGHTS
Catches of shad have greatly declined, but the species is still sought by anglers, despite its bony meat. It can be hot-smoked, which helps to soften the bones. Commercial netting of the fish as it enters spawning rivers has virtually ceased.

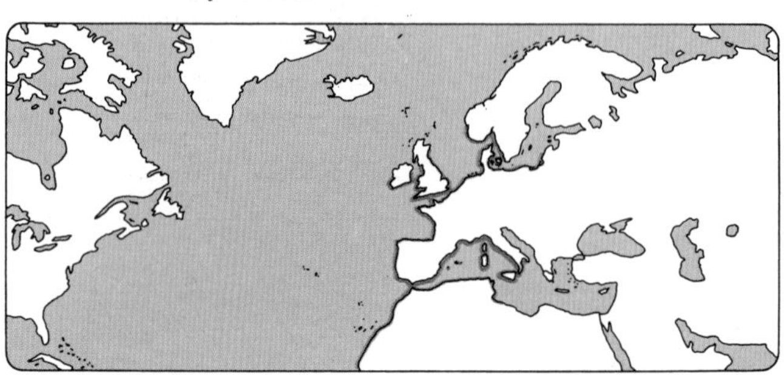

DK: Kraftigt fald i kommercielle fangster, men stadig en eftertragtet sportsfisk trods det benfyldte kød. Er lettere at udbene efter varmrygning

D: Starker Rückgang der Industriefänge, aber trotz des grätenreichen Fleisches gefragt bei Sportfischern. Ist nach dem Warm-räuchern einfacher zu entgräten

E: Han bajado mucho las capturas comerciales pero es un pescado deportivo popular, a pesar de sus muchas espinas, que son más fáciles de sacar ahumado en caliente

F: Les pêches commerciales sont en déclin, mais il reste très recherché par la pêche sportive. Les arêtes s'enlèvent plus facilement si la chair est fumée à chaud

I: Forte calo delle pesca commerciali, ma un pesce sportivo ricercato nonostante la grande presenza di spine. Si pulisce più facilmente dopo l'affumicatura a caldo

AMERICAN SMELT

Scientific name:
Osmerus mordax

Family: Osmeridae — smelts
Typical size: 15 to 20 cm

Smelts are caught at sea and in the Great Lakes and other major freshwater bodies in northern North America. They can be caught through holes in the ice (with hooks or box-nets) in winter.

FISHING METHODS

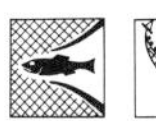
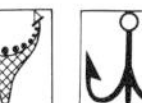

MOST IMPORTANT FISHING NATIONS
Canada, United States

PREPARATION

USED FOR

EATING QUALITIES
Smelts are characterized by delicate, pearly-white meat and an odour and flavour usually likened to freshly cut cucumber. Sea smelts are larger, more oily and have more flavour than those from freshwater, which are more valuable. Graded by size, smelts are cooked whole; the bones soften and can be eaten.

COMMON NAMES
D : Amerikanischer Stint
DK: Amerikansk smelt
E : Eperlano arco iris
F : Eperlan d'Amérique
GR: Iridízon eperlános
I : Sperlano, eperlano
IS: Vatnalodna, tannlodna
J : Kyuri-uo
N : Krøkle
NL: Spiering
P : Eperlano arco-íris
RU: Koriushka
S : Amerikansk nors
SF: Amerikankuore
US: Rainbow smelt

Nutrition data:
(100 g edible weight)

Water	78.8 g	Total lipid	
Calories	92 kcal	(fat)	2.4 g
Protein	17.6 g	Omega-3	0.7 mg

HIGHLIGHTS
Smelt fisheries fluctuate wildly from year to year. The American smelt is almost identical to the European smelt, *Osmerus eperlanus*, and to a Pacific smelt which is important in Siberia. Smelts were introduced to the Great Lakes and have thrived in the region.

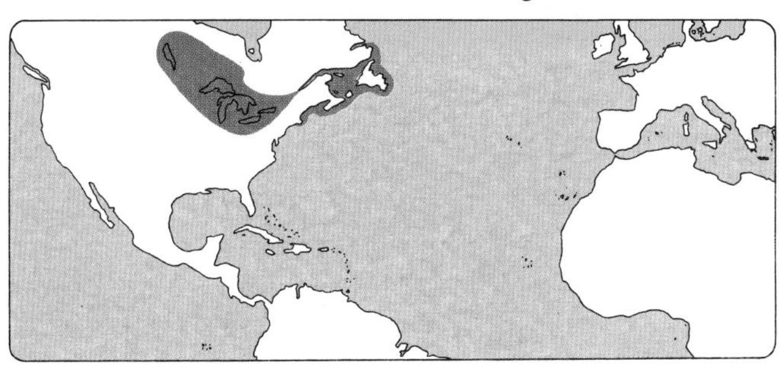

DK: Havarten er større, mere fedtrig og har bedre smag end ferskvandsarten. Tilberedes hel; bløde spiselige ben

D: Die Meeresart ist grösser, fetthaltiger und geschmacklich besser als die Süsswasserart. Wird im ganzen Stück zubereitet; hat weiche essbare Gräten

E: La especie de mar es más grande, contiene más grasa y tiene mejor sabor que la especie de agua dulce. Se cocina entero. Tiene espinas blandas y comestibles

F: L'éperlan de mer est plus grand, plus gras et il a meilleur goût que l'éperlan de lac. Il est préparé en entier et les arêtes tendres se mangent

I: La specie di mare è più grande, più grasso e di sapore più squisito della specie d'acqua dolce. Viene preparato intero e sono commestibili le spine tenere

ANGLERFISH/MONKFISH

Scientific name:
Lophius piscatorius

Synonyms: Sea devil, abbot, allmouth, fishing frog, frogfish
Family: Lophiidae — goosefishes
Typical size: up to 1.5 m, 30 kg

Anglerfish are increasingly popular for their firm, mild, boneless meat. Only the tail section is used. This is covered with many layers of skin. Some product is skinned before freezing, some is sold with the skins on. Skinned product is often cut from the central cartilage, giving two large fillet pieces.

DESCRIPTION

The monkfish is one of the ugliest marine fishes, with an enormous head, much of which is mouth. Partly because of the potential danger of being severely bitten by a dying monkfish, most fishermen lop off the tail section and throw the head and belly section of the fish overboard. Chefs and retail markets wanting whole monkfish for displays have to place special orders.

Anglerfish are found in small numbers in the Mediterranean and Black Seas, but most of the catch is taken in the Atlantic, where the species ranges from the Straits of Gibraltar as far north as the Barents Sea. The American goosefish ranges from New Brunswick to northern Florida. Both species are found from inshore waters to depths of 500 m and more.

COMMON NAMES

D: Seeteufel, Angler
DK: Havtaske, bredflab
E: Rape, sapo, rana pescadora
F: Baudroie, lotte
GR: Vatrochópsaro
I: Rana pescatrice, rospo
IS: Skötuselur
J: Anakou
N: Breiflabb
NL: Zeeduivel, hozemond
P: Tamboril, tamboril branco
RU: Moskoj chert
S: Marulk
SF: Merikrotti
TR: Fener baligi
US: Monkfish, goosefish

LOCAL NAMES

PO: Nawed zabnica
EG: Kott
IL: Shed yam
MO: Zaal ifoundou
TU: Bescatris

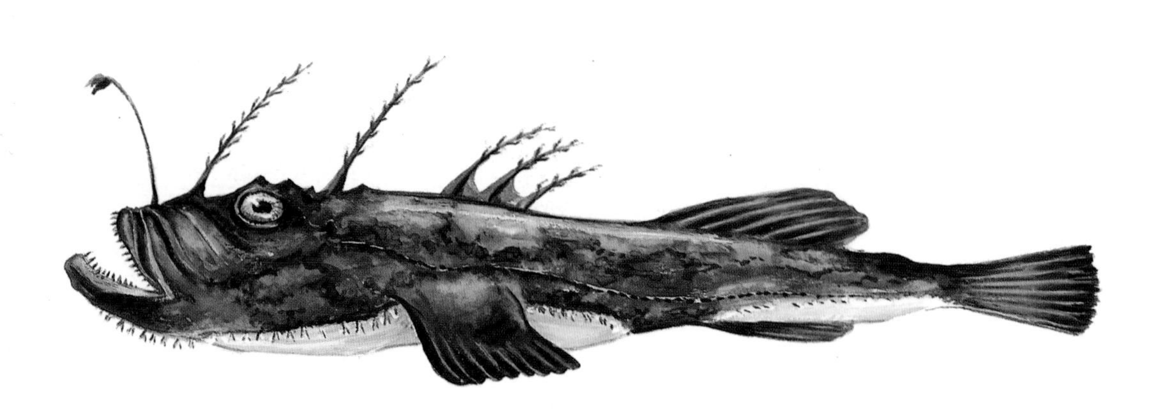

HIGHLIGHTS

Anglerfish are named for what appears to be a baited hook and line extending from the head and dangling over the mouth. This is the first ray of the dorsal fin, modified to include a flap of lure-like skin for the purpose of attracting prey, which are swallowed by the huge mouth when they come to investigate the lure.

The fish lie partly hidden in the sea bed, attracting prey with little effort. Growing to sometimes great sizes, they provide large fillets even though more than two thirds of the fish is discarded.

Nutrition data:
(100 g edible weight)

Water	83.3 g
Calories	66 kcal
Protein	15.7 g
Total lipid (fat)	0.4 g
Omega-3	0.1 mg

ANGLERFISH/MONKFISH *Lophius piscatorius*

FISHING METHODS

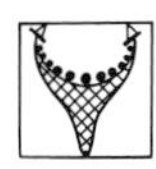

MOST IMPORTANT FISHING NATIONS

France, Spain, United Kingdom

PREPARATION

USED FOR

EATING QUALITIES

The meat from the anglerfish's tail section is firm, sweet and white. It has a texture almost like lobster tail meat, lacking the flake associated with most fish species. It is sometimes described as "poor man's lobster." This is a reference to the unfortunate and dishonest practice of some restaurants of substituting chunks of monkfish for lobster in certain well-sauced recipes, where the true nature, as well as the flavour, of the meat can be easily disguised.

Anglerfish meat is always boneless, an advantage in many markets. These large fish provide very big, meaty fillets which can be used in a wide variety of fish and shellfish recipes.

IMPORTANCE

Approximately 50,000 tons of anglerfish are caught yearly in European waters. Markets also have access to a further 12,000 tons or more of the closely related American monkfish (*L. americanus*). There are similar species in other parts of the world. Note that monkfish offered from New Zealand is actually a stargazer, a member of a different family. This fish has flakier meat and some small pinbones.

Anglerfish has been finding increasing acceptance in North America and Europe in recent years. At one time a trash fish in New England, often discarded, it is now a valuable resource that is specifically targeted for markets in the United States and Europe.

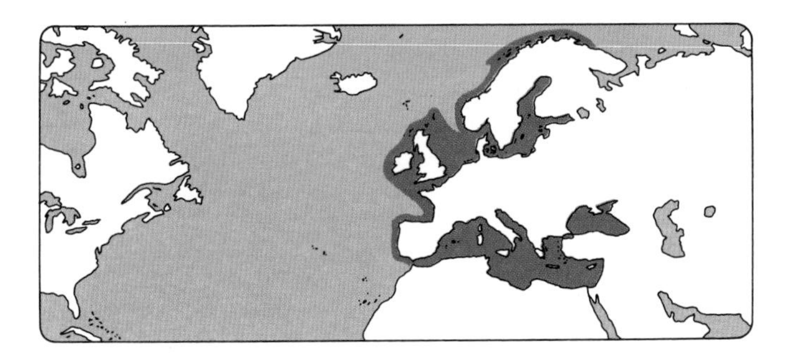

DK: Havtasken ligger på havbunden, hvor den lokker byttet ved at vifte med den forreste rygfinnestråle. Kan nå en betragtelig størrelse. Selvom 2/3 af fisken ikke anvendes, fås der store filetter

D: Der Seeteufel liegt am Meeresboden, wo er seine Beute durch das Wedeln mit dem ersten Strahl der Rückenflosse anlockt. Erzielt eine beträchtliche Grösse. Obwohl zwei Drittel nicht verwendet werden, ergeben sich grosse Filets

E: El rape se esconde en el fondo del mar, donde atrae la presa agitando la espina delantera de la aleta dorsal. Puede alcanzar un tamaño considerable. Aunque no se usan 2/3 del pescado, se obtienen grandes filetes

F: La lotte se dissimule au fond de la mer d'où elle attire sa proie en agitant l'épine de devant de sa nageoire dorsale. Elle peut devenir assez grande. Les 2/3 du poisson ne sont pas utilisés, mais les grands filets sont excellents

I: La rana pescatrice si nasconde sul fondo marino, attirando la preda movendo il raggio della pinna dorsale anteriore. Può giungere grandezze notevoli. Fornisce dei bei filetti grandi, anche se 2/3 del pesce non sono utilizzati

ARCTIC CHAR

Scientific name:
Salvelinus alpinus

Synonyms: Salmon trout, mountain trout, ilkalupik
Family: Salmonidae — salmonids
Typical size: 4 kg, 30 cm; may reach 15 kg

An anadromous circumpolar fish of cold northern waters, the char or charr is seldom seen commercially, except from tiny fisheries in Labrador, Iceland and Siberia and newly developing farms. It is considered one of the finest salmon-like fish to eat.

COMMON NAMES
D : Wandersaibling
DK: Fjeldørred
E : Salvelino, trucha alpina
F : Omble chevalier, omble
GR: Arktosalvelínos
I : Salmerino artico
IS: Bleikja
N : Arktisk røye
NL: Beekridder
P : Salvelino árctico
RU: Golets
S : Röding
SF: Nieriä
US: Arctic char

FISHING METHODS

MOST IMPORTANT FISHING NATIONS
Iceland, Canada, Russia, Norway

PREPARATION

USED FOR

EATING QUALITIES
Arctic char have white, amber or red flesh, well-flavoured, moist and firm, somewhat similar to brook trout. They are particularly good to eat if taken in the sea or shortly after they migrate into fresh water. For commercial use, the red-meated populations are preferred.

Nutrition data:
(100 g edible weight)

Water	71.0 g	Total lipid (fat)	8.1 g
Calories	154 kcal		
Protein	20.2 g	Omega-3	1.6 mg

Ocean form

HIGHLIGHTS
Arctic char are being farmed successfully in Iceland, Norway and Canada. So far, quantities are small, supplying high-priced gourmet markets. The fish is important to people of the far north for subsistence and has considerable value as a sport fish.

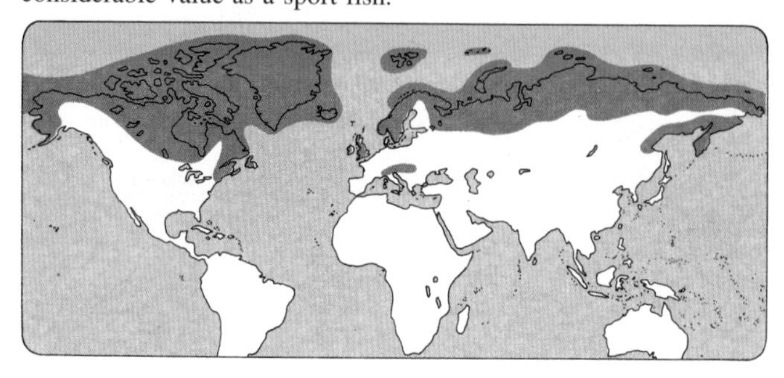

DK: Lille kommercielt fiskeri, men stigende interesse for at opdrætte denne fisk, som anses for en af de mest velsmagende ørredarter

D: Unbedeutender kommerzieller Fischfang, aber wachsendes Interesse an der Aufzucht dieses Fisches, der als eine der delikatesten Saiblingarten gilt

E: Pequeña pesca comercial, pero existe un interés cada vez mayor en criar este pescado, que es considerado una de las especies más sabrosas

F: La pêche commerciale est peu importante mais l'intérêt pour l'élevage augmente, car c'est l'espèce la plus délicate des truites

I: Scarca pesca commerciale, ma interesse crescente per l'allevamento di questo pesce che è considerato uno dei più gustosi fra le specie di trote

ATLANTIC CROAKER

Scientific name:
Micropogonias undulatus

Synonyms: Crocus, hardhead
Family: Sciaenidae — drums
Typical size: 25 cm

Atlantic croakers yield about 60,000 tons a year to the fisheries of Brazil and Uruguay. Argentina and the United States take smaller quantities. They are mostly taken during heavy seasonal runs when the fish shoal inshore to feed.

FISHING METHODS

   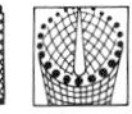

MOST IMPORTANT FISHING NATIONS

Brazil, Uruguay, Argentina

PREPARATION

USED FOR

EATING QUALITIES

Atlantic croaker has tasty but rather dark meat and plenty of bones. Because of its small size, it is usually dressed and trimmed, then pan-fried. Fillets are sometimes processed from the largest fish, which may reach over 1500 g, although most are about 500 g.

COMMON NAMES

D: Atlantischer Umberfisch
DK: Atlantisk trommefisk
E: Corvinón brasileño
F: Tambour brésilien
GR: Mylokópi tis Vrazilías
I: Ombrina
IS: Atlantsbaulari
NL: Atlantische ombervis
P: Rabeta brasileira
RU: Volnisty gorbyl
SF: Aaltorumpukala
US: Atlantic croaker

Nutrition data:
(100 g edible weight)

Water	78.0 g	Total lipid	
Calories	100 kcal	(fat)	3.2 g
Protein	17.8 g	Omega-3	0.2 mg

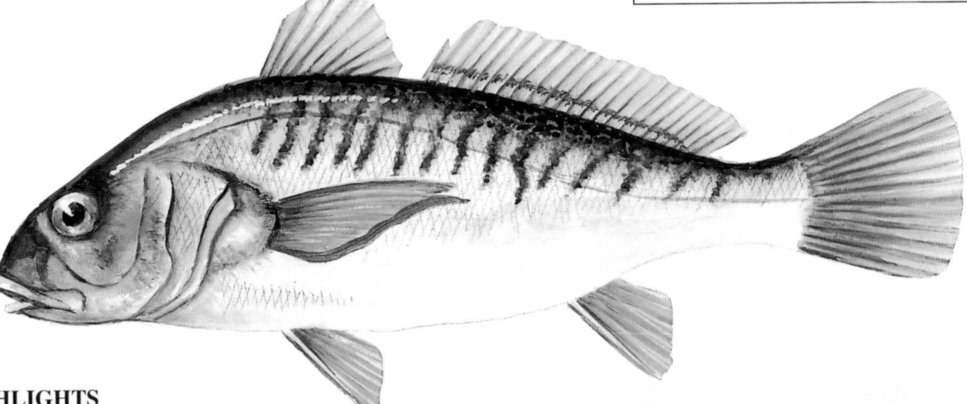

HIGHLIGHTS

North American resources of Atlantic croaker have greatly declined, especially in the Chesapeake Bay where they were once abundant. There are underexploited stocks in the Gulf of Mexico; South American resources appear to be substantial.

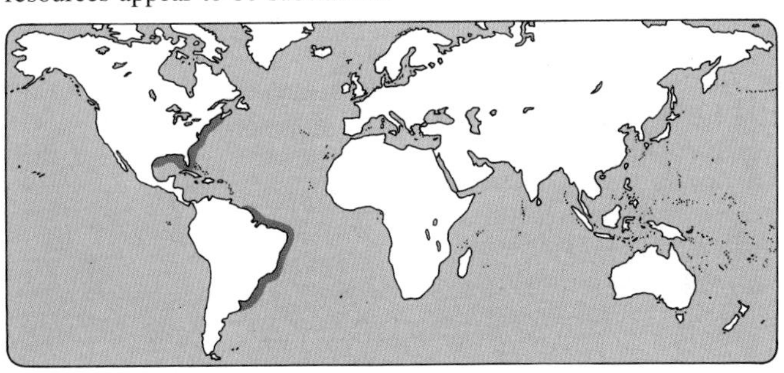

DK: Arten er på kraftig tilbagegang i Nordamerika, specielt i Chesapeake Bugten med den førhen så talrige bestand

D: Starker Rückgang der Art in Nordamerika, besonders in der Chesapeake-Bucht, die früher so grosse Bestände auswies

E: Han bajado mucho las capturas de estaespecie en Norteamérica, especialmente en la Bahía de Chesapeake, donde había abundancia en el pasado

F: Cette espèce est en fort déclin en Amérique du Nord, en particulier dans la Baie de Chesapeake qui, autrefois, abritait une population très nombreuse

I: Le risorse di questa specie stanno calando drasticamente nell'America Settentrionale, in particolare nel Chesapeake Bay, dove una volta era abbondante

ATLANTIC MACKEREL

Scientific name:
Scomber scombrus

Family: Scombridae — tunas and mackerels
Typical size: 30 cm, 750 g

Most of the landings of this temperate Atlantic species are made in Europe, principally by Norway and Scotland from the North Sea and adjacent waters. These two countries account for about 60 percent of world landings of 500,000 to 700,000 tons. Although the USA and Canada produce only small quantities, a century ago catches of over 100,000 tons were recorded. The resource in the western Atlantic is currently estimated by US government scientists to be capable of supporting a fishery of over 200,000 tons.

DESCRIPTION

Atlantic mackerel are generally small, but can reach 60 cm and over 2 kg when older (in 1925, a mackerel of 3.4 kg was reported from the Gulf of Maine). Fat content varies greatly through the year. For most uses, the fish is preferred when it has the highest fat content, which may exceed 22 percent. In some populations, the oil can fall to less than 3 percent immediately after the fish have spawned.

Although salted mackerel used to be an important product in North America, that market has disappeared. The species remains highly regarded in Europe, where it is used for many different recipes.

COMMON NAMES

D: Makrele, gemeine Makrele
DK: Makrel
E: Caballa del Atlántico
F: Maquereau commun
GR: Skoumbrí
I: Maccerello, sgombro
IS: Makríll
J: Saba, hirasaba, marusaba
N: Makrell
NL: Makreel
P: Sarda
RU: Atlanticheskaya makrel
S: Makrill
SF: Makrilli
TR: Uskumru
US: Atlantic mackerel

LOCAL NAMES

PO: Makrela
TU: Sqoumri
MO: Kabaila
CH: Kau yu, faa gau
TH: Pla thu
IN: Kembung

HIGHLIGHTS

Over the centuries, catches of mackerel from both sides of the North Atlantic have fluctuated enormously. The resource is clearly cyclical; at the present time, the stocks appear to be at the lower point of a cycle in the Eastern Atlantic, but comparatively abundant on the Western side. However, because of fears that over-harvesting has led to reduced numbers, there are strict controls on fishing throughout the range of the species. Many scientists believe that the health of the stock is related to the survival of juvenile fish; the factors that govern this are not clearly understood.

Nutrition data:
(100 g edible weight)

Water	64.0 g
Calories	220 kcal
Protein	18.7 g
Total lipid (fat)	16.1 g
Omega-3	1.8 mg

ATLANTIC MACKEREL *Scomber scombrus*

FISHING METHODS

   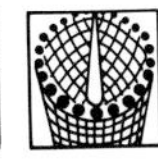

MOST IMPORTANT FISHING NATIONS

Norway, United Kingdom (Scotland)

PREPARATION

USED FOR

EATING QUALITIES

The brownish meat and medium-sized flake of the Atlantic mackerel has excellent flavour and texture when fresh, but the high oil content turns the meat rancid in a very short time. Fish for the fresh and frozen markets must be handled with great care and speed. The flavour intensifies as the fish ages; there is a noticeable difference in fish that is less than perfectly fresh.

Smoked and cured mackerel products are widely used, especially in Europe. Such preparations have good shelf life and add considerable value to what is still an inexpensive raw material. Hot smoked mackerel can be compared with smoked trout for delicate texture and taste. Many cured products remain popular with European consumers.

IMPORTANCE

The huge schools of mackerel which once provided Europeans with a major staple food have all but gone. Nevertheless, the species is still important and well regarded by consumers. Like other pelagic fish which school heavily near the coast, mackerel are easy to catch in large numbers, but this makes it difficult to handle them with sufficient speed to ensure top quality.

Despite the decline in stocks in European waters, supplies of mackerel could be increased from North American countries if over-restrictive fishery management regimes were relaxed.

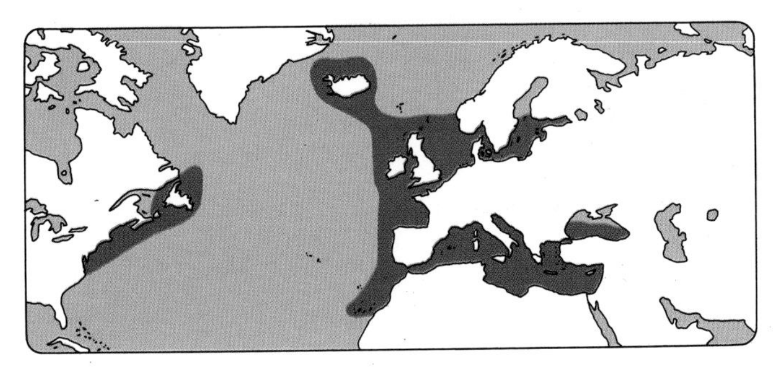

DK: En lille, fedtrig fisk som altid har været et vigtigt næringsmiddel i Europa. Fangsterne er langt lavere end tidligere, men arten er fortsat vigtig med mange anvendelsesmuligheder

D: Ein kleiner, fetthaltiger Fisch, der in Europa immer ein Grundnahrungsmittel gewesen ist. Die Fänge sind viel geringer als früher, aber die Art ist mit ihren vielen Anwendungsmöglichkeiten weiterhin wichtig

E: Pescado pequeño y rico en grasa, que siempre ha sido un alimento importante en Europa. Las capturas son mucho más bajas que antes, pero la especie sigue siendo importante gracias a sus múltiples usos

F: Petit poisson riche en matière grasse, qui depuis toujours constitue un aliment important en Europe. Les pêches sont en forte baisse, mais il conserve son importance du fait de ses multiples utilisations

I: Un piccolo pesce grasso e da secoli un genere alimentare importante in Europa. Pescato in quantità molto inferiori a quelle di prima, ma la specie rimane importante per le svariate possibilità di impiego

ATLANTIC SALMON

Scientific name:
Salmo salar

Family: Salmonidae — salmonids
Typical size: 65 cm, 9 kg

Once, Atlantic salmon teemed into the rivers of the North Atlantic as the fish returned to their natal streams to spawn. Now, Atlantic salmon has almost disappeared from the wild, but has been so successfully domesticated that it is again a major source of high quality seafood.

Commercial catches of wild salmon are minimal and angling is severely restricted, but the lore and mystique of this fish continues to attract attention, as intense efforts to restore salmon runs into hundreds of rivers continue on both sides of the Atlantic.

DESCRIPTION
Atlantic salmon are silver-skinned fish with small, cross-like marks on the back and upper parts of the sides. The shape of these spots is distinctive and helps to distinguish this species from other salmon.

Wild Atlantic salmon can grow to enormous size: a fish of 45 kg was netted in Finnmark (Norway). Such fish have spawned and returned to the ocean several times. Farmed fish are not usually permitted to reach sexual maturity before they are harvested, but are grown to market size requirements.

COMMON NAMES
D: Lachs, atlantischer Lachs
DK: Laks
E: Salmón
F: Saumon atlantique
GR: Solomós tou Atlantikoú
I: Salmone atlantico
IS: Lax
N: Laks
NL: Zalm
P: Salmao do Atlântico
RU: Atlantichesky losos
S: Lax
SF: Lohi
TR: Alabalik
US: Atlantic salmon

LOCAL NAMES
PO: Losós

HIGHLIGHTS
The enormous expansion of salmon farming, especially in northern Europe, has made Atlantic salmon one of the world's most important marine foods after more than a century of decline in natural stocks. Once abundant throughout the estuaries and rivers of the North Atlantic, Atlantic salmon had almost disappeared when farmers began to raise it successfully in the late 1970s.

A premium quality fish for eating and smoking, Atlantic salmon has earned new markets for the hugely increased production of the farms.

Nutrition data:
(100 g edible weight)

Water	67.2 g
Calories	180 kcal
Protein	20.2 g
Total lipid (fat)	11.0 g
Omega-3	1.8 mg

ATLANTIC SALMON *Salmo salar*

FISHING METHODS

 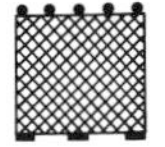 

MOST IMPORTANT FISHING NATIONS

Norway, United Kingdom, Chile, Canada

PREPARATION

USED FOR

EATING QUALITIES

Fine flavour, firm texture and excellent consumer acceptability mark the advantages that high quality salmon bring to the fish business. Atlantic salmon, thanks to sophisticated aquaculture, is consistently high quality, earning large and loyal markets throughout the world. The meat is pink to red, less intensely coloured than sockeye, similar to coho and chinook. Breeders are developing fish with redder meat.

Atlantic salmon is the species of choice for smokers in Europe. Smoked salmon is the most important product made from this species, which, unlike Pacific salmon, is seldom canned. Cold-smoked salmon is an important item in France, Germany, Italy, the United Kingdom and other European nations.

IMPORTANCE

Farmers have made this species one of the world's most valuable. Output exceeds 300,000 tons a year, with Norway the dominant producer. Top quality farmed Atlantic salmon is available fresh year round. It is now grown far from its natural range (see map), on the Pacific coast of Canada and the United States, in Chile and even in New Zealand and Australia.

Wild salmon is making a very slow return to its former habitats in North Atlantic rivers, where, it is hoped, it will again become a major attraction for recreational fishermen.

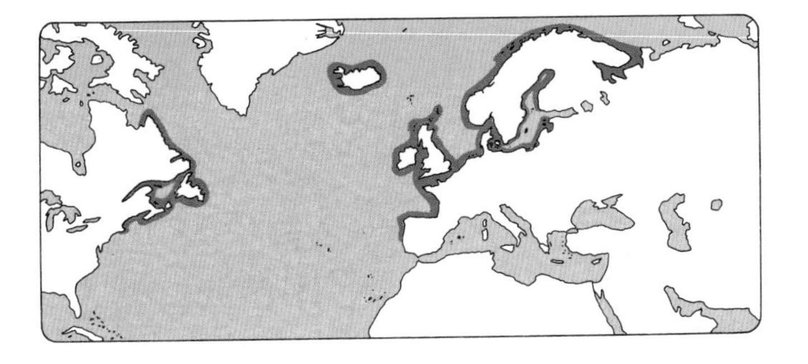

DK: En førsteklasses spisefisk i såvel fersk som røget form. Laksen har vundet nye markeder som følge af det voldsomt stigende opdræt og er nu igen en af verdens vigtigste fisk

D: Frisch und geräuchert ein erstklassiger Speisefisch. Der Lachs hat als Folge der massiv erweiterten Aufzucht neue Märkte erobert und gehört jetzt wieder zu den wichtigsten Nutzfischen der Welt

E: Pescado comestible de primera calidad tanto fresco como ahumado. El salmón ha ganado nuevos mercados como consecuencia de la cría cada vez más intensiva, y ha vuelto a ser uno de los pescados más importantes del mundo

F: Un poisson de très grande classe tant à l'état frais que fumé. Grâce à l'élevage en énorme progression, le saumon a conquis de nouveaux marchés et il compte de nouveau parmi les plus importants poissons du monde

I: Un pesce commestibile di altissima qualità sia fresco che affumicato. Il salmone ha guadagnato nuovi mercati in seguito all'allevamento in forte aumento ed è ritornato ad essere uno dei pesci più importanti del mondo

BELUGA STURGEON

Scientific name:
Huso huso

Synonym: Great sturgeon
Family: Acipenseridae — sturgeons
Typical size: up to 5 m, 800 kg

The largest of all the sturgeon family, the beluga also provides the most expensive caviar: its greyish-black eggs are the most highly prized by caviar connoisseurs. The largest beluga documented was caught in the Volga River near Astrakhan in 1736 and weighed 2,072 kg. There are ancient reports of even bigger fish. The giants still exist: a specimen caught in the Volga in 1989, now stuffed and displayed in the Astrakhan Museum, weighed 978 kilos.

DESCRIPTION
The beluga is an anadromous and freshwater species found in the Caspian, Black and Azov Seas and the major rivers of this region. A few may still inhabit the eastern Mediterranean as far west as the River Po in Italy. Beluga may live for more than 100 years and may be over 5 m long. The females mature slowly and carry eggs only once every five to seven years.

The closely related kaluga (*Huso dauricus*) is a freshwater sturgeon found in rivers and lakes in the Azov basin, along the border between Siberia and China. The kaluga may grow even larger than the beluga and provides caviar of equal quality.

COMMON NAMES
D: Europäischer Hausen
DK: Hus, beluga
E: Esturión beluga, husio
F: Béluga, grand esturgeon
GR: Mouroúna, stourióni
I: Storione ladano
IS: Hússtyrja
NL: Huso, beluga
P: Esturjao beluga
RU: Beluga
S: Husen
SF: Kitasampi
TR: Mersin morinasi
US: Beluga sturgeon

HIGHLIGHTS
Beluga caviar is packed traditionally in tins with blue lids. In recent times, the term beluga has come to be applied to all top quality grayish-black caviar, irrespective of species; the word beluga on the label often refers to a quality rating rather than being a guarantee that the eggs come from the beluga species.

Although the economics of the beluga fisheries are based almost entirely on the value of the caviar, the meat is also edible, especially if smoked or made into soup in traditional Russian manner.

Nutrition data:
(100 g edible weight)

Water	78.7 g
Calories	97 kcal
Protein	16.2 g
Total lipid (fat)	3.6 g
Omega-3	0.7 mg

BELUGA STURGEON *Huso huso*

FISHING METHODS

MOST IMPORTANT FISHING NATIONS
Russia, Kazakhstan, Iran

PREPARATION
Caviar

USED FOR

EATING QUALITIES
Beluga caviar, especially the lower-salt or malasol variety, is the most expensive and most highly regarded of all sturgeon caviar. Traditionally, most caviar was exported in bulk packs and repacked in market countries into containers small enough to be affordable for the consumer. There is a developing trend towards consumer packing in the supplying countries, including both Russia and Kazakhstan. This trend will help to improve the quality of the product bought by the end-user and will, perhaps, counteract some of the problems mentioned in the following paragraphs.

IMPORTANCE
The beluga fishery was tightly controlled during the Soviet era to protect the limited numbers of these slow-growing fish. Since the breakup of the USSR, the species has suffered from increasing poaching. Experts believe that numbers are dwindling. Controls over harvesting, as well as over the quality and export of caviar, no longer exist in practice. Caviar buyers throughout the world have found the once reliable quality of Russian caviar exports is a thing of the past. Importers now have to make detailed inspections to ensure that the product is fresh, properly processed and accurately labelled.

Although beluga have not yet been successfully domesticated, a hybrid of a female beluga and a male sterlet (*Acipenser ruthenus*), called bester, is being successfully farmed for its high quality eggs.

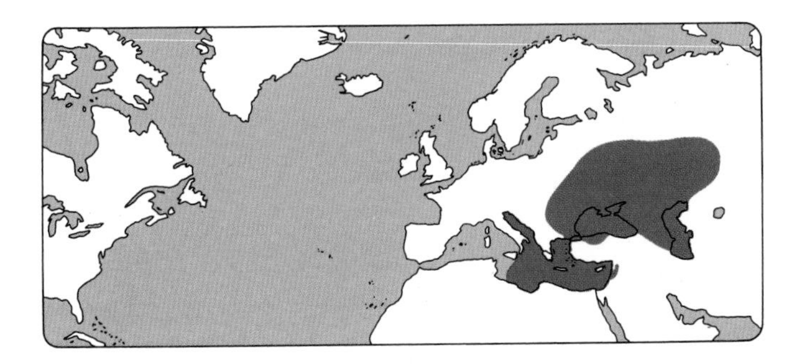

DK: Beluga kaviar, og i særdeleshed den saltfattige 'malosol' variant, er den dyreste og højest værdsatte kaviar. Der er i dag en tendens til at levere kaviaren som et færdigpakket produkt

D: Beluga und insbesondere die wenig gesalzene Variante 'malosol' ist der teuerste und begehrteste Kaviar. Der Kaviar wird heute zunehmend als Fertigprodukt geliefert

E: El caviar beluga, y en particular la variante 'malosol', pobre en sal, es el caviar más caro y más apreciado. Hoy día hay una tendencia a suministrar el caviar como producto envasado

F: Le caviar beluga, et en particulier la variante 'malosol' qui est pauvre en sel, est le caviar le plus cher et le plus apprécié dans le monde entier. La tendence va vers la vente du caviar conditionné sous emballage 'consommateur'

I: Il caviale Beluga, e in particolare la variante poco salata 'malosol', è il caviale più costoso e più pregiato. Oggi i paesi produttori lo forniscono, sempre più spesso, come prodotto preconfezionato

BIG SKATE

Scientific name:
Raja binoculata

Family: Rajidae — skates
Typical size: 1.8 m, 91 kg

COMMON NAMES
RU: Skat
US: Big skate

Big skate is one of the largest skates in the world, reaching 2.4 m in length and a similar size across the wings. It is marketed in small quantities, but often regarded as a nuisance fish by commercial harvesters, who target more valuable species.

FISHING METHODS

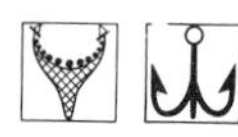

MOST IMPORTANT FISHING NATIONS
Canada

PREPARATION

USED FOR

EATING QUALITIES
Reddish coloured when raw, the meat of the big skate turns white when cooked. It is moist and well flavoured, with a moderately firm texture. Because they are so large, the wings of this species are usually filleted and skinned.

Nutrition data:
(100 g edible weight)

Water	83.1 g	Total lipid (fat)	0.4 g
Calories	64 kcal		
Protein	15.0 g	Omega-3	0.1 mg

HIGHLIGHTS
The horny egg case of the big skate is up to 30 cm long and may contain seven eggs. The large pectoral fins, known as wings or flaps in the seafood industry, are the only part of the fish used; the rest is discarded or turned into fishmeal.

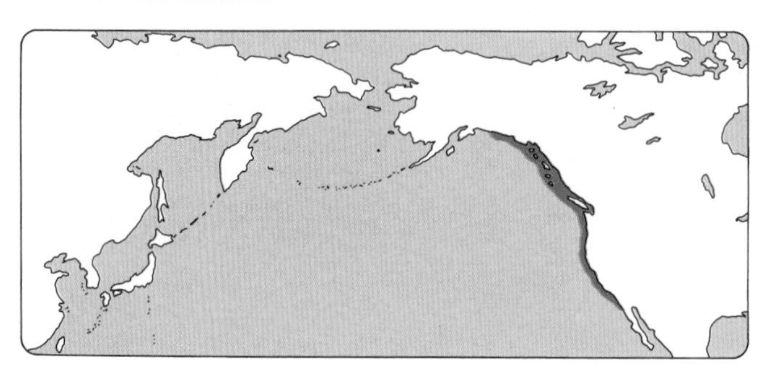

DK: I rokkens indtil 30 cm lange, hornede ægkapsel kan ligge 7 æg. De store brystfinner spises, mens resten destrueres eller anvendes til fiskemel

D: In der bis zu 30 cm langen Eierkapsel aus Horn finden 7 Eier Platz. Die Brustflossen werden gegessen; der Rest wird vernichtet oder zu Fischmehl verarbeitet

E: La cápsula córnea de la raya, de hasta 30 cm de longitud, puede contener 7 huevos. Las aletas pectorales se comen, el resto se destruye o aprovecha para harina

F: La capsule ovulaire cornée de la raie peut atteindre 30 cm de long et contenir 7 oeufs. Seules les grandes nageoires pectorales sont comestibles

I: La capsula cornea lunga fino a 30 cm può contenere 7 uova. Le grandi pinne pettorali sono commestibili, mentre il resto viene distrutto o usato per farina

BLACK SEA BASS

Scientific name:
Centropristis striata

Synonym: Sea bass
Family: Serranidae — sea basses
Typical size: 55 cm, 2.75 kg

Black sea bass ranges from Florida to Cape Cod. A reef fish, it is landed mainly from Long Island to Carolina, though supplies are available from Florida in winter. It has poisonous spines on its back, so must be handled with care.

FISHING METHODS

MOST IMPORTANT FISHING NATIONS
United States

PREPARATION

USED FOR

EATING QUALITIES
Black sea bass has firm, white meat and excellent flavour. It is versatile in use, particularly popular for cooking whole in Chinese cuisine. Because it often contains parasitic worms, the fish should be thoroughly bled when caught.

COMMON NAMES
D: Schwarzer Sägebarsch
DK: Sort havaborre
E: Serrano estriado
F: Fanfre noir
GR: Mavróperka
I: Perchia striata
IS: Svarti gaddborri
NL: Zwarte zeebaars
P: Serrano estriado
RU: Cherny morskoj okun
S: Svart havsabborre
SF: Kalliomeriahven
US: Black sea bass

Nutrition data:
(100 g edible weight)

Water	78.3 g	Total lipid (fat)	2.0 g
Calories	92 kcal	Omega-3	0.6 mg
Protein	18.4 g		

HIGHLIGHTS
Despite poisonous spines, parasites and a reputation for poor shelf life, the black sea bass is popular with commercial users and recreational anglers because of the fine quality of its meat. Limited supplies are all sold fresh; there is not enough caught to freeze.

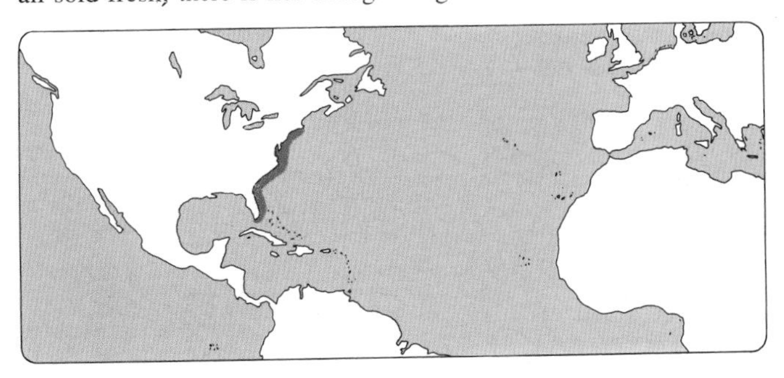

DK: En vigtig konsum- og sportsfisk selvom den har giftige pigge, parasitter og er letfordærvelig

D: Ein wichtiger Speise- und Sportfisch, obwohl er giftige Stacheln und Parasiten hat und leicht verderblich ist

E: Importante pescado deportivo y de consumo, aunque tiene espinas venenosas, parásitos, y se deteriora fácilmente

F: En dépit de ses épines vénéneuses, ses parasites et une chair périssable, le bar est très recherché par les pêcheurs tant professionnels que sportifs

I: Un pesce di consumo e sportivo importante anche se coperto di spine velenose e parasiti e per di più deperibile

BLACKBACK FLOUNDER

Scientific name:
Pleuronectes americanus

Family: Pleuronectidae — right-eye flounders
Typical size: up to 60 cm, 3 kg
Also called: *Pseudoparalichthys americanus*

COMMON NAMES
RU: Belobokaya kambala
US: Lemon sole

Scientific names of the genus have changed from Pseudopleuronectes to Pseudoparalichthys and now to Pleuronectes, but the fish remains an important small inshore flounder of the Atlantic coast of North America, in shallow water from Labrador to Georgia.

FISHING METHODS

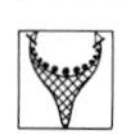 

MOST IMPORTANT FISHING NATIONS
United States, Canada

PREPARATION

USED FOR

EATING QUALITIES
Blackback flounder has a bland taste, but good texture. The meat is greyish, especially from the dark skin side, but it is whiter after cooking. The dark skin is usually removed from the fillets before they are sold, to improve the appearance.

Nutrition data:
(100 g edible weight)

Water	77.9 g	Total lipid	
Calories	k87cal	(fat)	1.0 g
Protein	19.6 g	Omega-3	0.2 mg

HIGHLIGHTS
Known as winter flounder because it follows colder water inshore, this species is locally important in the central part of its range. Fish over 3.5 lb (1.58 kg) are sold in the United States (legally) as lemon sole, underlining the lack of distinction between flounders and soles.

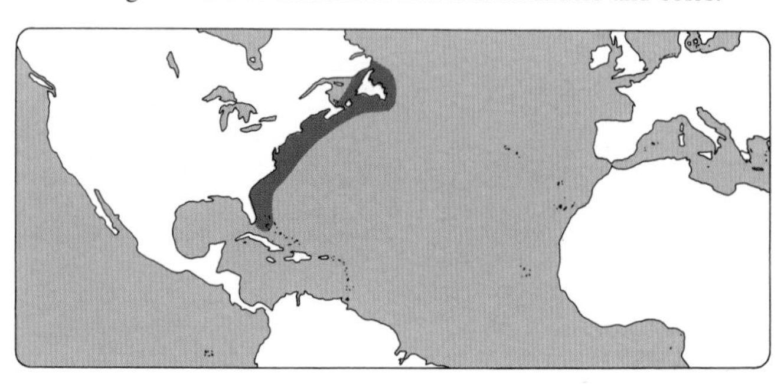

DK: Eksemplarer på over 3,5 lb (1,58 kg) sælges legalt i USA som rødtunge, hvilket understreger den manglende skelnen mellem flyndere og tunger

D: Exemplare von mehr als 3 Pfund werden legal als Rotzunge in den USA verkauft, woraus die fehlende Unterscheidung zwischen Flundern und Zungen erkennbar ist

E: Ejemplares de más de 1,58 kg se venden legalmente en EE.UU. como falsa limanda, lo que muestra que no se hace distinción entre las peludas y los lenguados

F: Des exemplaires de plus de 1,58 kg sont vendus légalement aux E.-U. sous le nom de limande sole ce qui illustre le manque de distinction entre les flets et soles

I: Esemplari di oltre 1,58 kg sono venduti legalmente negli USA come sogliola limanda, confermando la scarsa distinzione tra passere e sogliole

BLACKTIP SHARK

Scientific name:
Carcharhinus limbatus

Synonyms: Blackfin shark, soupfin shark
Family: Carcharhinidae — requiem sharks
Typical size: 150 cm

Blacktip sharks are found throughout the world's warm and temperate waters. The species is caught close inshore, but is also found in the open ocean. Not favoured as a game fish, it is nevertheless capable of fighting hard, sometimes leaping out of the water when hooked. This generally small shark can grow as large as 255 cm and 122 kg.

FISHING METHODS

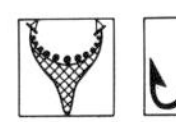

MOST IMPORTANT FISHING NATIONS

Data not available

PREPARATION

USED FOR

EATING QUALITIES

Light-coloured meat with mild flavour makes this species a favourite among shark consumers. Blacktip has earned a market niche in the United States in recent years. The skins are used for leather.

COMMON NAMES

D: Kleiner Schwarzspitzenhai
DK: Atlantisk sorttippet haj
E: Tiburón macuira
F: Requin bordé, requin blanc
I: Squalo pinne nere
IS: Svartuggi
N: Atlantisk svarttipped hai
NL: Zwartpuntrifhaai
P: Tubarao de pontas negras
RU: Chernoperaya akula
S: Svartfenad haj
SF: Mustapilkkahai
US: Blacktip shark

LOCAL NAMES

MA: Yu nipah

Nutrition data:
(100 g edible weight)

Water	74.0 g	Total lipid	
Calories	91 kcal	(fat)	0.3 g
Protein	22.0 g	Omega-3	0.1 mg

HIGHLIGHTS

Blacktip is regarded as one of the best quality sharks to eat. It inhabits coastal waters, including estuaries, where it remains at or near the surface. It is often caught by sport fishermen. Commercial landings often result from accidental capture in nets.

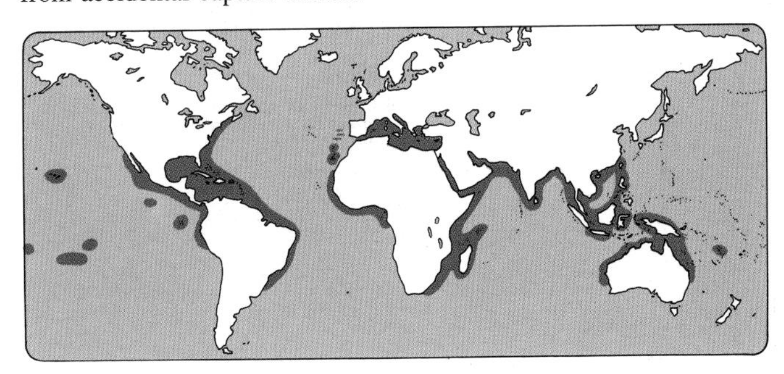

DK: En af de foretrukne, spiselige hajer. Den holder til langs kysten og er derfor en yndet sportsfisk

D: Einer der bevorzugten, essbaren Haie. Kommt in Küstennähe vor und ist deshalb bei Sportfischern beliebt

E: Uno de los tiburones comestibles de preferencia. Vive a lo largo de las costas, y por eso es un pescado muy popular entre los pescadores deportivos

F: L'un des requins comestibles recherché pour sa bonne chair. Il vit le long de la côte, où il est souvent pêché par les pêcheurs sportifs

I: Uno degli squali commestibili di preferenza. Vive lungo le coste ed è quindi un pesce sportivo popolare

BLUE CATFISH

Scientific name:
Ictalurus furcatus

Family: Ictaluridae — bullhead catfishes
Typical size: 50 cm, 3 kg

COMMON NAMES
RU: Sinyaya zubatka
US: Blue catfish

One of the largest N. American catfishes, the blue catfish can reach 1.5 m and 45 kg. Specimens of 68 kg were recorded in the 19th century. It is a favourite recreational fish throughout its range, which covers large rivers from Minnesota to Mexico. It is farmed in small quantities, both for meat and for stocking rivers for anglers.

FISHING METHODS

MOST IMPORTANT FISHING NATIONS
United States

PREPARATION

USED FOR

EATING QUALITIES
Blue catfish has excellent meat: firm, white and sweet. It is quite easy to remove the bones from the large fillets. Like other catfishes, it has thick skin which should be removed before cooking.

Nutrition data:
(100 g edible weight)

Water	76.5 g	Total lipid (fat)	3.0 g
Calories	103 kcal		
Protein	19.0 g	Omega-3	0.6 mg

HIGHLIGHTS
Unlike other bullheads, the blue catfish prefers clear, strongly flowing water. A predatory species, it is a strong fighter, popular with anglers for sport, its large size and its fine meat. The species has been widely stocked, so its current range is much greater than its original one.

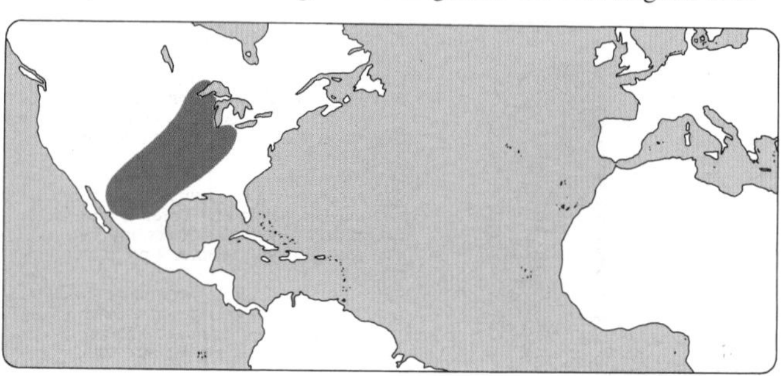

DK: Med en størrelse på indtil 1,5 m og 45 kg er dette en af Nordamerikas største dværgmaller. I 1800-tallet fangedes eksemplarer på op til 68 kg

D: Bei einer Länge von bis zu 1,5 m und einem Gewicht von 45 kg ist dies einer der grössten Zwergwelse Nordamerikas. Im 19. Jh. wurden Exemplare von 68 kg gefangen

E: Es uno de los más grandes bagres de Norteamérica, que puede alcanzar 1,5 m y 45 kg. En el siglo XIX se capturaron ejemplares de 68 kg

F: L'un des plus grands ictaluridae de l'Amérique du Nord pouvant atteindre 1,5 m et 45 kg. Des exemplaires de 68 kg ont été pêchés au 19e siècle

I: Con i suoi 1,5 m. e 45 kg, questo è uno dei pesci gatto più grandi dell' America Settentrionale. Nel secolo scorso se ne catturarono esemplari di 68 kg

BLUE WAREHOU

Scientific name:
Seriolella brama

Synonyms: Striped trevally, warehou, snotgall trevally
Family: Centrolophidae — medusafishes
Typical size: 40 to 60 cm, 4 kg; up to 76 cm

Blue warehou, named for their deep purple backs, are targeted around southern Australia and New Zealand's South Island by trawlers and gillnetters. Young fish are also sought by anglers. In Australia it is caught with the spotted warehou (*S. punctata*). The two species are sold together and not normally distinguished.

COMMON NAMES

D: Seriolella
J: Warefu
RU: Leschevidnaya seriolella
US: Blue warehou

FISHING METHODS

MOST IMPORTANT FISHING NATIONS

New Zealand, Australia

PREPARATION

USED FOR

    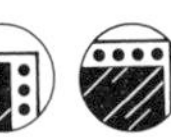

EATING QUALITIES

Blue warehou has pinkish or pale brown meat, which whitens when it is cooked. It has moderate oil content and a fairly soft texture, but holds together well through most cooking methods.

Nutrition data:
(100 g edible weight)

Water	73.5 g	Total lipid	
Calories	123 kcal	(fat)	4.4 g
Protein	20.8 g	Omega-3	1.1 mg

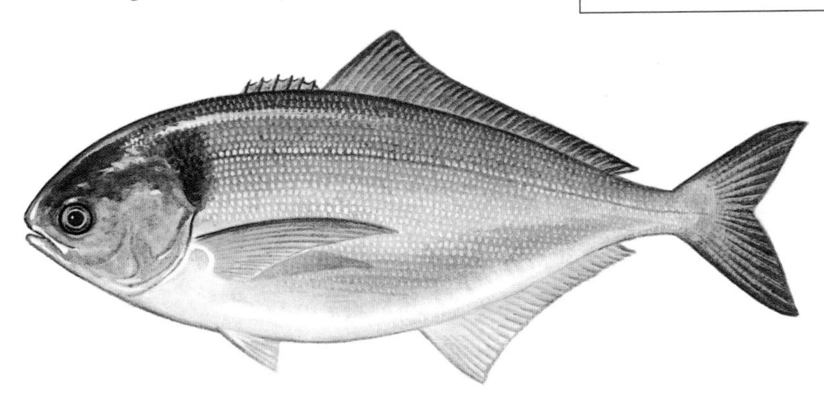

HIGHLIGHTS

This fast growing fish reaches about 25 cm in its first year and may live for ten or more years. Although the size of the resource is not known, warehou catches have recently increased substantially without apparent effect.

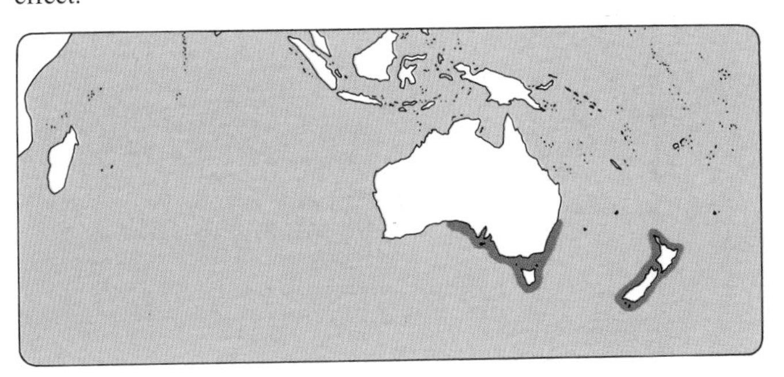

DK: Stigende fiskeri efter denne hurtigt voksende fisk, som i sit første leveår bliver omkring 25 cm lang. Lever i minimum 10 år

D: Zunehmende Fänge dieses schnell wachsenden Fisches, der in seinem ersten Lebensjahr ungefähr 25 cm lang wird. Wird 10 Jahre und älter

E: Sigue aumentando la pesca de esta especie de rápido crecimiento que llega a ser de unos 25 cm de longitud durante su primer año de vida. Vive como mínimo 10 años

F: Ce poisson de croissance rapide atteint une longueur de 25 cm la première année et il peut atteindre plus de dix ans. La pêche au warehou continue à progresser

I: E' aumentata sensibilmente la pesca di questo pesce a crescita rapida che nel primo anno di vita raggiunge i 25 cm in lunghezza. Vive almeno per dieci anni

BLUE WHITING/POUTASSOU

Scientific name: *Micromesistius poutassou*

Synonyms: Poutassou, couch's whiting
Family: Gadidae — cods
Typical size: 15 to 30 cm, up to 50 cm

Harvests of 500,000 tons and more of blue whiting are taken mainly in deep waters of the northeast Atlantic, with some fishing also in the Mediterranean. Some experts believe the resource can support fishing of more than 1 million tons yearly.

FISHING METHODS

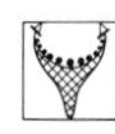

MOST IMPORTANT FISHING NATIONS
Norway, Russia, Faroe Islands, Denmark, Spain

PREPARATION

USED FOR

EATING QUALITIES
The blue whiting's poor keeping quality means most of the catch of this abundant species is relegated to low value uses such as fishmeal. It is reported to have little flavour, with very soft, whitish meat and sometimes parasitic infestation which affects meat quality.

COMMON NAMES
D: Blauer Wittling
DK: Sortmund, blåhvilling
E: Bacaladilla, perlita
F: Merlan bleu, poutassou
GR: Prosfygáki, tsimbláki
I: Melu, potassolo
IS: Kolmunni
N: Kolmule, blagunnar
NL: Blauwe wijting
P: Verdinho, pichelim
RU: Severnaya putassu
S: Kolmule, blåvitling
SF: Mustakitaturska
TR: Mezgit, bakayaro
US: Blue whiting

LOCAL NAMES
EG: Nazelli
IL: Shibbut albin
MO: Abadekho
PO: Blekitek
TU: Nazalli azraq

Nutrition data:
(100 g edible weight)

Water	80.0 g	Total lipid	
Calories	72 kcal	(fat)	0.4 g
Protein	17.1 g	Omega-3	0.1 mg

HIGHLIGHTS
This abundant species may in the future be used for the manufacture of blocks of surimi, used for processing into imitation crab and other products. It is not often eaten fresh: most of the catch is processed, much of it for animal feed.

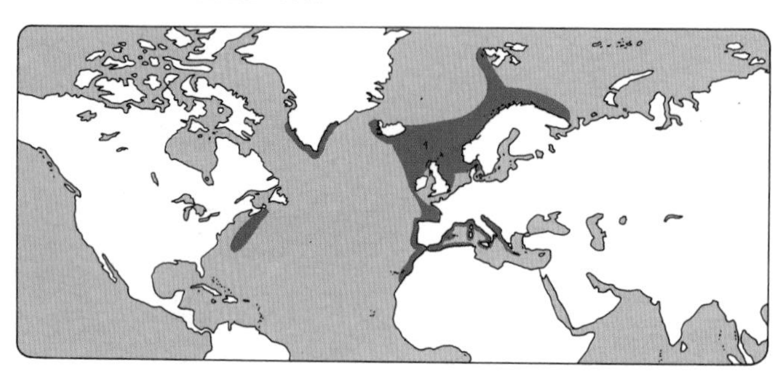

DK: En talrig og potentiel art til fremstilling af frosne blokke, der senere forarbejdes til surimi (krabbeerstatning)

D: Eine Zahlenmässig starke Art mit Zukunft für die Herstellung gefrorener Blöcke, die später zu 'Surimi' (Krabbenersatz) verarbeitet werden

E: Especie abundante que puede usarse para la fabricación de bloques congelados para 'surimi' (imitación de cangrejos)

F: Cette espèce abondante sera tout indiquée pour être congelée en blocs qui, par la suite, sont transformés en surimi (imitation de crabes)

I: Una specie abbondante che in futuro potrà essere usata per la produzione di blocchi congelati di surimi da utilizzare per granchio imitato e prodotti simili

BLUENOSE WAREHOU

Scientific name:
Hyperoglyphe antarctica

Synonym: Antarctic butterfish
Family: Centrolophidae — medusafishes
Typical size: 60 to 80 cm, 6 kg

Fishing for bluenose in southern Australian waters has intensified. It is popular domestically and is exported to Japan for sashimi and for cooking. The fishery is increasingly regulated, partly because of large catches of other species incidentally to the targeted bluenose.

FISHING METHODS

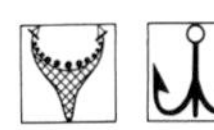

MOST IMPORTANT FISHING NATIONS
New Zealand, Australia

PREPARATION
Sashimi

USED FOR

EATING QUALITIES
Bluenose has firm but moist meat with a mild flavour. Brownish when raw, it whitens when cooked. It can be used in most recipes suitable for white-meated fish. Warehou may grow as large as 20 kg; big fish yield very thick, meaty fillets.

COMMON NAMES
D: Antarktischer Schwarzfisch
DK: Antarktisk sortfisk
E: Pez nariz azul antártico
F: Rouffe à nez bleu
I: Ricciola di fondale austr.
J: Nankyokumedai
NL: Antarctische botervis
P: Liro antárctico
RU: Giperoglif
US: Bluenose

LOCAL NAMES
AU: Blue eye

Nutrition data:
(100 g edible weight)

Water	76.8 g	Total lipid (fat)	2.7 g
Calories	100 kcal		
Protein	19.0 g	Omega-3	1.0 mg

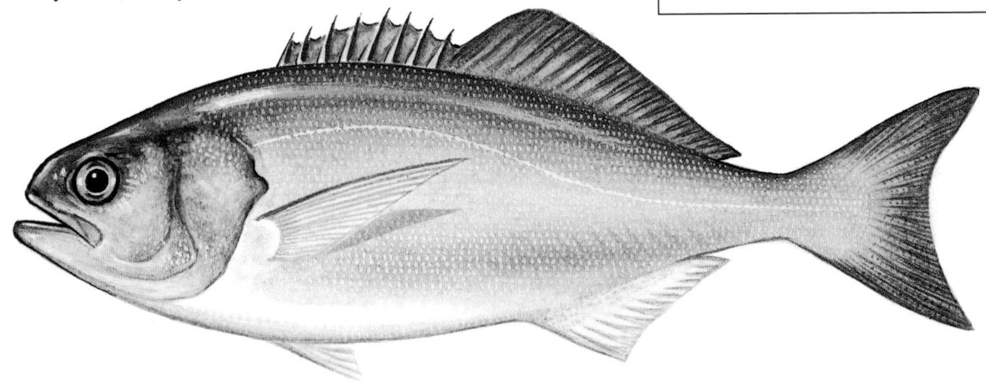

HIGHLIGHTS
There is a small resource of bluenose around New Zealand, yielding about 1,000 to 1,500 tons a year. The species likes rocky sea bed, so is mostly caught on longlines, with a few taken as by-catch in trawl fisheries.

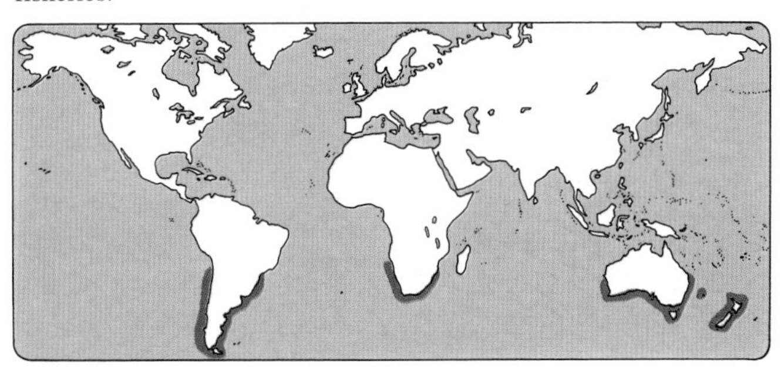

DK: Er populær i Australien, men exporteres også til Japan til 'sashimi'-forarbejdning samt kogning. Fiskes oftest med langline, da den foretrækker klippebund

D: In Australien beliebt, wird aber auch in Japan zu 'Sashimi' verarbeitet und gekocht. Wird meistens mit Langleine gefangen, da er Felsboden bevorzugt

E: Es popular en Australia, pero se exporta también a Japón, donde sirve para 'sashimi' y para cocer. Se captura con sedal porque prefiere los fondos rocosos

F: Très apprécié en Australie et au Japon où il est préparé en 'sashimi' ou en plats cuits. Il préfère les fonds rocheux et est le plus souvent pêché aux cordes

I: Popolare in Australia, ma esportato anche nel Giappone per la trasformazione in sashimi. Preferisce il fondo roccioso ed è normalmente preso con il palamito

BREAM

Scientific name:
Abramis brama

Family: Cyprinidae — carps and minnows
Typical size: 60 cm, 5 kg

The common bream is a popular angling fish in western Europe, where it has been propagated in many lakes to enhance the sporting quality of the water.

Bream shoal by size, so an angler who strikes a large fish (they may grow as large as 9 kg) can expect to find more in the immediate area. Since the largest fish are found in the smallest shoals, it is probable that the fish stay with others of their year throughout their lives.

DESCRIPTION
Frequently called bronze bream because of the dark golden colour often seen on its sides, the common bream varies considerably in appearance according to its age and location. Young fish are silvery while older fish tend to develop the typical dark green backs and bronze sides. Specimens from some lakes are much lighter.

The bream is found through most of northern and central Europe into Russia and the Caspian basin. It has a very deep, compressed body and prefers slow moving, even stagnant water, where it feeds on organisms extracted from the muddy bottom.

COMMON NAMES
D: Blei, Brachsen, Brassen
DK: Brasen
E: Brema común, brema
F: Brème commune, brème
GR: Lestiá
I: Brama, abramide
IS: Leirslabbi
N: Brasme
NL: Brasem
P: Brema
RU: Lesch
S: Braxen
SF: Lahna
TR: Tahta baligi
US: Bream

LOCAL NAMES
PO: Leszcz

HIGHLIGHTS
Although it is not an exceptional fighter, the breams's persistence accounts for its popularity among anglers. In Russia, it is an important food fish.

Bream often live in turgid waters which are low in oxygen and have the capacity to survive out of water for extended periods. This makes them particularly suitable for use in areas which lack ice, refrigeration chains and other amenities of modern distribution systems.

Nutrition data:
(100 g edible weight)

Water	75.0 g
Calories	122 kcal
Protein	18.0 g
Total lipid (fat)	5.5 g
Omega-3	0.4 mg

BREAM *Abramis brama*

FISHING METHODS

 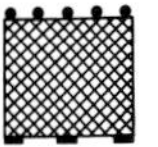 

MOST IMPORTANT FISHING NATIONS
Russia

PREPARATION

USED FOR

EATING QUALITIES
Bream provide firm meat with a reasonable flavour. The species is mainly eaten in areas which are not well supplied with marine fishes. In Germany, it is considered to be similar to carp, and an acceptable substitute for carp in many dishes. The flesh is a little dark and, like many freshwater species, may contain parasites which should be removed before the fish is cooked.

Small bream can be cleaned and trimmed, ready for pan-frying or broiling. Larger fish may be dressed and baked, but are usually filleted before cooking. Bream is generally cooked with the skin on, which helps to hold the meat flakes together.

IMPORTANCE
Commercial catches of 60,000 to 80,000 tons a year are made mostly by Russia, with smaller quantities from other republics of the former Soviet Union, including Khazakstan and Azerbaijan. It is an important source of animal protein in some areas.

In Ireland, the bream was introduced and has flourished, supporting the tourist industry by attracting anglers to remote parts. The species has also been economically successful in eastern England, where it has been stocked heavily in the Norfolk Broads. Young fish are reared and introduced into suitable water. Presumably, bream could be farmed if sufficient demand for the could be meat developed.

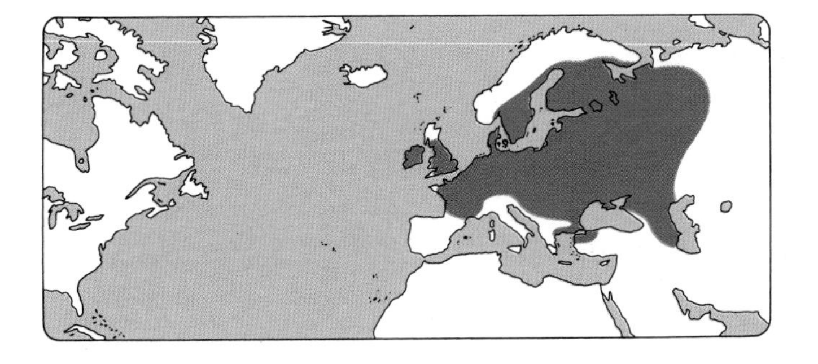

DK: En vigtig spisefisk i Rusland. Evnen til at kunne overleve uden vand i længere tid, gør den specielt velegnet til distribution i områder uden is- og kølefaciliteter

D: Ein wichtiger Speisefisch in Rusland. Wegen seiner Fähigkeit, längere Zeit ohne Wasser auszukommen, ist er besonders gut geeignet zum Vertrieb in Gegenden ohne Gefrier- und Kühlmöglichkeiten

E: Importante pescado comestible en Rusia. Su aguante fuera del agua durante bastante tiempo lo hace especialmente apropiado para distribución donde no hay hielo ni instalaciones frigoríficas

F: Un poisson comestible important en Russie. Comme il est capable de survivre assez longtemps hors l'eau, il se prête particulièrement bien à la distribution dans des régions dépourvues d'installations de glace et de réfrigération

I: Un importante pesce commestibile in Russia. La sua capacità di sopravvivere fuori dall'acqua per lunghi periodi lo rende particolarmente adatto alla distribuzione in zone prive di ghiaccio e impianti frigoriferi

BRILL

Scientific name: *Scophthalmus rhombus*

Synonyms: Kite, brett, brit, pearl
Family: Bothidae — left-eye flounders
Typical size: up to 60 cm

Unlike European turbot, which is covered in small tubercles, brill has regular scales, enabling quick distinction to be made between these similar species.

FISHING METHODS

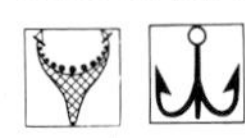

MOST IMPORTANT FISHING NATIONS
Not available

PREPARATION

USED FOR

EATING QUALITIES
Brill is similar to the European turbot but not considered as good. The meat is white, firm and tasty. It may be substituted for turbot in some restaurants, especially if well sauced. Nevertheless, brill is a fine fish in its own right, better than many other flatfish. Its reputation has suffered by comparing it to its close relative, the turbot.

COMMON NAMES
D: Glattbutt, Kleist, Tarbutt
DK: Slethvarre
E: Rémol, rapante, corujo
F: Barbue, turbot lisse
GR: Pissí, rómvos
I: Rombo liscio, soaso
IS: Slétthverfa
N: Slettvar
NL: Griet
P: Rodovalho
RU: Gladkij bril
S: Slätvar
SF: Silokampela
TR: Civisiz kalkan
US: Flounder, brill

Nutrition data:
(100 g edible weight)

Water	78.7 g	Total lipid	
Calories	95 kcal	(fat)	2.7 g
Protein	17.0 g	Omega-3	0.5 mg

HIGHLIGHTS
Brill is found only on the eastern side of the North Atlantic, especially in the Mediterranean and nearby Atlantic coasts. An inshore species, it is targeted by anglers for its meat and by commercial fishermen for local sale.

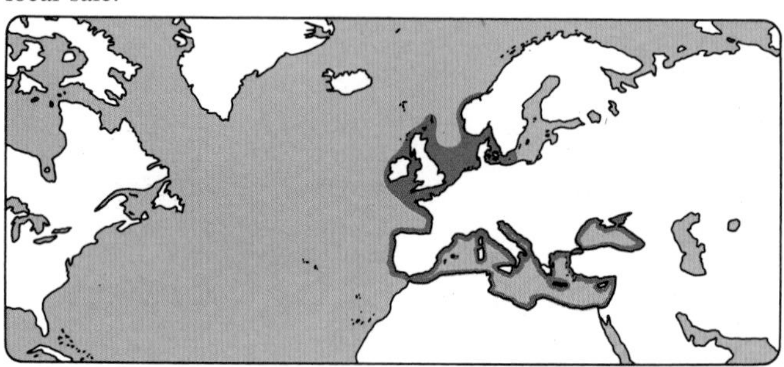

DK: I modsætning til pighvarre, som er dækket af små benknuder, har denne rigtige skæl, og forveksling er derfor umulig

D: Im Gegensatz zum Steinbutt, der von kleinen Knochenhöckern bedeckt ist, hat dieser Schuppen, und Verwechslung ist deshalb ausgeschlossen

E: A diferencia del rodaballo, que tiene pequeñas protuberancias en la piel, tiene escamas, lo que evita confundirlo con otras especies

F: A la différence du turbot, qui est couvert de petites tubercules d'os, la barbue est couverte d'écailles, et il n'est ainsi pas possible de les confondre

I: Contrariamente al rombo chiodato che è coperto di piccoli nodi ossei, questo pesce ha squame vere e proprie ed è quindi impossibile confonderli

Scientific name:
Scomberomorus semifasciatus

BROADBARRED KING MACKEREL

Synonym: Korean broadbarred king mackerel
Family: Scombridae — tunas and mackerels
Typical size: 50 cm, 2 kg

This species is restricted to northern waters around Australia and southern parts of Papua New Guinea. It can grow to a very large size, sometimes exceeding 120 cm and 10 kg. It is commonly called grey mackerel in Australia, as the colours fade quickly after death.

COMMON NAMES
D: Tigermakrele
DK: Bredbåndet kongemakrel
E: Carite tigre
F: Thazard tigre
GR: Skoumbrí tígris
I: Maccarello reale austral.
NL: Tijgermakreel
P: Serra de faixas
RU: Korejskaya korolev. makrel
US: King mackerel

FISHING METHODS

 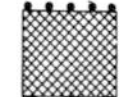 

MOST IMPORTANT FISHING NATIONS
Australia

PREPARATION

USED FOR

 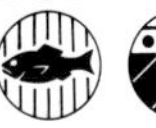

EATING QUALITIES
Spanish mackerels are among the best eating fish. The flavour is delicate and the finely-textured meat has a small flake. The brownish colour of the raw meat turns white when cooked. Excellent for fish and chips and smoking; very versatile in most fish recipes.

Nutrition data:
(100 g edible weight)

Water	75.5 g	Total lipid	
Calories	107 kcal	(fat)	3.0 g
Protein	20.0 g	Omega-3	0.6 mg

HIGHLIGHTS
Locally important in Queensland and the Northern Territories of Australia, the broadbarred king mackerel is mostly sold fresh, either as steaks or fillets with the skin on. There is some recreational fishing, especially for larger fish and for small specimens for bait.

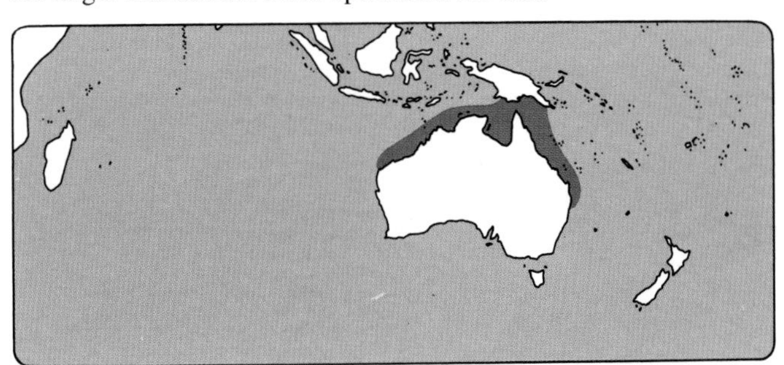

DK: Populær art fisket af erhvervs- og lystfiskere. Sælges primært frisk (skiver/filetter) på det australske marked

D: Beliebte Fischart bei Industrie- und Sportfischern. Wird hauptsächlich frisch (Scheiben/Filets) auf dem australischen Markt verkauft

E: Especie popular entre los pescadores profesionales y deportivos. En primer lugar, se vende fresco, en rodajas o filetes, en el mercado australiano

F: Espèces populaire, pêchée par les pêcheurs professionnels et sportifs. Est surtout vendu frais (tranches/filets) sur le marché australien

I: Specie ricercata dai pescatori professionali e sportivi. Venduto soprattutto fresco, in fette e filetti, sul mercato australiano

BROOK TROUT

Scientific name: *Salvelinus fontinalis*

Synonyms: Speckled char, brook char, speckled trout, squaretail
Family: Salmonidae — salmonids
Typical size: 15 to 30 cm, 0.5 to 5 kg

Native to eastern North America from the far north to the mountain streams of the Carolinas, brook trout have been successfully transplanted to Europe, Australia, New Zealand, South America and other parts. Freshwater and anadromous forms are both sought eagerly by anglers. The species is not normally sold commercially.

FISHING METHODS

MOST IMPORTANT FISHING NATIONS
France, Canada

PREPARATION

USED FOR

EATING QUALITIES
Brook trout has excellent flavour and moist texture, with a small flake. The meat colour varies from white to bright orange, according to the age and diet of the fish.

COMMON NAMES
D: Bachsaibling
DK: Kildeørred
E: Trucha de fontana
F: Saumon de fontaine
GR: Salvelínos, limnopéstrofa
I: Salmerino di fontana
IS: Lindableikja
J: Kawamasu
N: Bekkerøye
NL: Bronforel
P: Truta das fontes
RU: Paliya, amerikanski goletz
S: Amerikansk bäckröding
SF: Puronieriä
TR: Alabalik türü
US: Brook trout

LOCAL NAMES
PO: Pstrag zródlany

Nutrition data:
(100 g edible weight)

Water	76.0 g	Total lipid	
Calories	102 kcal	(fat)	2.5 g
Protein	20.0 g	Omega-3	0.3 mg

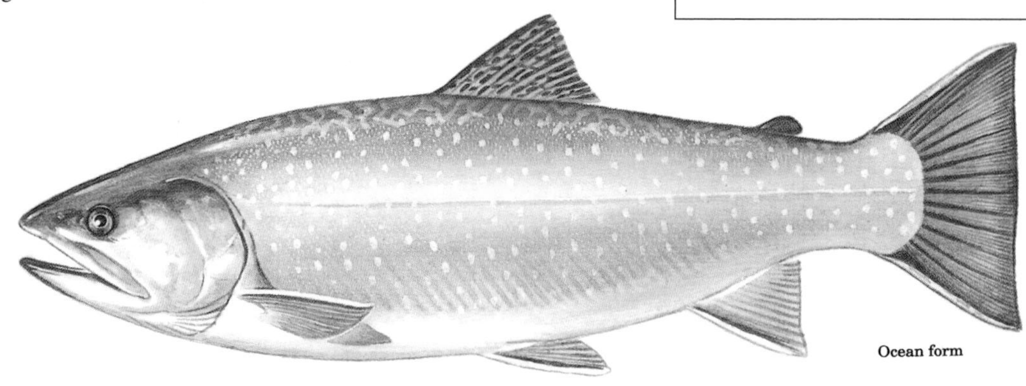

Ocean form

HIGHLIGHTS
Brook trout is greatly prized as a sport fish, especially by fly fishermen. It is domesticated, but is not raised commercially for meat because of its slow growth. Farmed fish are stocked in streams to attract anglers and the money they bring to the surrounding area.

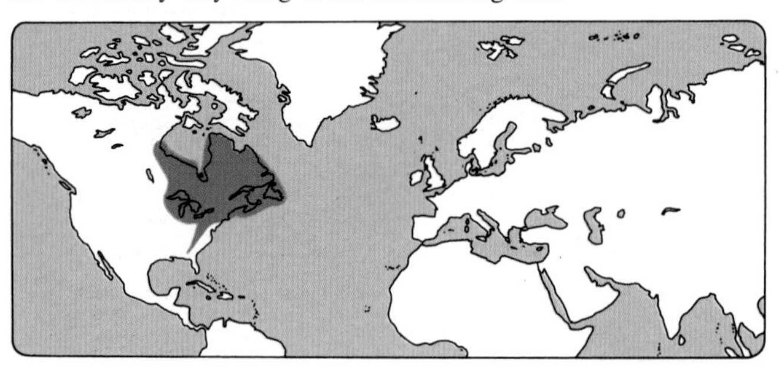

DK: En meget eftertragtet ørred, der primært fanges med flue. Denne sportsfisk opdrættes ikke kommercielt, da den er langsomt voksende

D: Ein sehr begehrter Saibling, der hauptsächlich auf Fliege gefangen wird. Dieser Sportfisch wird nicht zu kommerziellen Zwecken aufgezogen, da er langsam wächst

E: Trucha muy popular que normalmente es capturada con mosca. Este pescado deportivo no sirve para la cría comercial porque crece lentamente

F: Truite très recherchée, principalement pêchée à la mouche. De croissance très lente, ce poisson sportif ne fait pas l'objet d'un élevage commercial

I: Una trota molto richiesta, pescata prevalentemente con la mosca. Questo pesce sportivo non viene allevato commercialmente, in quanto cresce lentamente

BROWN TROUT

Scientific name:
Salmo trutta

Synonym: River trout
Family: Salmonidae — salmonids
Typical size: 50 cm, 6 kg

Brown trout are freshwater fish. Anadromous races are called sea trout and may grow as large as 10 kg; a specimen of 21 kg was recorded in Finland. Some freshwater races grow to enormous sizes: one from the Caspian basin was reported at over 50 kg. Originally native to Europe and Asia from Iceland to Afghanistan, the species has been successfully introduced to many temperate regions.

COMMON NAMES
D: Bachforelle
DK: Bækørred
E: Trucha común
F: Truite commune
GR: Péstrofa
I: Trota, trota fario
IS: Lækjarsilungur
J: Brown trout
N: Ørret
NL: Beekforel, rivierforel
P: Truta comum
RU: Rutschjevnaja forel
S: Bäcköring
SF: Purotaimen
TR: Alabalik
US: Brown trout

LOCAL NAMES
PO: Pstrag potokowy

FISHING METHODS

MOST IMPORTANT FISHING NATIONS
Denmark, Portugal, France

PREPARATION

USED FOR

EATING QUALITIES
Brown trout has delicately flavoured, sweet meat with a small flake. Colour varies from red, especially in sea trout, to yellowish-white.

Nutrition data:
(100 g edible weight)

Water	74.1 g	Total lipid	
Calories	112 kcal	(fat)	3.8 g
Protein	19.4 g	Omega-3	0.8 mg

HIGHLIGHTS
Brown trout is the most important European freshwater gamefish. It is also popular in many areas where it has been introduced, including North America. There is no commercial fishing for freshwater browns, but the limited populations of sea trout are targeted.

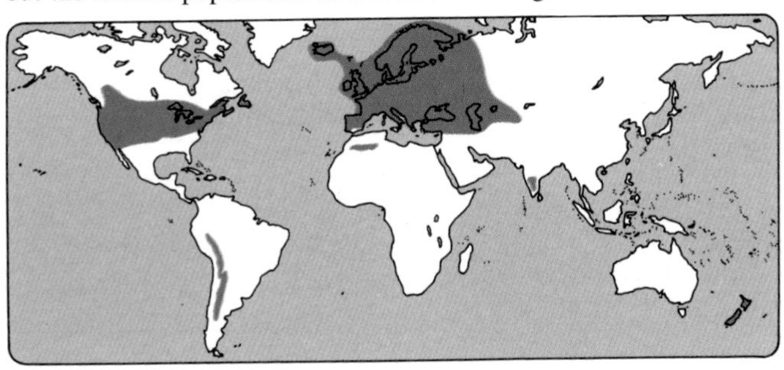

DK: Den vigtigste ferskvands-sportsfisk i Europa. I modsætning til den anadrome havørred, er der intet kommercielt fiskeri efter ferskvandsarten

D: Der wichtigste Süsswasser-Sportfisch in Europa. Im Gegensatz zur anadromen Meerforelle ist die Süsswasserart ohne wirtschaftliche Bedeutung

E: El pescado deportivo de agua dulce más importante en Europa. Contrariamente a la trucha anádroma de mar, esta especie no tiene importancia para la pesca comercial

F: Le poisson sportif d'eau douce le plus important en Europe. Contrairement à la truite de mer anadrome, l'espèce d'eau douce n'est pas pêchée commercialement

I: Il pesce d'acqua dolce più importante per la pesca sportiva. Contrariamente alla trota di mare anadroma, non ha importanza per la pesca professionale

BUTTERFISH

Scientific name:
Peprilus triacanthus

Synonyms: Dollar fish, American butterfish, shiner
Family: Stromateidae — butterfishes
Typical size: up to 24 cm, 500 g

Butterfish is frozen for export sale, and must be carefully graded, 80 to 105 g, 105 to 150 g, 150 to 200 g and 200 g up (which are called extra large). The fish is packed whole, including the guts. Inferior packs, sometimes as large as 50 kg, are prepared for domestic use, most of which is for animal feed.

DESCRIPTION

Butterfish are small, silvery, laterally compressed fish which shoal in large numbers. It is one of the few fish which feed on jellyfish and sea gooseberries, organisms which provide only small amounts of nutrition.

Although it ranges from Newfoundland to Florida, the commercial harvest is largely made off the mid-Atlantic states, from New York to Virginia.

At one time, butterfish were heavily infested with parasites, but this problem has been reduced with increasing fishing activity, which reduces the fish population's pressure on its food supply, so allowing the remaining fish to be better fed, healthier and more resistant.

COMMON NAMES

D: Dollarfisch, Butterfisch
DK: Dollarfisk
E: Pez mantequilla americano
F: Stromaté à fossettes
GR: Psevdolítsa tou Atlantikoú
I: Fieto americano
IS: Smjörfiskur, lummari
N: Dollarfisk
NL: Amerikaanse botervis
P: Peixe manteiga americano
RU: Maslyannaya ryba baterfish
S: Dollarfisk
SF: Amerikanvoikala
US: Butterfish

HIGHLIGHTS

Caught in large numbers during its summer migration to the shores of the east coast of the United States, the butterfish supports an export trade but is not highly regarded in areas where it is caught. Recreational fishermen often catch them for sport, but most are discarded; Americans are particularly offended by bony fish and the butterfish is one of the boniest.

The resource appears to be substantial and catches could be increased if markets were developed. The species might be better canned.

Nutrition data:
(100 g edible weight)

Water	74.1 g
Calories	141 kcal
Protein	17.3 g
Total lipid (fat)	8.0 g
Omega-3	1.6 mg

BUTTERFISH *Peprilus triacanthus*

FISHING METHODS

MOST IMPORTANT FISHING NATIONS
United States

PREPARATION

USED FOR

EATING QUALITIES
Butterfish is oily and moist, with an excellent flavour. The flesh is greyish, but turns almost white when it is cooked. The biggest problem with butterfish is the bone structure: there are numerous small bones which make the fish extremely awkward to eat.

Butterfish is excellent hot smoked; the process softens the bones and makes them sufficiently consumable that the product becomes acceptable to a wider market. However, smoked butterfish, though a delicacy, is not regularly distributed in the United States.

IMPORTANCE
Butterfish catches vary widely between 2,000 and 10,000 tons a year. It is probable that recreational catches account for a considerable quantity of unrecorded landings in addition. Most of the catch is frozen and exported to Japan. Catches depend partly on the seasonal availability of the species and partly on the fluctuating price patterns of the Japanese market, which is very specific about the grading, handling and packing of the product it buys.

Peprilus burti is a virtually identical species that ranges into the Gulf of Mexico. In practice, the two species are not distinguished. *Peprilus simillimus* is a very similar species in the north Pacific, mainly caught off California. This butterfish is available throughout the year and is sometimes sold as Pacific pompano.

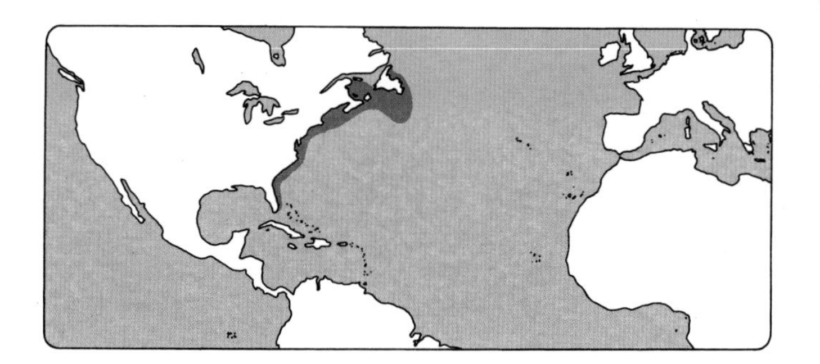

DK: Fanges i store mængder under sommervandringen til USA's østkyst. Størstedelen af fangsten exporteres, da den ikke er særlig højt estimeret som spisefisk i de områder, hvor den fanges

D: Grosser Fang während der Sommerwanderung an die Ostküste der USA. Der grösste Teil wird exportiert, da er in den Fanggebieten als Speisefisch nicht sonderlich geschätzt wird

E: Se captura en grandes cantidades cuando migra en verano a la costa este de EE.UU. Se exporta la mayor parte de las capturas, porque no es estimado como pescado comestible en las zonas donde es capturado

F: Pêchée en grandes quantitées pendant la migration d'été vers la côte Est des Etats-Unis. N'étant pas très appréciée comme poisson comestible dans les régions où elle est pêchée, la majorité de la pêche est exportée

I: Catturato in grandi quantità durante l'estate quando si sposta verso la costa orientale degli Stati Uniti. Viene quasi esclusivamente esportato, non essendo molto stimato come pesce commestibile nelle zone in cui vive

CALIFORNIA PILCHARD

Scientific name:
Sardinops caeruleus

Synonym: Monterrey Pacific sardine
Family: Clupeidae — herrings, sardines
Typical size: 25 cm, up to 36 cm
See also Chilean pilchard

Often called the Pacific sardine, the species is considered by some experts to be the same as the Chilean pilchard, *S. sagax*. It is returning to California, with a limited fishery under 20,000 tons, and migrating northwards, but Mexican catches have declined.

FISHING METHODS

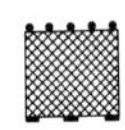 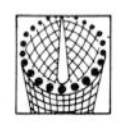

MOST IMPORTANT FISHING NATIONS

Mexico, United States

PREPARATION

 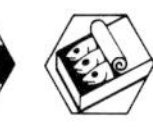 

USED FOR

EATING QUALITIES

These small pilchards are excellent canned. They can also be used like European sardines. If properly handled and carefully cooked, the bones lift easily from the meat, which is sweet and delicate, with a small flake and grey to whitish colour.

COMMON NAMES

D: Kalifornische Sardine
DK: Californisk sardin
E: Sardina de California
F: Sardinops de Californie
GR: Sardéla tis Kalifórnias
I: Sardina di California
IS: Sardína
J: Iwashi
N: Kalifornisk sardin
NL: Californische pelser
P: Sardinopa da Califórnia
RU: Kalifornijskaya sardina
S: Kalifornisk sardin
SF: Kaliforniansardiini
US: Pacific sardine

Nutrition data:
(100 g edible weight)

Water	68.3 g	Total lipid	
Calories	174 kcal	(fat)	12.0 g
Protein	16.4 g	Omega-3	1.6 mg

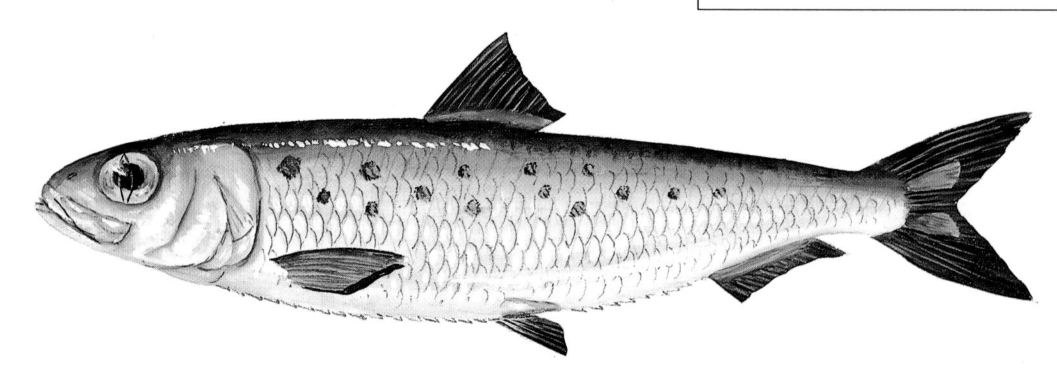

HIGHLIGHTS

The huge resource of the 1930s crashed as the ocean water cooled slightly. Scientists have identified long-term cycles of about 60 years of abundance and shortage. Sardines were once canned or reduced along the entire Pacific coast of N. America.

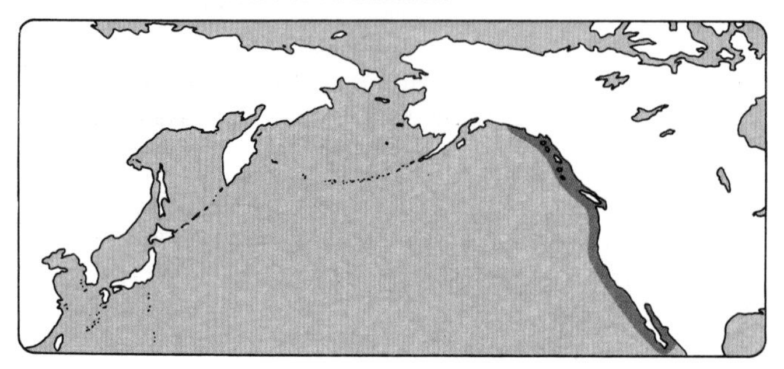

DK: Klimatiske ændringer reducerede katastrofalt 30'ernes enorme forekomster. Er igen stigende som følge af den naturlige cyklus

D: Klimaänderungen haben den Riesenbestand der 30er Jahre katastrophal dezimiert. Ist infolge des natürlichen Zyklus wieder im Anstieg begriffen

E: Cambios climáticos han reducido gravemente las enormes existencias de los años treinta. Vuelven a aumentar gracias al ciclo natural

F: Des changements climatiques ont réduit les énormes ressources des années 30 de façon dramatique. Est de nouveau en hausse grâce au cycle naturel

I: I cambiamenti climatici hanno ridotto drasticamente le risorse enormi degli anni '30. Sta aumentando nuovamente grazie al ciclo naturale favorevole

CAPE HAKE

Scientific name:
Merluccius capensis

Synonyms: Stockfish, shallow-water hake
Family: Merlucciidae — merluccid hakes
Typical size: 40 to 60 cm, up to 120 cm

The combined resource of Cape hake and the deepwater Cape hake (*M. paradoxus*) produce catches of 400,000 to 500,000 tons yearly. The potential sustainable catch is estimated to be 620,000 tons, although in the 1970s landings of over 1 million tons were recorded.

FISHING METHODS

MOST IMPORTANT FISHING NATIONS

Spain, South Africa, Namibia, Zaire

PREPARATION

USED FOR

EATING QUALITIES

Cape hake is considered one of the best of the small hakes or whitings, with quite firm, sweet, white meat and a medium flake. It is a good raw material for hot smoking in headless dressed form. Blocks are widely available and versatile.

COMMON NAMES

D: Kap-Hecht
DK: Sydafrikansk kulmule
E: Merluza del Cabo
F: Merlu blanc du Cap
GR: Bakaliáros tis N. Afrikís
I: Nasello del Capo
IS: Höfdalysingur
J: Merulûsa
NL: Kaapse heek
P: Pescada da Africa do Sul
RU: Kapskaya meduza
SF: Kapinkummeliturska
US: South African whiting

LOCAL NAMES

AN: Marmota
SO: Shallow-water hake

Nutrition data:
(100 g edible weight)

Water	80.2 g	Total lipid	
Calories	84 kcal	(fat)	1.2 g
Protein	18.3 g	Omega-3	0.3 mg

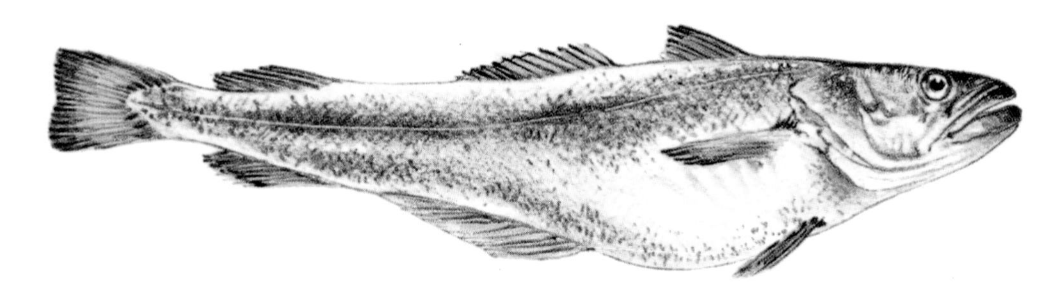

HIGHLIGHTS

Cape hake is a fine quality, white-meated fish, versatile in use. Spanish factory trawlers are major harvesters, but coastal states in southern Africa are increasingly entering the fishery for domestic consumption as well as for export.

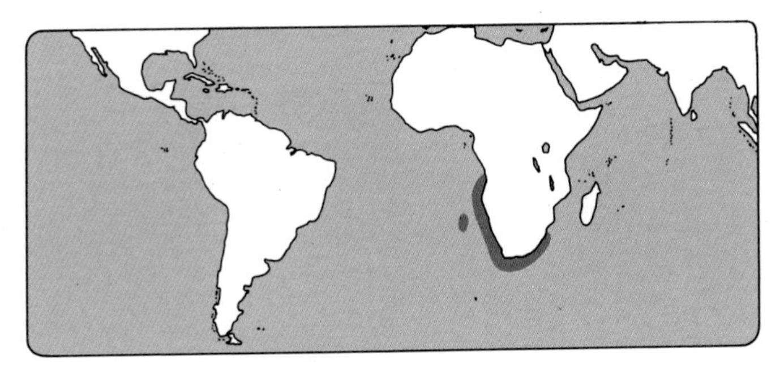

DK: Anses for en af familiens fineste med ret fast, sødt, lyst kød med middelstore flager. Nedfryses normalt i blokke

D: Gilt als einer der delikatesten der Familie mit recht festem, weissem Fleisch in mittelgrossen Lagen. Wird meistens in Blöcken gefroren

E: Uno de los más finos de la familia de carne bastante firme, dulce y blanca, con lonchas transversales de tamaño medio. Se congela normalmente en bloques

F: Compte parmi les plus fins de la famille. Chair blanche et ferme en lamelles de taille moyenne, de goût assez doux. Est généralement congelé en blocs

I: Considerato uno dei migliori della famiglia con carni sode, dolci e chiare con fiocchi medi. Normalmente è congelato in blocchi

CAPELIN

Scientific name:
Mallotus villosus

Family: Osmeridae — smelts
Typical size: 13 to 20 cm, 40 g

Capelin provides catches of over 1 mn tons a year, although levels have declined substantially from earlier peaks. The species is a vital food for cod and other fish. Decline in capelin stocks may be reducing resources of more valuable fishes throughout northern seas.

FISHING METHODS

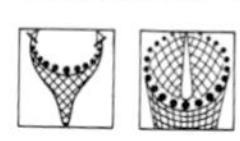

MOST IMPORTANT FISHING NATIONS
Norway, Iceland

PREPARATION
Roe  

USED FOR

EATING QUALITIES
Capelin have been a staple food for inhabitants of polar regions, but are seldom eaten further south, although canned capelin can be palatable. Female fish, separated from males by high-speed machines, are valued for their roe. Males are used for animal feed and fishmeal.

COMMON NAMES
D: Lodde, Polarstint, Kapelan
DK: Lodde
E: Capelán
F: Capelan
GR: Kapelános
I: Capelin
IS: Lodna
J: Karafuto-shishamo
N: Lodde
NL: Lodde
P: Capelim
RU: Moiva
S: Lodda
SF: Villakuore
US: Capelin

Nutrition data:
(100 g edible weight)

Water	65.6 g	Total lipid	
Calories	212 kcal	(fat)	17.5 g
Protein	13.6 g	Omega-3	3.5 mg

HIGHLIGHTS
The species is found in cold northern waters of the Atlantic and Pacific. Although palatable products can be made from oily, pre-spawning fish (which may have fat content as high as 23 percent), capelin is eaten mainly dried, in Japan.

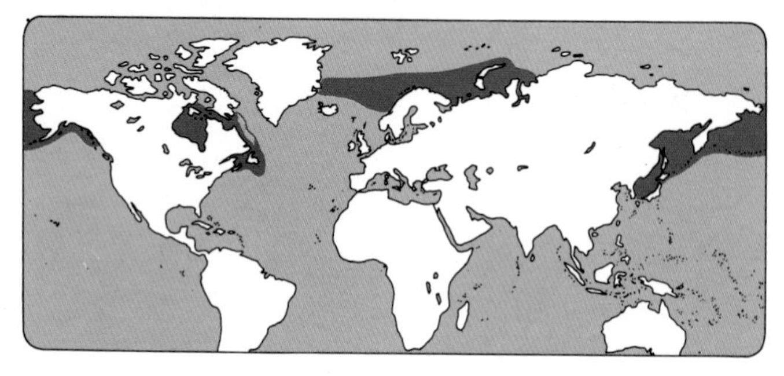

DK: Automatik sørger for hurtig sortering af de to køn. Hunnerne giver en fortrinlig rogn, hvorimod hannerne mest anvendes til dyrefoder og fiskemel

D: Die beiden Geschlechter werden maschinell voneinander getrennt. Die Weibchen liefern vorzüglichen Rogen; die Männchen werden zu Tierfutter und Mehl verarbeitet

E: Las hembras, separadas a máquina de los machos, son apreciadas por sus exquisitas huevas. Los machos sirven de alimento para animales y harina de pescado

F: Les femelles, séparées mécaniquement des mâles, sont très appréciées pour leurs oeufs, les mâles n'étant généralement transformés en pâtée et en farine

I: Le femmine, separate dalle macchine automatiche secondo il sesso, forniscono delle ottime uova, mentre i maschi vengono usati per foraggio e farina di pesce

CATFISH/WOLFFISH

Scientific name:
Anarhichas lupus

Synonym: Rockfish
Family: Anarhichadidae — wolffishes
Typical size: 150 cm, 18 kg

The Atlantic ocean catfish or wolffish is a solitary species, taken mainly as a by-catch of fishing directed at cod, haddock and other bottom-dwelling fishes. Wolffish eat shellfish, including clams and lobsters, which perhaps accounts for their fine flavour and certainly explains the massive teeth, used for crushing shells.

FISHING METHODS

 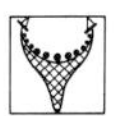 

MOST IMPORTANT FISHING NATIONS
Iceland, Russia, Scotland

PREPARATION

USED FOR

 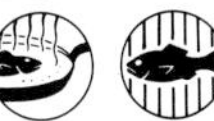 

EATING QUALITIES
Wolffish have white, firm, sweetly flavoured meat which can be substituted for the finest Dover sole. It is versatile, usable in many recipes and methods of preparation.

COMMON NAMES
D: Gestreifter Seewolf
DK: Stribet havkat
E: Perro del Norte
F: Loup de l'Atlantique
GR: Lykópsaro
I: Lupo di mare, bavosa lupa
IS: Steinbítur
J: Taiseiyo-namazu
N: Gråsteinbit
NL: Zeewolf
P: Peixe lobo riscado, gata
RU: Zubatka
S: Havskatt
SF: Merikissa
US: Ocean catfish

Nutrition data:
(100 g edible weight)

Water	80.1 g	Total lipid	
Calories	96 kcal	(fat)	2.8 g
Protein	17.6 g	Omega-3	0.7 mg

HIGHLIGHTS
Boneless fillets of ocean catfish are among the finest food from the North Atlantic. Substitutes are sometimes offered: the presence of pinbones, which are absent from properly prepared fillets of wolffish, are a frequent indicator that an inferior species has been used.

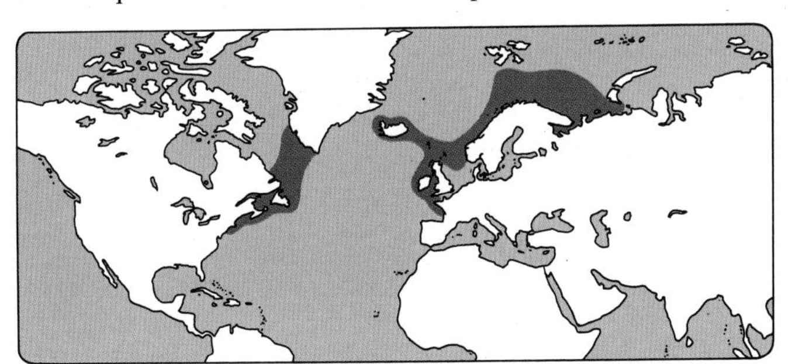

DK: Fiskens benløse filetter regnes blandt Nordatlantens fineste spise. Fisken tages normalt som bi-fangst

D: Die grätenlosen Filets dieses Fisches werden zu den auserlesensten Gerichten gezählt, die der Nordatlantik zu bieten hat. Der Fisch ist normalerweise Beifang

E: Sus filetes sin espinas son considerados uno de los manjares más exquisitos del Atlántico Norte. Normalmente, se coge como captura secundaria

F: Les filets sans arêtes de ce poisson de l'Atlantique Nord qui est principalement pêché en prise secondaire, comptent parmi les mets les plus fins

I: I filetti senza spine di questo pesce sono considerati uno dei cibi più pregiati dell'Atlantico del Nord. Normalmente è preso come cattura secondaria

CHANNEL CATFISH

Scientific name: *Ictalurus punctatus*

Family: Ictaluridae — bullhead catfishes
Typical size: 35 cm, 2 kg

Channel catfish supports the largest aquaculture industry in the United States and has largely displaced the cheaper but less consistent Amazonian catfish from Brazil (*Brachyplatysoma vaillanti*) in the U.S. market.

A wide range of product forms is available throughout the year, chilled or frozen. Many farmers are organised into processing and marketing groups which have developed strong brand-name identities. Catfish marketers have learned from the chicken industry, successfully applying modern quality control and marketing techniques.

DESCRIPTION
Channel catfish require clean, well oxygenated water. Wild fish are found mostly in rivers and streams; ponds used by farmers have to be aerated and carefully maintained.

The species has been widely introduced as an angling fish throughout North America. Small numbers have also been taken to Europe, often for aquariums; it is doubtful if buyers realized that their pets might reach 1.25 m in length and a weight of 25 kg.

COMMON NAMES
D: Getüpfelter Gabelwels
DK: Plettet dværgmalle
E: Coto punteado
F: Poisson-chat tacheté
GR: Gatópsaro
I: Pesce gatto puntado
IS: Kanalgrani, skurdgrani
N: Plettet dvergmalle
NL: Kanaalmeerval
P: Peixe gato pontuado
RU: Kanavnaya zubatka
SF: Pilkkupiikkimonni
US: Channel catfish

HIGHLIGHTS
Channel catfish is raised in ponds and delivered in special tankers alive to the processing plant. It is killed and processed quickly: it takes only ten minutes to convert live fish into processed products entering the freezer. As a result of these techniques, consumers rely on the quality and consistency of farmed catfish products.

In the wild, channel catfish ranges from the St. Lawrence to northern Mexico and is a popular angling species. Commercial fisheries on the Great Lakes have now given way to competition from farmed fish.

Nutrition data:
(100 g edible weight)

Water	76.4 g
Calories	112 kcal
Protein	18.2 g
Total lipid (fat)	4.3 g
Omega-3	0.4 mg

FISHING METHODS

MOST IMPORTANT FISHING NATIONS

United States

PREPARATION

USED FOR

EATING QUALITIES

Channel catfish has firm, white meat with a delicate flavour and a small flake. Dressed fish, fillets and steaks are most often coated and fried, but the fish can be used in a wide range of recipes, including some suitable for good quality marine fish such as sole. The thick skin and lateral fat line of red meat are always removed by processors.

Wild catfish sometimes taste sour, a condition caused by its feed. Similar problems with farmed catfish are avoided, as growers sample the ponds before harvesting them. Tainted fish are not put into the market.

IMPORTANCE

Farmers in the United States, mostly in Mississippi, Alabama and Arkansas, now produce over 200,000 tons of catfish yearly; they have the capacity to increase production rapidly. Catfish is one of the few fish to be successfully farmed for sale at a low price: most aquaculture is based on premium species such as shrimp and salmon. Catfish is now being farmed on a small scale in Cuba, Mexico and other countries.

American catfish growers demonstrated the potential of successful marketing programmes for fish. Markets were developed quickly as production rose in the 1980s. Catfish, offered in many product forms from simple skinned and dressed fish to highly sophisticated consumer ready products, is now a staple food as well as an example, to the whole seafood industry, of modern marketing techniques.

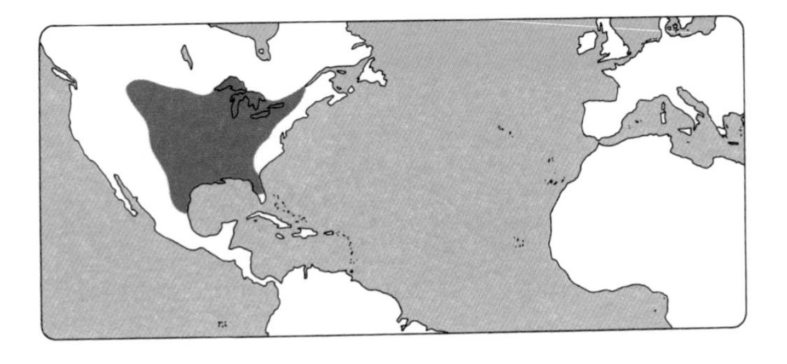

DK: Fisken opdrættes i dambrug og transporteres levende i specielle tanke til fabrikken, hvor den slagtes og omgående forarbejdes: det tager kun 10 min. at omdanne den levende fisk til et forarbejdet produkt

D: Der Fisch wird in Teichen gezüchtet und in Spezialbehältern lebend zur Fabrik transportiert, wo er sofort geschlachtet und verarbeitet wird; dieser Prozess dauert nur 10 Minuten

E: El pescado se cría en piscifactorías y se transporta vivo en tanques especiales a la fábrica, donde es sacrificado y preparado inmediatamente: la transformación del pescado vivo en productos elaborados se realiza en sólo 10 min.

F: Le poisson est cultivé en viviers et ensuite transporté à l'état vivant dans des réservoirs spéciaux directement à l'usine, où il est abattu et transformé immédiatement. En 10 minutes, le poisson vivant est transformé en produit fini

I: Viene allevato in vivai e trasportato vivo in apposite cisterne alla fabbrica, dove viene sottoposto immediatamente alla lavorazione. Ci vogliono solo 10 minuti per trasformare il pesce vivo in prodotto finito

CHERRY SALMON

Synonyms: Masu salmon, salmon trout, Japanese salmon
Family: Salmonidae — salmonids
Typical size: 45 cm, 2 kg

Scientific name:
Oncorhynchus masou

COMMON NAMES

D: Masu-Lachs, Japan-Lachs
DK: Japanlaks, masulaks
E: Salmón cherry
F: Saumon japonais
GR: Solomós
I: Salmone giapponese
J: Sakuramasu, yamame
N: Japansk laks
NL: Japanse zalm, masouzalm
P: Salmao japonês
RU: Sima, mazu
S: Japansk lax
SF: Masulohi
US: Cherry salmon

Masou, cherry or Japanese salmon is found only in limited areas of northern Japan. A landlocked form is found in southern Japan. It is very similar to coho, though generally smaller. Catches average about 3,000 tons yearly, of which about 25 percent is from fresh water. Cherry salmon is now farmed in Chile and supplies are expected to increase in the future as a result.

FISHING METHODS

MOST IMPORTANT FISHING NATIONS

Japan, Chile

PREPARATION

USED FOR

EATING QUALITIES

Masou has pink to reddish meat, a little paler than coho, but very similar in flavour and texture, not unlike sea-grown rainbow trout.

Nutrition data:
(100 g edible weight)

Water	66.7 g	Total lipid	
Calories	195 kcal	(fat)	13.6 g
Protein	18.2 g	Omega-3	2.7 mg

Ocean form

HIGHLIGHTS

Cherry salmon is the only Pacific salmon not found on the eastern side of the ocean. Valued in its native Japan, it is quite rare, not often seen in markets. The fish occasionally reaches 10 kg, but most anadromous and freshwater individuals are much smaller.

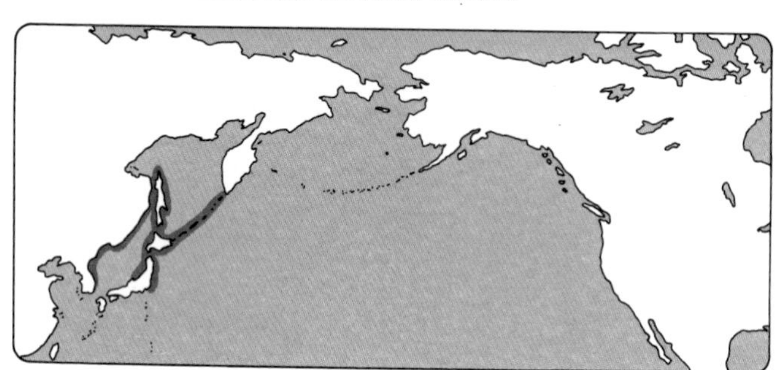

DK: Minder om sølvlaks i smag og konsistens. Findes kun i et begrænset omfang i det nordlige Japan. Opdrættes på forsøgs-basis i Chile

D: Erinnert im Geschmack und von der Konsistenz her an Silberlachs. Kommt in beschränktem Umfang in Nordjapan vor. Aufzuchtversuche in Chile

E: Similar en sabor y consistencia al salmón plateado. Sólo se encuentra en algunas zonas del norte de Japón. Se cría experimentalmente en Chile

F: Ressemble en goût et en consistance au saumon argenté. Existe en nombre limité au Japon du Nord. Au Chili, on fait des élevages à titre d'essai

I: Simile al salmone argentato come sapore e consistenza. La sua diffusione è limitata al Giappone del Nord. Viene allevato su base sperimentale in Cile

Scientific name:
Sardinops sagax

CHILEAN PILCHARD

Synonyms: Chilean Pacific sardine, South American pilchard
Family: Clupeidae — herrings, sardines
Typical size: 20 cm
See also California pilchard

Landings of this important species fluctuate between 2 and 5 million tons. With the anchoveta, it is the basis of the fishmeal industry in Peru and Chile. Attempts over many years to utilize more of the harvest for human consumption are slowly having an effect.

COMMON NAMES

D: Chilenische Sardine
DK: Chilensk sardin
E: Sardina chilena
F: Sardine chilienne
GR: Sardéla tou Peroú
I: Sardina del Cile
IS: Kyrrahafssardína
J: Iwashi
NL: Chileense pelser
P: Sardinopa chilena
RU: Chilijskaya sardina
S: Chilesardin
SF: Perunsardiini

LOCAL NAMES

CL: Sardina española
EC: Sardina
PE: Sardina

FISHING METHODS

MOST IMPORTANT FISHING NATIONS

Peru, Chile

PREPARATION

Fishmeal  

USED FOR

EATING QUALITIES

Chilean pilchards make a palatable and inexpensive canned fish which goes well in tomato and other strongly flavoured sauces. The small bones are softened and made edible by the canning process, adding calcium to the already high nutritional value of the fish.

Nutrition data:
(100 g edible weight)

Water	69.7 g	Total lipid	
Calories	154 kcal	(fat)	8.2 g
Protein	20.0 g	Omega-3	1.6 mg

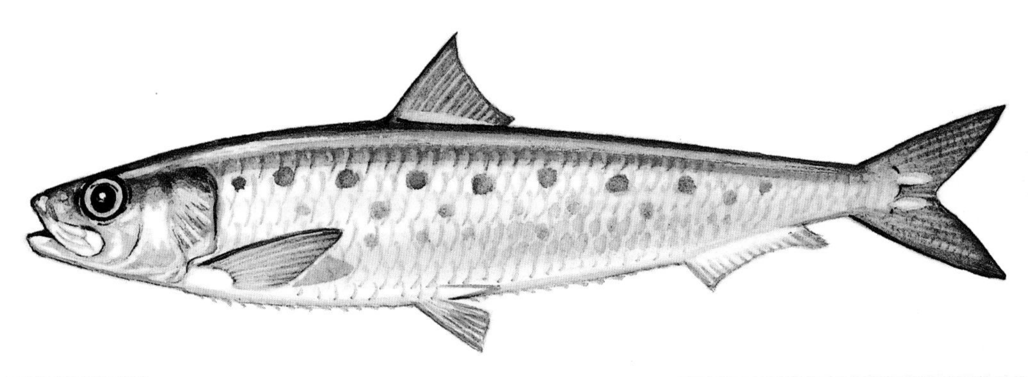

HIGHLIGHTS

Although used primarily for fishmeal, the pilchard is excellent canned and it can also be frozen. However, the huge quantities caught at one time and the industrial methods used in handling do not produce very high quality material for purposes other than fishmeal and canning.

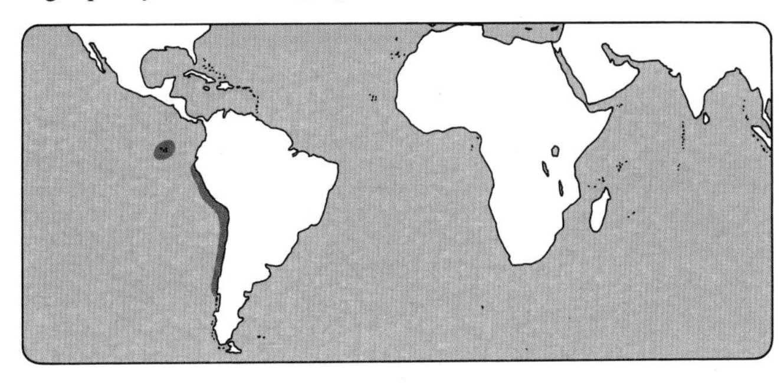

DK: De enorme, nærmest maste, fangster forringer kvaliteten i en sådan grad, at fisken udelukkende anvendes til fiskemel og konserves

D: Die riesigen, fast zerquetschten Fänge beeinträchtigen die Qualität in dem Masse, dass der Fisch ausschliesslich zu Fischmehl oder Konserven verarbeitet wird

E: Las enormes capturas y los métodos industriales de pesca deterioran la calidad, sirviendo el pescado únicamente para harina de pescado y enlatado

F: Les énormes quantités pêchées et les méthodes brutales de pêche abîment le poisson au point qu'il est utilisé uniquement pour farine de poisson et conserves

I: Le grosse quantità pescate ed i metodi usati deteriorano la qualità al punto tale che il pesce serve unicamente per farina di pesce e conserva

CHILEAN WHITING/HAKE

Scientific name:
Merluccius gayi

Synonyms: South Pacific hake, Pacific silver hake
Family: Merlucciidae — merluccid hakes
Typical size: 50 cm, up to 115 cm

Long used for food in Chile, the species was harvested in Peru mainly for fishmeal until the late 1960s, when its value as an exportable food product became apparent. It is known as whiting in the United States, where large quantities of dressed fish and blocks are used for fish portions, sticks and other ready-to-use products.

FISHING METHODS

 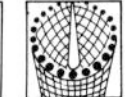

MOST IMPORTANT FISHING NATIONS
Peru, Chile, Russia

PREPARATION

USED FOR

  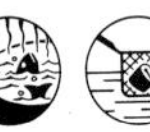 

EATING QUALITIES
This hake has soft, white meat with little flavour and a small flake. It is not suitable for smoking. Fillets are deep-skinned to remove the lateral fat line to extend shelf life and improve colour.

COMMON NAMES
D: Chilenischer Seehecht
DK: Chilensk kulmule
E: Merluza chilena
F: Merlu du Chili
GR: Bakaliáros tou Peroú
I: Nasello del Cile
IS: Sílelysingur
J: Chiri-heiku
N: Lysing
NL: Chileense heek
P: Pescada do Chile
RU: Chilijskaya meduza
SF: Perunkummeliturska
US: Chilean whiting

LOCAL NAMES
CL: Merluza
PE: Huaycuya, merlango

Nutrition data:
(100 g edible weight)

Water	85.0 g	Total lipid	
Calories	81 kcal	(fat)	1.6 g
Protein	16.6 g	Omega-3	0.5 mg

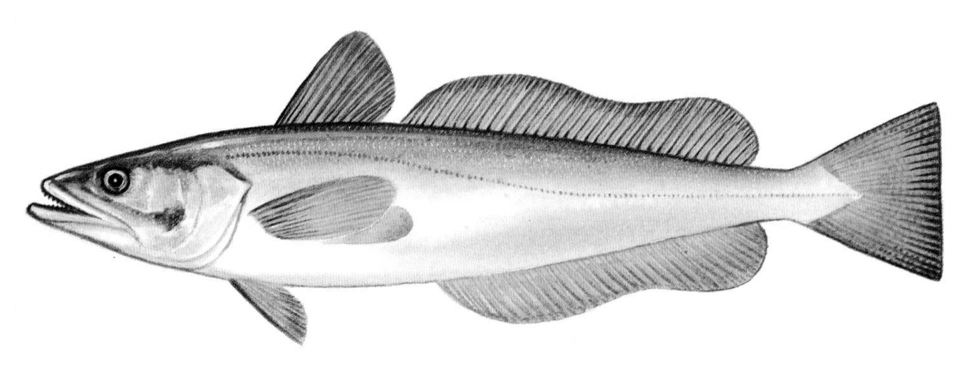

HIGHLIGHTS
Catches have greatly increased recently to over 200,000 tons yearly as factory trawlers have been attracted to the resource. Although the meat is soft, it can be made into inexpensive fillets and blocks which are acceptable in North America and Europe.

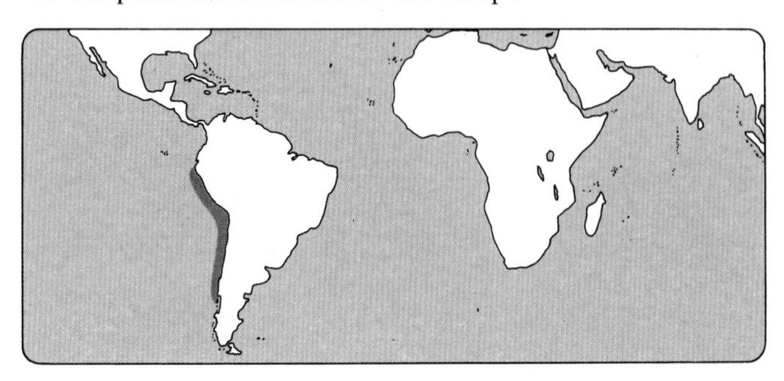

DK: Til trods for den bløde konsistens, anvendes kødet til filetter og blokke. Eksporteres til Nordamerika og Europa som lavprisprodukter

D: Trotz seiner weichen Konsistenz wird das Fleisch zu Filets und Blöcken verarbeitet, die als Billigware nach Nordamerika und Europa exportiert werden

E: A pesar de su consistencia blanda, la carne es usada para filetes y bloques. Se exporta a Norteamérica y Europa como productos de precios bajos

F: Malgré sa consistance molle, la chair est utilisée en filets et en blocs. Exporté vers les Etats-Unis et l'Europe sous forme de produits de discount

I: Nonostante la sua consistenza molle, la carne si adopera per filetti e blocchi. Esportato nell'America Settentrionale ed in Europa come prodotto a basso prezzo

CHILIPEPPER ROCKFISH

Scientific name:
Sebastes goodei

Family: Scorpaenidae — scorpionfishes
Typical size: up to 56 cm

COMMON NAMES
E: Chancharro pimienta
F: Sébaste piment
US: Chilipepper rockfish

Chilipepper rockfish is targeted by recreational anglers as well as commercial fishermen. It frequents rocky reefs as well as sandy and muddy bottoms, where it can be trawled.

FISHING METHODS

MOST IMPORTANT FISHING NATIONS
United States

PREPARATION

USED FOR

EATING QUALITIES
This red-skinned rockfish has pinkish meat which whitens when cooked. It has little flavour and a delicate texture. It is best used as a base for spices, stuffings or other enhancers, as it lacks the character to stand on its own. It is a popular species in North America, widely sold in supermarkets with other rockfish.

Nutrition data:
(100 g edible weight)

Water	79.3 g	Total lipid	
Calories	86 kcal	(fat)	1.6 g
Protein	18.0 g	Omega-3	0.3 mg

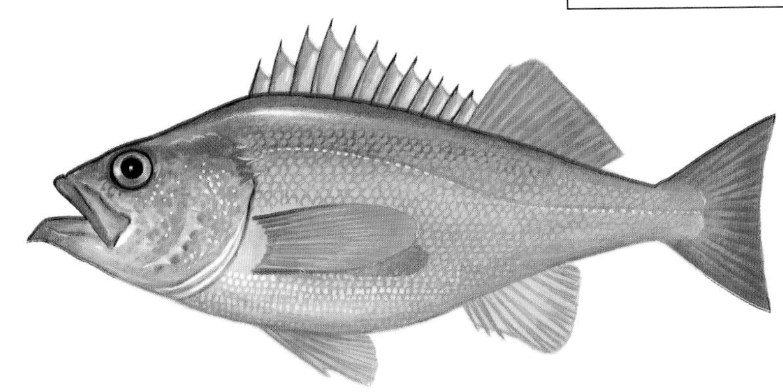

HIGHLIGHTS
Chilipepper rockfish is commercially important in California. Unlike most of its relatives in the Pacific, it does not range into the Gulf of Alaska. It is similar in appearance to the Pacific ocean perch (*S. alutus*) and often confused with it in trade.

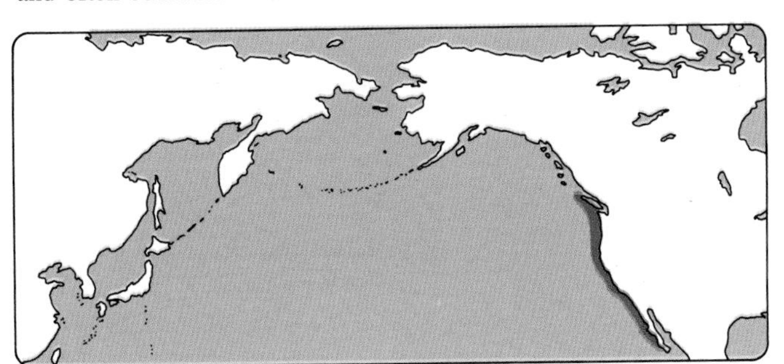

DK: En kommercielt vigtig fisk i Californien. I modsætning til størsteparten af familien, findes den ikke nordpå i Alaskabugten

D: Ein wichtiger Wirtschaftsfisch in Kalifornien. Im Gegensatz zum grössten Teil der Familie kommt er nicht im Golf von Alaska vor

E: Pescado de importancia comercial en California. A diferencia de la mayor parte de la familia, no vive al norte del Golfo de Alaska

F: Un poisson de grand intérêt commercial en Californie. Contrairement à la majeure partie de la famille, il n'est pas repéré jusque dans les eaux du Golfe d'Alaska

I: Un pesce commerciale importante della California. A differenza delle altre specie della famiglia è assente nel Golfo dell'Alasca

CHINOOK/KING SALMON

Scientific name:
Oncorhynchus tshawytscha

Synonyms: Spring salmon, quinnat salmon
Family: Salmonidae — salmonids
Typical size: 100 cm, 14 kg

Chinook, king or spring salmon are the largest Pacific salmon and the most valued by the market. They are also an important recreational fish in Canada and the United States, although declining stocks in those countries mean there are fewer opportunities for anglers to hook one of these desirable fish.

DESCRIPTION
Chinooks can be easily recognized by their black gums, but their size can also be a distinguishing feature. Mature fish vary greatly, but the record is nearly 57 kg and there are many fish which weigh in at 20 kg or more. One reason that kings grow so large is that some individuals in certain runs are capable of spawning more than once, like the Atlantic salmon, *Salmo salar*. Other salmon species typically die immediately after spawning. The king is the only Pacific species which may spawn more than once, giving it the opportunity to grow larger each year it returns to the ocean.

Populations along the Asian coast of the Pacific are quite small. The species has been introduced into the Great Lakes, where it has become established and helps to support a recreational industry.

COMMON NAMES
D: Königslachs, Quinnat
DK: Kongelaks
E: Salmón real
F: Saumon royal, chinook
GR: Solomós tou Irinikoú
I: Salmone reale
IS: Kóngalax
J: Masunosuke
N: Kongelaks
NL: Chinook zalm
P: Salmao real
RU: Chavycha
S: Kungslax
SF: Kuningaslohi
US: King salmon, chinook

HIGHLIGHTS
Some populations of chinook salmon swim 3000 km or more from the sea to their spawning grounds. Since the fish do not eat during their migration, they enter the river with a great deal of energy, stored as fat, for the long journey. Such fish are prime for eating, especially moist and full of flavour. The Columbia and Yukon Rivers have such runs starting upriver in spring. Fish from these and other long runs are especially targeted and fetch a very high price. Autumn runs of chinooks tend to be shorter and the fish less fat.

Nutrition data:
(100 g edible weight)

Water	73.2 g
Calories	174 kcal
Protein	20.1 g
Total lipid (fat)	10.4 g
Omega-3	1.4 mg

CHINOOK/KING SALMON *Oncorhynchus tshawytscha*

FISHING METHODS

MOST IMPORTANT FISHING NATIONS

Canada, United States

PREPARATION

USED FOR

EATING QUALITIES

King salmon has red meat, not so red as sockeye but redder than Atlantic salmon. The meat is moist, full of flavour, with a large flake. It smokes well. Both hot smoking and cold smoking produce excellent results. Salted sides of chinook salmon were once shipped from the Pacific to markets in eastern North America as well as Europe. These pickled sides were then smoked to local recipes and tastes.

Some kings have white meat. This is not an indication of poor quality, but is a genetic trait. White meated kings taste the same as their red meated cousins.

IMPORTANCE

Chinook salmon catches are far lower than they were in the middle of the 19th century, when the species supported the rapid development of the then new canned salmon industry. Now, catches are less than 30,000 tons a year and in decline. Canada produces about half the total, including some farmed fish.

Although chinooks are domesticated and have been farmed widely in Canada, the United States and as far away from their natural range as Chile and New Zealand, many farmers have switched to Atlantic salmon or coho, which give more reliable results. Kings have proved quite hard to farm profitably, although the high market price for the species means that a number of farmers are persevering.

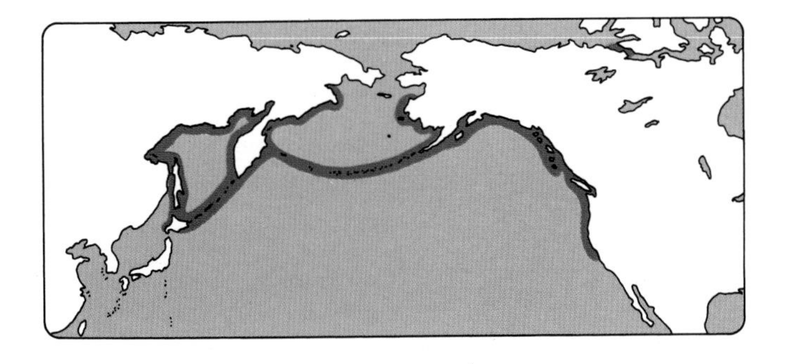

DK: Nogle bestande svømmer ca. 3.000 km fra havet til gydestedet. Fisken er specielt delikat og velsmagende ved vandringen op ad floderne, når energireserverne i form af fedt er store

D: Einige Bestände schwimmen rund 3000 km vom Meer weg zu ihren Laichplätzen. Der Fisch ist besonders delikat und schmackhaft, während er die Flüsse hinaufsteigt und seine Energiereserven in Form von Fettpolstern gross sind

E: Algunas especies recorren unos 3.000 km desde el mar hasta el lugar de reproducción. Este pescado es particularmente exquisito cuando sube por los ríos, con reservas de energía muy grandes en forma de grasa

F: Certaines populations parcourent plus de 3000 km depuis la mer pour frayer. Le saumon en fraie est spécialement savoureux et délicat lorsque, plein de grandes réserves énergie, il commence la remontée des rivières

I: Alcune popolazioni percorrono circa 3.000 km dal mare al luogo di riproduzione. Questo pesce è particolarmente squisito nei momenti in cui risale i fiumi, quando le riserve di energia sotto forma di grasso sono grandi

CHUB MACKEREL

Scientific name: *Scomber japonicus*

Synonyms: Spanish mackerel, Pacific mackerel
Family: Scombridae — tunas and mackerels
Typical size: 40 cm, 750 g

A cosmopolitan temperate species, this important mackerel is fished by over 40 countries worldwide. Generally smaller and less oily than Atlantic mackerel, recent catches of 1 to 1.5 million tons are far lower than the record 2.8 million tons reported in 1978.

FISHING METHODS

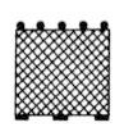 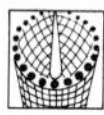

MOST IMPORTANT FISHING NATIONS

Japan, Russia, China, Korea, Chile, Peru, Ecuador

PREPARATION

USED FOR

EATING QUALITIES

Chub mackerel has dark coloured meat with a medium flake and a fairly strong flavour. It is excellent canned or smoked. Shelf life is limited: the fish must be handled well and frozen promptly after capture to avoid deterioration, especially rancidity.

COMMON NAMES

D: Japanische Makrele
DK: Spansk makrel
E: Estornino, caballa
F: Maquereau espagnol
GR: Koliós
I: Lanzardo, sgombro cavallo
IS: Spánskur makríll
J: Masaba
N: Spansk makrell
NL: Spaanse makreel
P: Cavala
RU: Japonskaya makrel
S: Spansk och japansk makrill
SF: Japaninmakrilli
TR: Kolyoz
US: Chub mackerel

LOCAL NAMES

BR: Cavalinha
EG: Scomber
ME: Cachorreta
MO: Kabaila
PO: Makrela kolias

Nutrition data:
(100 g edible weight)

Water	69.7 g	Total lipid	
Calories	154 kcal	(fat)	7.3 g
Protein	22.1 g	Omega-3	1.5 mg

HIGHLIGHTS

Chub mackerel schools in huge numbers in coastal waters, where it is easily caught. It migrates seasonally to remain in cool water; catches tend to be concentrated into short seasons when the shoals are present. Sport fishermen are the main users of the California resource.

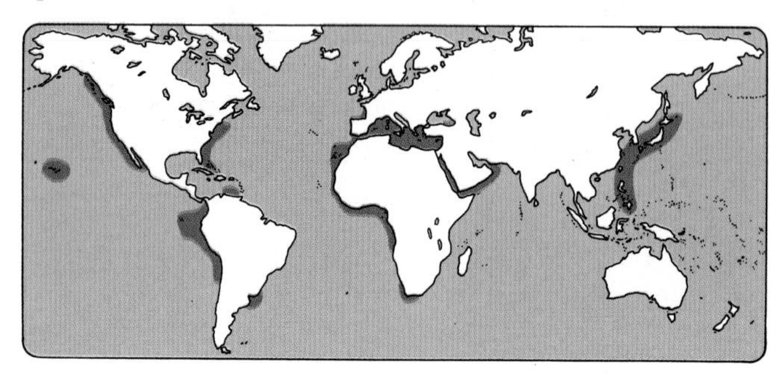

DK: Trods stort fald i fangster, stadig en af verdens vigtigste arter. Velegnet til konserves og rygning. Anvendes endvidere til fiskemel

D: Trotz gesunkener Fänge immer noch eine der wichtigsten Fischarten der Welt. Gut geeignet zum Konservieren und Räuchern. Ausserdem Verarbeitung zu Fischmehl

E: Sigue siendo una de las más importantes especies del mundo, aunque han bajado mucho las capturas. Excelente, enlatada o ahumada. Sirve para harina de pescado

F: Toujours l'une des plus importantes espèces du monde malgré une pêche en forte baisse. Excellent en conserve ou fumé. Est aussi utilisé pour farine

I: Ancora una delle specie più importanti del mondo, anche se le catture sono diminuite notevolmente. Eccellente conservata o affumicata. Usata anche per farina

Scientific name: *Oncorhynchus keta*

CHUM SALMON

Synonyms: Keta salmon, dog salmon, qualla
Family: Salmonidae — salmonids
Typical size: 3.5 to 7 kg, up to 15 kg

Chum is the second most important Pacific salmon species in volume, especially valued in Japan, where it is ranched (by hatchery production of fry) in large quantities. The roe is the premium material for sujiko and for salmon caviar (ikura). Japan accounts for about two thirds of the world's production of the species.

FISHING METHODS

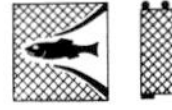 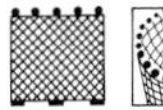 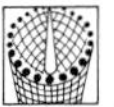 

MOST IMPORTANT FISHING NATIONS

Japan, United States, Russia, Canada

PREPARATION

Caviar    

USED FOR

    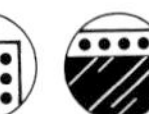

EATING QUALITIES

Pink to red meat with excellent flavour characterize chum salmon from the ocean. Once the fish begin their spawning migration, the meat quality deteriorates rapidly, losing colour, oil, taste and texture.

COMMON NAMES

D: Keta-Lachs, Hundslachs
DK: Ketalaks
E: Keta, salmón chum
F: Saumon keta, saumon chien
GR: Solomós kéta
I: Salmone keta
IS: Hundlax
J: Sake, shiro-zake
N: Ketalaks
NL: Chumzalm, ketazalm
P: Salmao cao
RU: Keta
S: Keta
SF: Koiralohi
US: Chum salmon

Nutrition data:
(100 g edible weight)

Water	75.4 g	Total lipid	
Calories	115 kcal	(fat)	3.8 g
Protein	20.1 g	Omega-3	0.6 mg

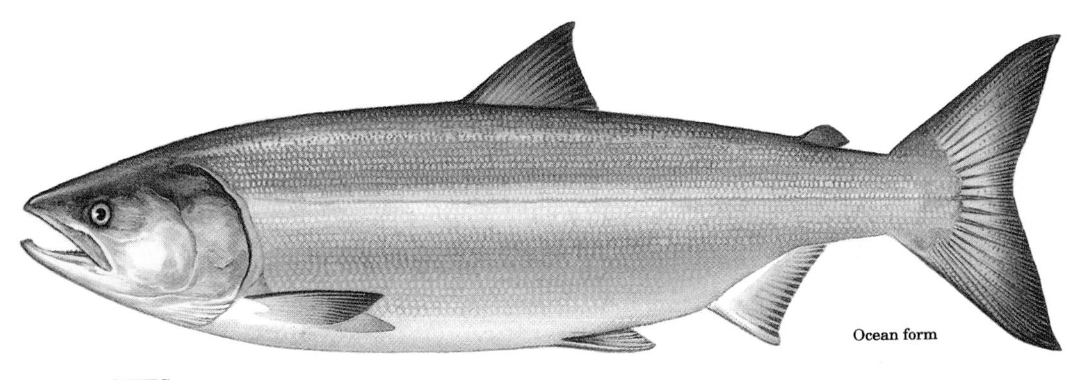
Ocean form

HIGHLIGHTS

Chum salmon from the ocean is a premium fish, excellent for eating and smoking. The roe is sometimes more valuable than the meat of sexually mature fish. World catches exceed 200,000 tons. Resources are declining in North America while being enhanced in Japan.

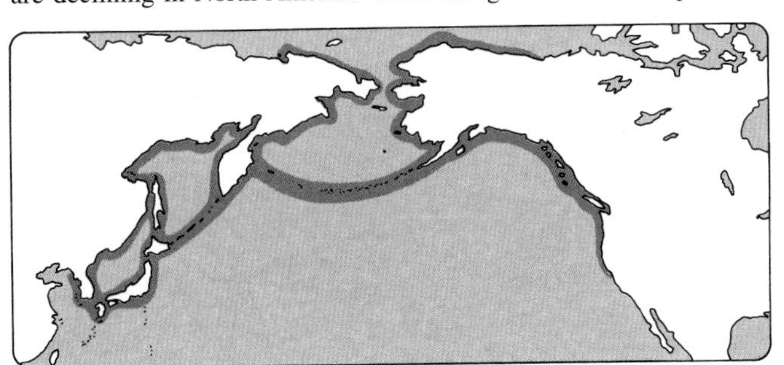

DK: I perioden, hvor den lever i havet, er den en førsteklasses spisefisk. Rognen er mere værdifuld end kødet fra de kønsmodne fisk

D: Während seines Aufenthalts im Meer ein ausgezeichneter Speisefisch. Bei geschlechtsreifen Fischen ist der Rogen wertvoller als das Fleisch

E: En la época en que vive en el mar, es un pescado comestible muy solicitado. Las huevas son más apreciadas que la carne de los ejemplares adultos

F: Pendant la période où il vit dans la mer, c'est un délicieux poisson de table. Les oeufs du poisson mature ont plus de valeur que la chair

I: All'epoca in cui vive nel mare è un pesce commestibile molto pregiato. Degli esemplari adulti sono più preziose le uova che non le carni

COD

Scientific name:
Gadus morhua

Synonym: Atlantic cod
Family: Gadidae — cods
Typical size: 60 cm, 2.5 kg

Atlantic cod has long been a staple food for Europeans: its bones have been identified in sites from the Mesolithic Age. Reports of its abundance in the New World tempted adventurous European fishermen to begin exploiting N. American stocks 500 years ago. The prosperity of many early settlements in Canada and New England depended on salting cod for European markets.

Cod stocks in the North Atlantic stabilized in the early 1980s but now are again hard pressed. Much European and Canadian fishing is severely curtailed. Only the far northern stocks, which are harvested mainly by Russia, appear strong.

DESCRIPTION
Grey-greenish to reddish brown fish, speckled on the back, Atlantic cod can be distinguished from haddock by its pale lateral line. Cod are cold-water fish, moving away from the warming shore in summer.

Most cod now landed are small by the standards of the past. The largest ever recorded (in Massachusetts in 1895) weighed 96 kilos and measured almost 2 m. Now, fish of 5 kg are considered large.

COMMON NAMES
D: Dorsch, Kabeljau
DK: Torsk
E: Bacalao
F: Morue, cabillaud
GR: Gádos, bakaliáros
I: Merluzzo bianco
IS: Thorskur
J: Tara, madara
N: Torsk, skrei
NL: Kabeljauw, gul
P: Bacalhau
RU: Atlanticheskaya treska
S: Torsk
SF: Turska
TR: Morina
US: Cod

LOCAL NAMES
PO: Dorsz

HIGHLIGHTS
The mighty cod resources of the N. Atlantic provided 30 percent of the world's groundfish in 1983. Since then, catches have declined and cod's dominant market position is threatened. Salt cod, once a staple, is now a European luxury. Atlantic cod is often priced too high for major products such as blocks and stockfish.

Higher prices and reduced supplies are attracting Norwegian fish farmers into growing cod, initially using young fish caught in the wild. Hatcheries are being developed to ensure a supply of fry.

Nutrition data:
(100 g edible weight)

Water	80.8 g
Calories	80 kcal
Protein	18.3 g
Total lipid (fat)	0.7 g
Omega-3	0.3 mg

FISHING METHODS

   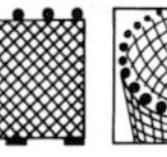  

MOST IMPORTANT FISHING NATIONS
Norway, Iceland, Denmark, United Kingdom, Russia, Canada, USA

PREPARATION

USED FOR

EATING QUALITIES
Cod has white, moist meat, a large flake and mild flavour, suitable for almost any cooking method. It is widely available smoked, salted or dried. It is available fresh or frozen dressed, filleted, as loins or portions and in numerous packs and styles. Fillet blocks are made for the production of portions, fingers and other processed products. Smoked cod, stockfish, bacalao and lutefisk are among familiar cured products. Cod roes are boiled, sometimes smoked; cheeks and tongues are used by maritime communities. Cod livers were once used as a source of vitamins A and D.

IMPORTANCE
Cod landings have declined by almost half from the 2.3 million tons reported in the early 1980s, but the species remains of vital importance for many fishing communities, especially in Iceland, Norway and Canada. Cod is still a very important fish, even though it is no longer the predominant species in markets in Europe and North America.

Cod have high fecundity, producing millions of eggs and reaching first maturity as young as two years. So, although cod stocks have suffered frequent collapses in the past, they have usually recovered within a fairly short time once fishing pressure was eased. However, humans are not the only predators: increasing numbers of highly protected seals have decimated some cod stocks and infected others with parasites, affecting the chances that the species will again recover.

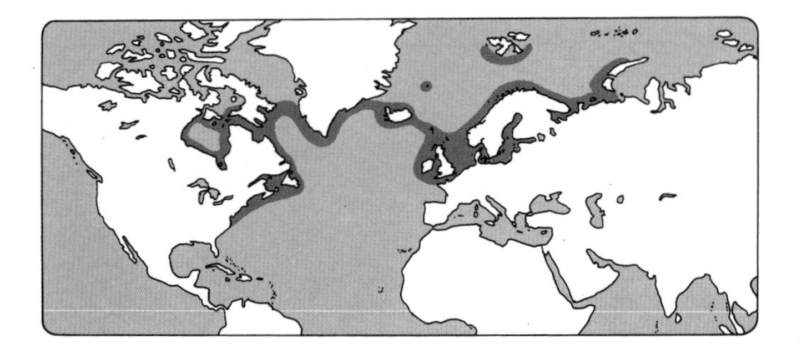

DK: Faldet i Nordatlantens førhen kolossale bestand truer nu fiskens førende markedsposition. Saltet torsk er blevet en luksusvare på det europæiske marked. Højere priser har øget interessen for opdræt

D: Der Rückgang des früher riesigen Bestands im Nordatlantik gefährdet jetzt die führende Marktposition des Fischs. Salzdorsch gilt auf dem europäischen Markt als Delikatesse. Höhere Preise haben das Interesse an Aufzucht gefördert

E: Es amenazada su posición predominante en el mercado por la caída de las existencias en el Atlántico Norte. El bacalao salado es hoy un producto de lujo en el mercado europeo. Los altos precios han aumentado el interés en su cría

F: La chute de la population autrefois colossale dans l'Atlantique nord menace maintenant sa position de leader du marché. La morue salée est devenue un produit de luxe en Europe, et les prix élévés ont augmenté l'intérêt de l'élevage

I: Il calo della popolazione di questo pesce, un tempo colossale nel Nordatlantico, minaccia ora la sua posizione leader sul mercato. Il baccalà è diventato un prodotto di lusso in Europa, aumentando l'interesse per l'allevamento

COHO/SILVER SALMON

Scientific name:
Oncorhynchus kisutch

Synonyms: Blueback, medium red salmon, jack salmon
Family: Salmonidae — salmonids
Typical size: 75 cm, 6 kg

COMMON NAMES
D: Kisutchlachs, Silberlachs
DK: Koho, sølvlaks
E: Salmón plateado
F: Saumon argenté
GR: Asiménios solomós, kócho
I: Salmone argentato
IS: Silfurlax
J: Gin-zake
NL: Cohozalm
P: Salmao prateado
RU: Kizhuch
S: Stillahavslax
SF: Hopealohi
US: Coho salmon, silver salmon

Coho or silver salmon (the names are interchangeable) are found on both sides of the North Pacific. Attempts to introduce them to the Atlantic have been mostly unsuccessful, although occasionally a coho is caught there. Landlocked races of the fish are found in a number of North American lakes and the stocks in Lake Michigan and Lake Superior support important recreational fisheries. Coho is an important recreational fish on the Pacific coast of North America. As these resources decline, the tendency is for the remaining fish to be reserved for anglers, excluding the commercial fishery.

Coho is a good raw material for cold-smoking. Frozen, dressed fish are sold to European markets for this purpose.

DESCRIPTION

Ocean run coho are silver fish with small black spots on the back and upper lobe of the tail. The lower jaw has white gums, which distinguishes coho from the black-gummed chinook (*O. tshawytscha*). The largest cohos weighed 14 kg for a sea-run fish in Canada and 15 kg for a landlocked coho in Lake Michigan.

HIGHLIGHTS

Coho salmon is highly valued for its tasty red meat and delicate texture. It is the most domesticated Pacific salmon as well as the one most like Atlantic salmon. It is being grown in increasing quantities, especially in Japan. Wild runs of coho have been declining and many appear to be in some danger. Nevertheless, the supply is being well maintained by farmers.

Pan-sized coho salmon are a popular alternative to trout. These can be grown entirely in fresh water.

Nutrition data:
(100 g edible weight)

Water	72.6 g
Calories	140 kcal
Protein	21.6 g
Total lipid (fat)	6.0 g
Omega-3	0.8 mg

COHO/SILVER SALMON *Oncorhynchus kisutch*

FISHING METHODS

 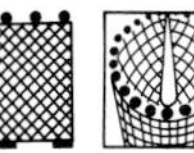 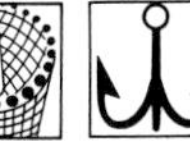

MOST IMPORTANT FISHING NATIONS

Japan, United States, Chile, Canada

PREPARATION

USED FOR

EATING QUALITIES

Coho salmon has red meat, lighter than chinook, similar to Atlantic salmon. The flavour is a little milder than Atlantic, the texture similarly firm and moist. Coho is especially valued in Japan. Coho is labelled "medium red" salmon when canned, which is nowadays quite rare.

The skin reddens as the fish matures, but the flesh colour usually remains the same, although the taste may deteriorate a little.

Trolled fish is more valuable than netted fish, because it is gutted on board the fishing boat and handled individually. Quality of the meat is invariably better, but production of trolled fish has fallen substantially, partly because fishery managers can enforce better control of the harvesting if the fish are caught in the rivers, where they are netted, rather than in the ocean, where they take hooks.

IMPORTANCE

World production of coho is under 100,000 tons a year. Catches have been falling in North America, but farmed production in Chile and Japan has increased significantly, leading to greater overall production, Farming of cohos has not been profitable in British Columbia, Canada, even though the fish is native to those waters. Japanese farmers have been successful, as have their competitors in Chile where the fish is unknown naturally (there are no fish of this genus naturally in the southern hemisphere).

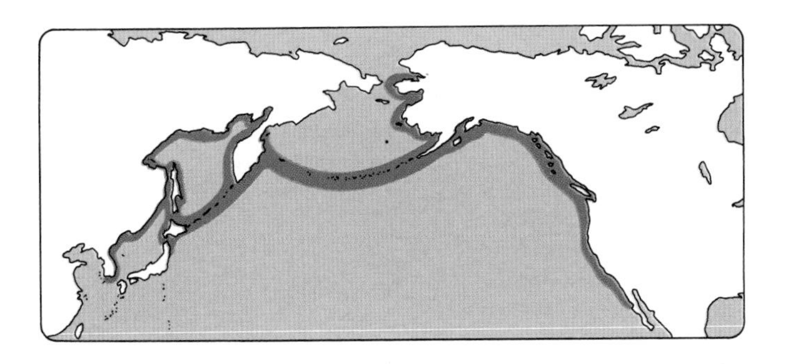

DK: Coho'ens velsmagende røde kød og delikate konsistens gør den til en højt estimeret spisefisk. Faldet i den oprindelige art i Nordamerika opvejes af et tiltagende opdræt i specielt Japan, som er det største marked

D: Das wohlschmeckende rote Fleisch und die delikate Konsistenz machen den Silberlachs zu einem hoch geschätzten Speisefisch. Der Rückgang der Art in Nordamerika wird durch zunehmende Aufzucht, besonders in Japan, aufgewogen

E: El salmón coho es muy apreciado por su sabrosa carne roja y su delicada consistencia. La caída de la especie primitiva en Norteamérica ha sido compensada por la cría, importante especialmente en Japón

F: La chair rouge très délicate fait du saumon argenté un poisson de table très apprécié. Le déclin en Amérique du Nord de l'espèce d'origine est compensé par un élevage croissant, surtout au Japon qui en est aussi le plus grand marché

I: Il salmone argentato è altamente stimato per le sue pregiate carni rosse e la consistenza delicata. Il calo della specie originaria dell'America del Nord è compensato dall'allevamento, importante specialmente nel Giappone

COMMON DAB

Scientific name: *Limanda limanda*

Family: Pleuronectidae — right-eye flounders
Typical size: 25 cm

Dabs are one of many small flatfish used in Europe mostly fresh, dressed. They are seldom large enough to fillet and the flavour is better if they are cooked on the bone.

DESCRIPTION
Dabs have brown, often speckled skin, without the distinctive orange spots that distinguish plaice from the other flounders, although sometimes the speckles are orange or rusty red in colour. It is very similar to the yellowtail flounder (*Limanda ferruginea*), a popular species in the United States and Canada, but it is smaller than yellowtail. Dabs grow slowly and live about 12 years.

Dabs can be distinguished from other, similar flounders by the curve in the lateral line around the pectoral fin.

It is found from the White Sea in the north through the North Sea and English Channel to the northern half of the Bay of Biscay. It is also found around Iceland. It lives in water between 20 and 150 m deep, depths that are easily accessible to fishing vessels. It prefers sandy bottoms and, like other flatfish, burrows into the sand to hide.

COMMON NAMES
D: Kliesche, Scharbe
DK: Ising, slette
E: Limanda nórdica, gallo
F: Limande commune
GR: Limánta, chomatída
I: Limanda
IS: Sandkoli
J: Karei
N: Sandflyndre
NL: Schar
P: Solha escura do Norte
RU: Limanda, ershovatka
S: Sandskädda
SF: Hietakampela
TR: Pisi baligi
US: Dab, flounder

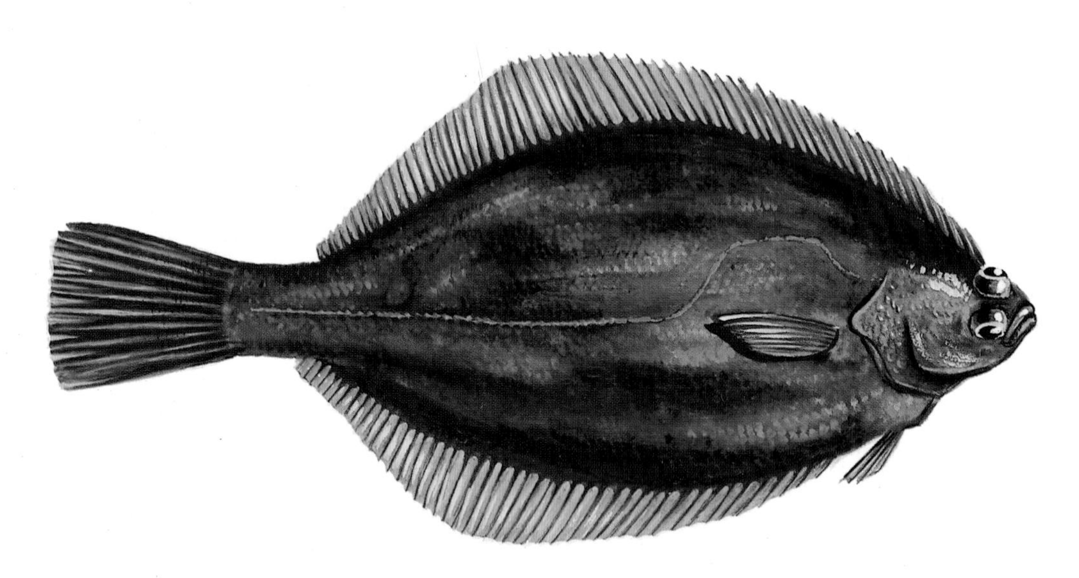

HIGHLIGHTS
Dabs, although small, are an important European flatfish. Mostly cooked whole because they are too small to fillet, the sweet meat continues to attract consumers as the resource and catches decline.

Dabs are frequently with other small flatfish, such as plaice (*Pleuronectes platessa*) and European flounder (*Platichthys flesus*). Fishermen and processors are not always more successful than consumers at identifying the different species; statistics, as well as market differences between these fish, may not always be significant.

Nutrition data:
(100 g edible weight)

Water	79.1 g
Calories	74 kcal
Protein	15.7 g
Total lipid (fat)	1.2 g
Omega-3	0.2 mg

COMMON DAB *Limanda limanda*

FISHING METHODS

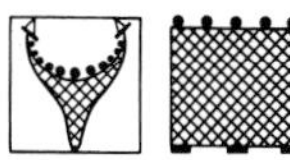

MOST IMPORTANT FISHING NATIONS
Denmark, France, Iceland, Scotland

PREPARATION

USED FOR

EATING QUALITIES
The common dab is a small flatfish with rather thick, rough skin which has to be removed before the meat can be eaten. Nevertheless, the meat is extremely sweet and tender. This is an excellent flounder and prized in many parts of northern Europe when it is avialable.

Dabs are usually cooked whole, on the bone. The meat flakes easily from the backbone and the tiny pinbones are generally undetectable.

Some experts consider that the taste is better if the fish is left for a couple of days to develop its full flavour. However, such beliefs are usually based on long familiarity with stale fish, which becomes the standard and therefore expected quality. Fresher fish may then not have the expected flavour and be considered inferior, although it is in fact better, fresher and more delicate.

In Denmark, Belgium and elsewhere in northern Europe, dabs are salted and dried. They are also sometimes smoked.

IMPORTANCE
Catches of common dab have declined dramatically in the last two decades and are now around 10,000 tons a year. Much of the fish that is now caught is taken as by-catch rather than from a directed fishery. Nevertheless, because it is a prized species, the dab continues to fetch a good price on European markets.

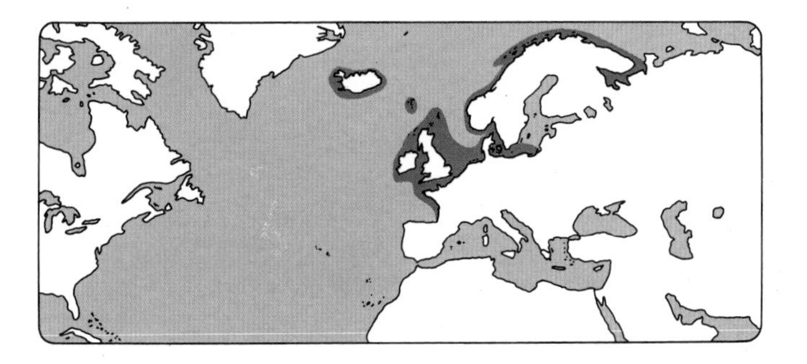

DK: En lille men vigtig europæisk fladfisk. Koges oftest hel da den størrelsesmæssigt ikke er velegnet til filettering. Sødt, delikat kød. Forveksles ofte med andre små fladfisk, f.eks. rødspætte

D: Kleiner, aber wichtiger europäischer Plattfisch. Wird meistens im ganzen Stück gekocht, da er zum Filetieren zu klein ist. Süsses, delikates Fleisch. Wird oft mit anderen kleinen Plattfischen, z.B. Scholle, verwechselt

E: Pequeño, pero importante pleuronecto europeo. Normalmente es cocido entero, porque es demasiado pequeño para filetear. Su carne es delicada y dulce. Se confunde a menudo con otros pleuronectos pequeños como la solla

F: Poisson plat européen, petit mais important. En raison de sa taille, il est le plus souvent préparé en entier. Chair délicate d'un goût très fin. Est parfois confondu avec d'autres poissons plats tels que la plie

I: Un pesce europeo piccolo, ma importante. Normalmente è bollito intero, perché troppo piccolo per farne i filetti. Carni dolci e pregiate. Spesso si confonde con altri piccoli pleuronettiformi come ad esempio la passera di mare

CONGER EEL

Scientific name:
Conger conger

Synonym: Conger
Family: Congridae — conger eels
Typical size: 100 to 175 cm

Congers may grow as large as 275 cm and 65 kg. Their ferocity and size makes them prime angling targets. Congers are found from Norway to Senegal in the eastern Atlantic, throughout the Mediterranean and in the Black Sea. Nowhere abundant, they live in rocky crevices and can only be caught with baited lines.

FISHING METHODS

MOST IMPORTANT FISHING NATIONS
France, Spain, Portugal, Morocco

PREPARATION

USED FOR

EATING QUALITIES
Conger eel is a fine fish to eat, with firm, sweet, white meat easily detached from the bones. The texture is meaty and the fish takes well to being roasted in the same way as beef.

COMMON NAMES
D: Gemeiner Meeraal, Conger
DK: Almindelig havål
E: Congrio europea
F: Congre commun
GR: Moungrí, dróngos
I: Grongo
IS: Hafáll
J: Anago
N: Havål
NL: Zeepaling, congeraal
P: Congro, safio
RU: Morskoj ugor, konger
S: Havsål
SF: Meriankerias
TR: Migri
US: Conger eel

LOCAL NAMES
MA: Malong
CH: Moon sin
IN: Pucok nipah, remang
PH: Pindanga
MO: Sanoure
TU: Gringou

Nutrition data:
(100 g edible weight)

Water	73.2 g	Total lipid	
Calories	114 kcal	(fat)	4.6 g
Protein	18.1 g	Omega-3	0.9 mg

HIGHLIGHTS
Not often sold commercially, the conger eel is a fine sport fish. A close relative, the American conger of the western Atlantic, is seldom targeted, although quite common: the word "eel" is not attractive to most American consumers, who regard eels rather as snakes.

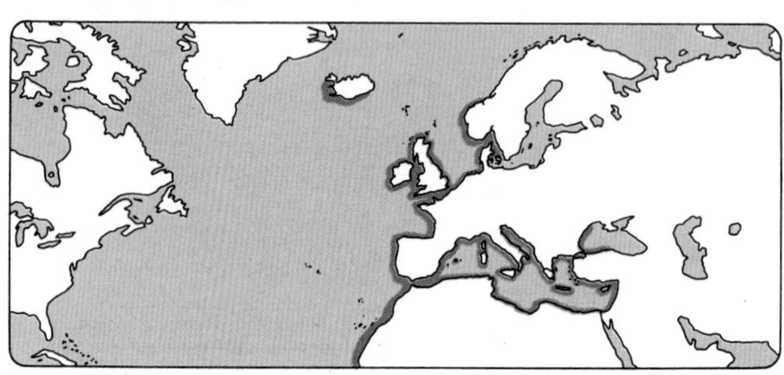

DK: Uden større økonomisk betydning, men en fin sportsfisk. Den tæt beslægtede amerikanske art fiskes kun i meget begrænset omfang

D: Ohne grössere wirtschaftliche Bedeutung, aber ein feiner Sportfisch. Die eng verwandte amerikanische Art wird nur in sehr begrenztem Umfang gefangen

E: No tiene importancia comercial, pero es un pescado deportivo apreciado. La especie afín americana es capturada con poca frecuencia

F: Le congre est sans importance commerciale mais c'est un excellent poisson sportif. Le congre américain étroitement apparenté, est rarement pêché

I: Senza grande valore economico, ma un ottimo pesce sportivo. La specie americana affine viene pescata solo in misura molto limitata

CORVINA

Scientific name:
Sciaena gilberti

Family: Sciaenidae — drums
Typical size: up to 60 cm

Corvina's firm meat can be used in the fiery marinades of lime juice and chili known as ceviche, in which the protein in the meat is denatured not by cooking but by the action of the acid. This dish is very popular.

COMMON NAMES
E: Corvina pampera, gringa
F: Courbine blonde
NL: Ombervis
RU: Stsiena
US: Drum

FISHING METHODS

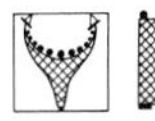  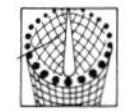

MOST IMPORTANT FISHING NATIONS
Peru

PREPARATION
Ceviche 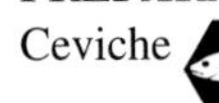

USED FOR

EATING QUALITIES
Corvina has white, firm meat with a mild flavour. It is versatile and can be used in many different fish recipes. It is an excellent quality fish, regarded as a prime table fish in South America, where the limited production is all consumed locally.

Nutrition data:
(100 g edible weight)

Water	78.9 g	Total lipid (fat)	0.7 g
Calories	81 kcal	Omega-3	0.1 mg
Protein	18.7 g		

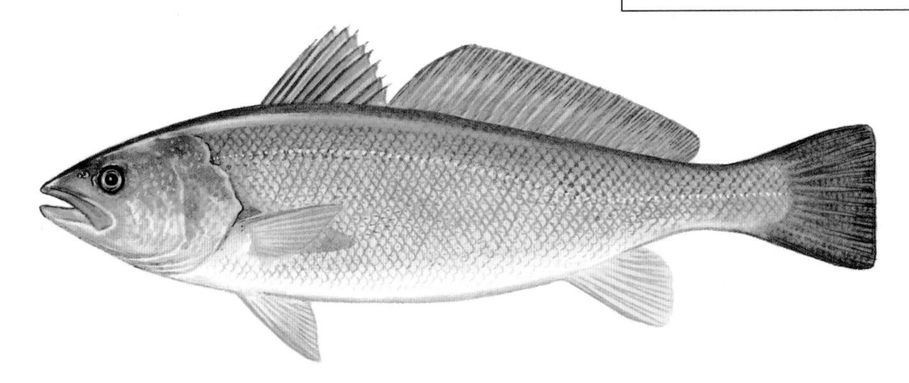

HIGHLIGHTS
A small resource, corvina is widely appreciated. Many other drums (and probably other families as well) are used as substitutes. Worldwide, there are around 200 species of drum, most of them providing good quality meat.

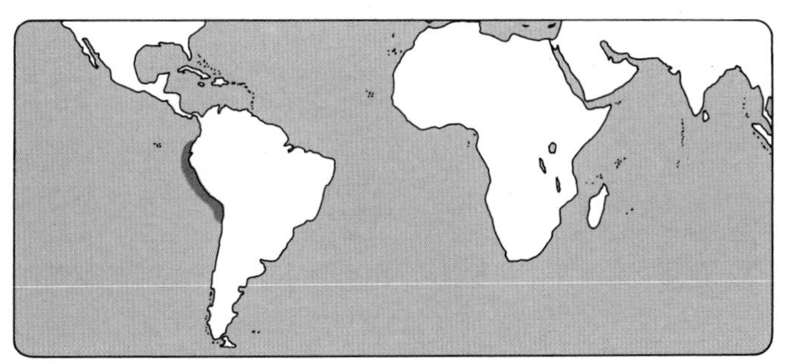

DK: Fiskens faste kød er velegnet til 'ceviche' - den stærkt krydrede marinade af limejuice og chili, hvor syren denaturerer kødets proteiner

D: Das feste Fleisch des Fisches eignet sich gut für 'Ceviche' - eine scharf gewürzte Marinade aus Limettensaft und Chili, die die Fleischproteine denaturiert

E: La carne firme del pescado es apropiada para 'ceviche', la salsa muy picante de zumo de lima y chili, en la que el ácido desnaturaliza las proteínas de la carne

F: La chair ferme convient très bien pour préparer la ceviche - marinade très épicée de jus de lime et piment rouge où l'acide dénature les protéines de la chair

I: Le sue carni sode sono adatte per ceviche - la marinata piccante a base di succo di limetta e peperoncino, di cui l'acido denatura le proteine della carne

CRUCIAN CARP

Scientific name:
Carassius carassius

Family: Cyprinidae — carps and minnows
Typical size: 15 cm, 150 g

Crucian carp is a small fish, but in the right conditions may reach 50 cm and 1.8 kg. It can survive with very little food for long periods. It is resistant to cold and to organic pollutants. It is also tolerant of low oxygen levels in the water, muddy water and other conditions which would destroy more sensitive fish.

In poor conditions it does not grow, staying small but surviving. It can be grown in marginal water that is unsuitable for most other kinds of edible finfish farming, but where conditions are good, the common carp grows faster and offers better market value.

DESCRIPTION
A deep bodied fish with a distinctively high back, the crucian carp lives in marshlands, weedy lakes and slow moving rivers, feeding on plants and insect larvae.

Its natural range extends from Spain across Europe and north-central Asia to northern China. Its range has been extended by man to most of China, and to Taiwan, Korea and Japan.

COMMON NAMES
D: Karausche, Giebel
DK: Karusse, søkarusse
E: Carpín, carpa gibel
F: Carassin commun, carouche
GR: Petaloúda
I: Carassio
IS: Grænkarpi
J: Funa
N: Karuss
NL: Kroeskarper, maankarper
P: Pimpao comum, serasmao
RU: Karas, kruglyi
S: Ruda
SF: Ruutana
TR: Kirmizi balik

LOCAL NAMES
PO: Karas

HIGHLIGHTS
Crucian carp is farmed on a large scale in China, where it is used widely for food. It is grown with other fish and can also be raised with rice in poly-culture operations.

In Europe, it is less well regarded as a food fish, but it is a part of some cuisines. The species hybridises readily with other cyprinids (especially the common carp) in the wild and in the laboratory, producing fish with the crucian carp's ability to survive in marginal conditions and the carp's superior growth rate.

Nutrition data:
(100 g edible weight)

Water	76.3 g
Calories	112 kcal
Protein	17.8 g
Total lipid (fat)	5.6 g
Omega-3	0.4 mg

CRUCIAN CARP

Carassius carassius

FISHING METHODS

 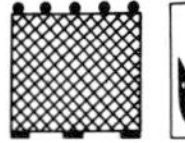 

MOST IMPORTANT FISHING NATIONS

China

PREPARATION

USED FOR

EATING QUALITIES

Crucian carp is similar to carp in eating characteristics, with tasty, pink to whitish meat. Like carp, it is better if the skin and underlying fat are removed before cooking.

Crucian carp is particularly susceptible to developing unpleasant flavours and odours from living in muddy or dirty water. However, the meat quality can be improved by re-stocking the fish in clean water for several weeks before it is eaten.

The meat is also improved by bleeding the fish, though with small fish that is not always practicable. It is not a problem to bleed fish harvested from ponds.

IMPORTANCE

China produces over 200,000 tons a year of crucian carp, all of it from aquaculture. Small quantities are also produced in Taiwan and Japan. The species is raised in ponds, in lakes and sometimes in rice paddies, so that protein and carbohydrates are grown together in the same water supply.

In eastern Europe and northwestern Asia, small numbers are raised in ponds, often with common carp (*Cyprinus carpio*). Because the crucian carp grows more slowly and eats food that the common carp could use, many growers prefer to remove the smaller fish from their ponds.

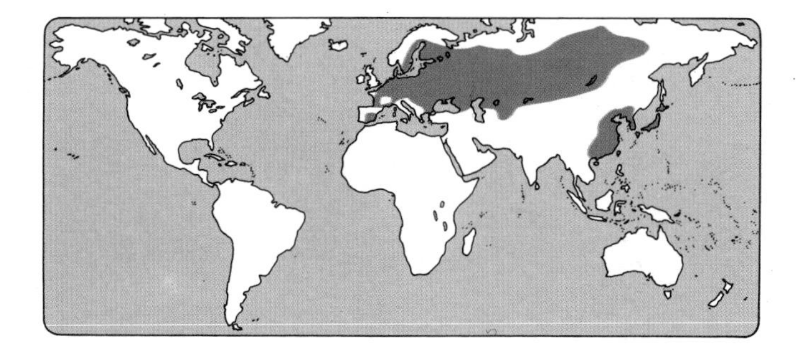

DK: Kina er den største karusseopdrætter. I modsætning til i Europa, hvor fisken ikke er særligt efterspurgt, anvendes den i Kina i talrige retter. Krydsning med karpe giver fine resultater

D: In China werden die meisten Karauschen gezüchtet und in zahlreichen Gerichten verwendet. In Europa dagegen ist der Fisch nicht sonderlich gefragt. Die Kreuzung mit Karpfen bringt gute Ergebnisse

E: China es el más importante criador de carpines. A diferencia de Europa, donde este pescado no es muy popular, es usado en China para la preparación de varios platos. El cruce con la carpa da buenos resultados

F: La Chine est la plus grande éleveuse du carassin. A l'opposé de l'Europe où le cyprin n'est pas très demandé, il entre dans de nombreux plats de la cuisine chinoise. Des croisements avec la carpe donnent de très bons résultats

I: Il carassio viene allevato su vasta scala in Cina, dove costituisce una importante componente di tanti piatti. In Europa, invece, non è molto richiesto. L'incrocio con la carpa comune dà ottimi risultati

CUTTHROAT TROUT

Scientific name:
Oncorhynchus clarki

Family: Salmonidae — salmonids
Typical size: 1 kg

This fine sport fish is found in the western parts of North America. There are small populations in Kamchatka and eastern Siberia. Large commercial fisheries once existed on American lakes, but the species is now quite rare and reserved for anglers. In California, one race of large cutthroat produced a specimen of 18.6 kg.

COMMON NAMES
D: Cutthroatforelle
DK: Cutthroat-ørred
E: Trucha
F: Truite
I: Trota
IS: Strandurridi, blódsilungur
N: Aure
NL: Purperforel
P: Truta
RU: Losos klarka
S: Strupsnitsöring
SF: Punakurkkulohi
US: Cutthroat trout

FISHING METHODS

MOST IMPORTANT FISHING NATIONS
United States, Canada

PREPARATION

USED FOR

EATING QUALITIES
Sea-run cutthroats are similar to steelhead, freshwater fish similar to brook trout. Either form offers delectable, delicate eating to the angler fortunate enough to catch one.

Nutrition data:
(100 g edible weight)

Water	76.7 g	Total lipid (fat)	5.2 g
Calories	125 kcal		
Protein	19.6 g	Omega-3	1.0 mg

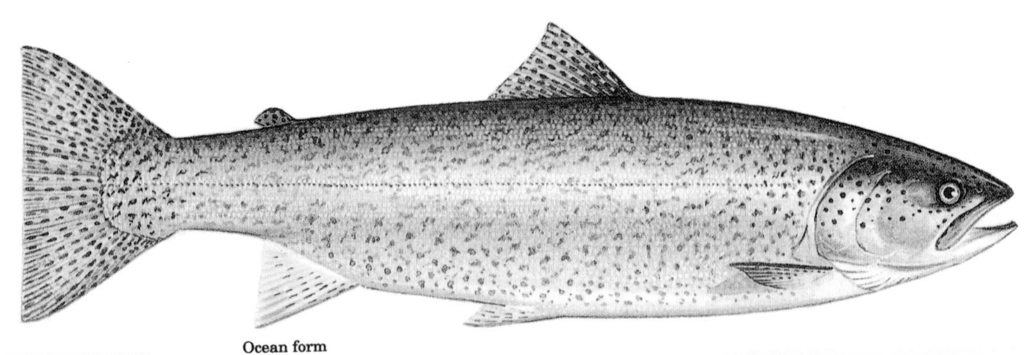

Ocean form

HIGHLIGHTS
Both sea-run and landlocked forms of cutthroat trout are reserved for sport fishermen. Both forms are highly regarded by anglers. The species interbreeds with rainbow trout (*O. mykiss*), making it difficult to use for stocking angling streams.

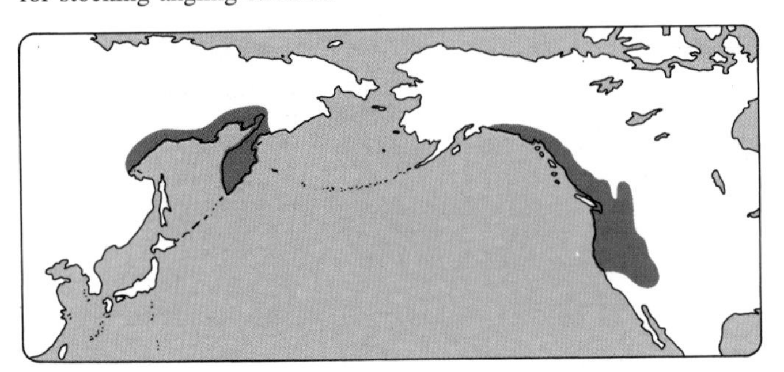

DK: Tidligere basis for et stort netfiskeri, nu kun forbeholdt sportsfiskere. Artens stamme er svær at bibeholde på grund af krydsning med andre arter

D: Früher ausgeprägte Netzfischerei, heute den Sportfischern vorbehalten. Die Stammart lässt sich wegen der Kreuzung mit anderen Arten schlecht bewahren

E: Base en el pasado de una gran pesca con redes ahora reservada para los pescadores deportivos. La especie es difícil de mantener debido al cruce con otras especies

F: Aujourd'hui, ce poisson est pratiquement réservée à la pêche sportive. L'espèce difficile à préserver car elle se croise avec d'autres truites

I: Prima oggetto di una larga pesca con le reti, ma ora riservato ai pescatori sportivi. La specie è difficile da mantenere, in quanto si incrocia con altre specie

Scientific name:

DOLLY VARDEN TROUT

Salvelinus malma

Synonyms: Bull trout, Dolly varden
Family: Salmonidae — salmonids
Typical size: 2 kg

The dolly varden trout or western char is found in rivers and lakes around the northern Pacific basin. There are anadromous and fresh water populations. Used as a source of food in Siberia, it is purely a sport fish in the United States and Canada. Freshwater populations sometimes grow very large: a western char of 16 kg was caught in a river in southern British Columbia.

FISHING METHODS

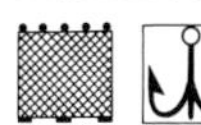

MOST IMPORTANT FISHING NATIONS
Russia, Canada

PREPARATION

USED FOR

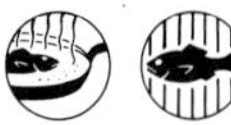

EATING QUALITIES
Dolly varden have firm, moist meat, coloured pink to red. The fish can be used in any trout recipe. Larger fish are like salmon.

COMMON NAMES
D: Malma-Saibling
DK: Malmaørred
E: Salvelino
F: Dolly Varden, omble malma
GR: Salvelínos tou Irinikoú
I: Salmerino
IS: Skrautbleikja
J: Miyabe-iwana, oshorokoma
NL: Malmaforel
P: Salvelino do Pacífico
RU: Tikhooke. kamennyj golets
SF: Härkänieriä
US: Dolly varden

Nutrition data:
(100 g edible weight)

Water	71.5 g	Total lipid	
Calories	153 kcal	(fat)	9.0 g
Protein	18.0 g	Omega-3	1.4 mg

HIGHLIGHTS
Very similar to the Arctic char, the dolly varden is primarily a game fish, although steelhead, coho and chinook found in the same waters are generally preferred by anglers for superior fighting qualities. Slow growing, the fish is not abundant.

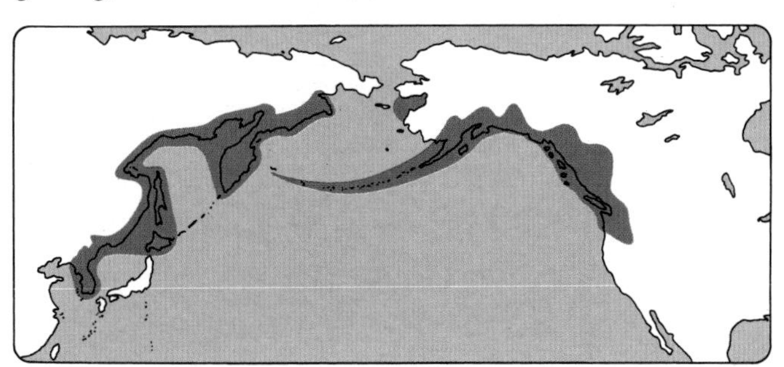

DK: Lever i floder og søer, der støder ud til det nordlige Stillehav. Hvor den i Sibirien er en spisefisk, er den i USA og Canada kun en sportsfisk

D: Lebt in Flüssen & Seen, die in den nördlichen Stillen Ozean einmünden. Ist in Sibirien ein Speisefisch, in den USA und Kanada nur ein Sportfisch

E: Vive en ríos y lagos que desembocan en el norte del Pacífico. En Siberia es un pescado comestible, pero en EE.UU. y Canadá sólo es un pescado deportivo

F: Vit dans les rivières et les lacs qui débouchent dans le nord du Pacifique. Comestible en Sibérie, il est seulement un poisson sportif aux E.-U. et au Canada

I: Vive nei fiumi e nei laghi che sboccano nel nord del Pacifico. Mentre è un pesce commestibile in Sibiria, negli Stati Uniti e nel Canada è solo un pesce sportivo

DOVER SOLE/COMMON SOLE

Scientific name:
Solea vulgaris

Family: Soleidae — soles
Typical size: up to 60 cm
Also known as *Solea solea*

The common or Dover sole is a highly rated species widely found in European coastal waters, although seldom taken by recreational fishermen.

DESCRIPTION

Soles are long, slender flatfish with oval bodies. They can be identified partly by their shape, which is more elliptical than that of most flounders, their nearest relatives. The body is quite thick, providing meaty fillets.

Common sole is found from the Mediterranean to the northern edges of the North Sea. It is most common in the North Sea and Bay of Biscay. These areas supply almost half the total catch in some years. Catches in the Mediterranean and in the Atlantic south of Biscay together account for about one third of the landings.

In winter, soles retreat to deeper water. There is a shallow trench in the North Sea called the Silver Pits by fishermen because in some winters it is teeming with soles seeking shelter from the cold. These fish are then easily caught in large numbers.

COMMON NAMES

D: Seezunge, Zunge
DK: Søtunge, tunge
E: Lenguado común
F: Sole commune, sole
GR: Glóssa
I: Sogliola
IS: Sólflúra
N: Tunge
NL: Tong
P: Linguado legítimo
RU: Morskoj yazyk
S: Sjötunga, äkta tunga
SF: Kielikampela, meriantura
TR: Dil
US: Dovèr sole

LOCAL NAMES

PO: Sola
EG: Samak moussa
IL: Sulit mezuya
MO: Hout-moussa
TU: Mdass

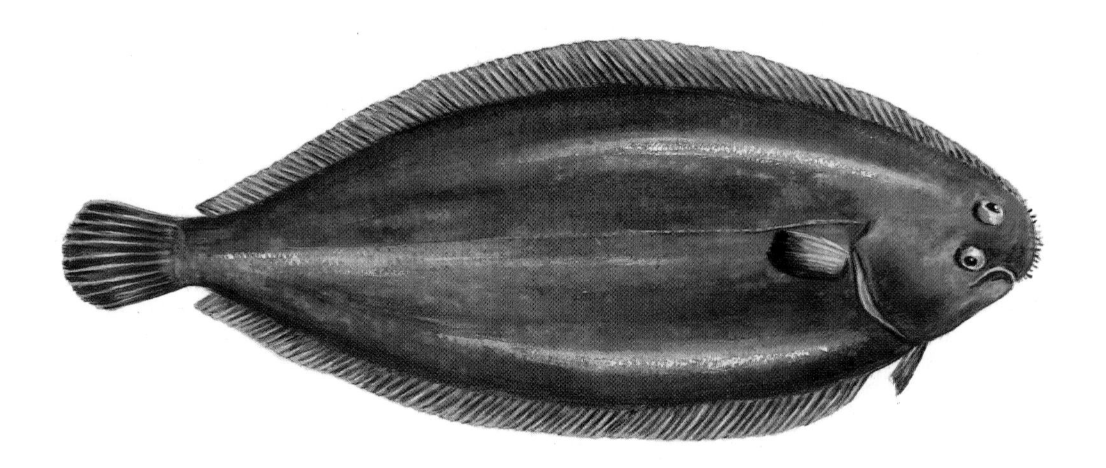

HIGHLIGHTS

The fine eating quality of the common sole has made the name "sole" so highly desirable that many species of fish not technically soles (that is, members of the Soleidae family) are given the name to improve their marketing appeal.

In addition to its fine meat, the sole has a reputation for keeping well without being frozen. Some experts say that its flavour improves after a day on ice, rather as beef improves with aging. Sole certainly freezes well, providing distant markets opportunity to enjoy it.

Nutrition data:
(100 g edible weight)

Water	80.0 g
Calories	89 kcal
Protein	18.1 g
Total lipid (fat)	1.8 g
Omega-3	0.4 mg

DOVER SOLE/COMMON SOLE *Solea vulgaris*

FISHING METHODS

MOST IMPORTANT FISHING NATIONS
Holland, Italy, United Kingdom, France, Denmark, Belgium

PREPARATION

USED FOR

EATING QUALITIES
The common sole is one of the finest flatfish in Europe, esteemed and costly wherever it is available. The meat is tender, white and sweet with a small, closely packed flake. The thick body provides a meaty fillet. Sole is the foundation of French fish cuisine, used as fillets or cooked on the bone in a huge variety of excellent dishes.

There is a story of a French chef in the Seventeenth Century who, when learning that his sole had not been delivered for an important feast, committed suicide (disreputably with a bacon slicer instead of a filleting knife). Apocryphal perhaps, but the tale indicates the importance of the species to the cuisine.

IMPORTANCE
Landings of sole are generally in excess of 50,000 tons a year, one of the most valuable European fisheries. Many countries record significant landings. No fishermen with sole available will miss the opportunity to profit from the resource.

Although often called Dover sole in English, the fish has no clear association with Dover. In the United States, a Pacific species also called Dover sole (*Microstomus pacificus*) is a grossly inferior flatfish with soft flesh. The occasional confusion caused by this nomenclature has not improved the reputation of the true Dover sole, which is found only in European waters.

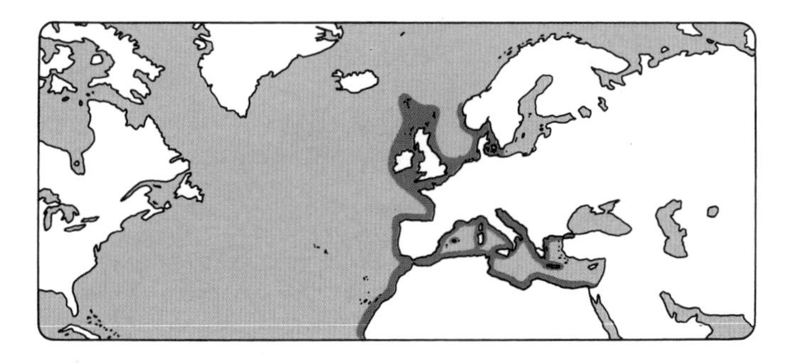

DK: En så delikat og eftertragtet spisefisk, at flere lignende arter, som ikke er af Soleidae-familien, bevidst markedsføres under navnet 'søtunge'

D: Ein so schmackhafter und gefragter Speisefisch, dass mehrere ähnliche Arten, die nicht zur Familie der Seezungen gehören, bewusst unter dem Namen 'Seezunge' auf den Markt kommen

E: Pescado comestible tan delicado y popular que varias especies similares, que no son de la familia de los lenguados, se comercializan intencionadamente bajo el nombre de 'lenguado común'

F: Un poisson de table tellement délicieux que plusieurs espèces qui lui ressemblent sans partenir à la famille des soleidae, sont commercialisées sciemment sous le nom de 'sole'

I: Un pesce commestibile cosi pregiato e richiesto che diverse specie simili, non appartenenti alla famiglia dei soleidi, vengono commercializzati sotto il nome di 'sogliola'

EUROPEAN CATFISH

Scientific name:
Silurus glanis

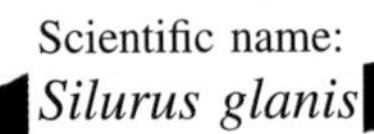

Synonyms: Danubian wels, wels catfish
Family: Siluridae — sheatfishes
Typical size: 1 to 3 m, 100 to 200 kg

The European catfish or wels is found in central and eastern Europe and across southern regions of the former Soviet Union to the basins of the Caspian and Aral Seas.

It is mainly found in large lakes and rivers, though it occasionally ventures into brackish water in the Baltic and Black Seas.

DESCRIPTION
The European catfish is distinguished by its flattened head with barbels around the mouth and the lack of an adipose fin.

A nocturnal feeder, this very large catfish is catholic in its dietary tastes. It is known to eat ducks, voles and crayfish as well as large and small fishes. Because of its size, there are few other organisms in its environment that it cannot attack and consume.

These catfish lack scales. Other members of the family, which is restricted to Europe and central and eastern Asia, are much smaller; some, reaching only 7 cm in length, are popular aquarium species in Europe.

COMMON NAMES
D: Wels, Waller, Flusswels
DK: Malle
E: Siluro
F: Silure d'Europe
GR: Goulianós
I: Siluro
IS: Fengrani
N: Malle
NL: Meerval
P: Siluro europeu
RU: Som
S: Mal
SF: Monni
TR: Yayin baligi
US: Catfish

HIGHLIGHTS
This is one of the largest catfish species: the record fish, taken in Russia's Dnieper River, was 5 m long and weighed 306 kg. Mature fish commonly reach 3 m and weigh 200 kg.

Widely used as a food fish, the species also yields glue, which is made from its swim bladder and bones.

Small quantities are farmed, but so far this fish has not adapted well to domestication.

Nutrition data:
(100 g edible weight)

Water	71.9 g
Calories	163 kcal
Protein	15.3 g
Total lipid (fat)	11.3 g
Omega-3	2.3 mg

EUROPEAN CATFISH *Silurus glanis*

FISHING METHODS

MOST IMPORTANT FISHING NATIONS
Russia, Khazakhstan

PREPARATION

USED FOR

EATING QUALITIES
Small specimens are tasty. They can be pan-fried if small enough. Fillets and steaks may be baked or grilled and the fish is versatile enough to be used in most recipes. The flavour is quite mild and the texture tender.

Large fish — and this species gets very large — tend to have coarse, darker meat which can develop a sour taste very quickly.

Like many other fishes, it is best if this is bled on capture. This lightens the colour of the meat, improves the flavour and extends shelf life.

Catfish is still canned in Russia and other countries of the former Soviet Union.

IMPORTANCE
The European catfish is a major commercial species in parts of eastern Europe and especially in Russia, Khazakhstan and other countries of the former U.S.S.R. Recorded production is over 10,000 tons yearly and it is thought that actual catches may be considerably higher. Because of its size, it is also targeted by anglers.

The species has been introduced beyond its natural range. It has survived, though apparently not prospered, in southern England. Small numbers are now farmed in France.

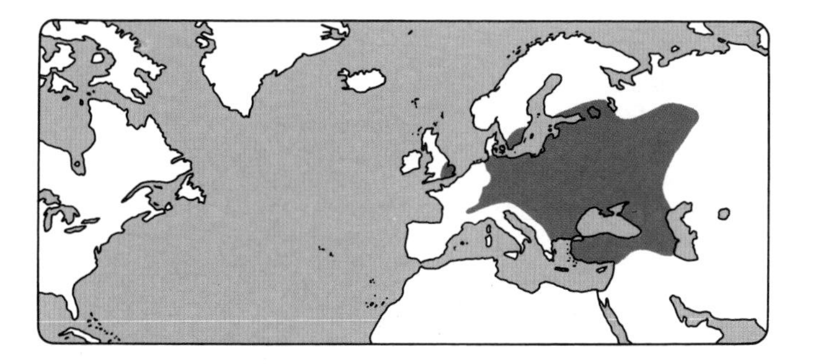

DK: En af de største mallearter: rekordfisken, fanget i Dnieper floden i Rusland, var 5 m lang og vejede 306 kg, mod normalt 3 m og 200 kg. Svømmeblæren og benene anvendes til limfremstilling

D: Eine der grössten Welsarten: der im russischen Dnepr gefangene Rekordfisch war 5 m lang und wog 306 kg, während ihre Normalgrösse bei 3 m und 200 kg liegt. Schwimmblase und Gräten werden zur Leimherstellung verwendet

E: Uno de los siluros más grandes: el pescado récord, capturado en el río Dnieper en Rusia, era de 6 m de longitud y pesaba 306 kg. Normalmente, tiene 3 m de longitud y pesa 200 kg. La vejiga natatoria y las espinas sirven para cola

F: L'un des plus grands des siluridae: le poisson record, pêché dans la Dnieper en Russie, mesurait 5 m et pesait 306 kg contre normalement 3 m et 200 kg. La vessie natatoire et les arêtes sont utilisées dans l'industrie de la colle

I: Una delle specie più grandi fra i Siluridae. L'esemplare record, catturato nel fiume Dnieper in Russia, era lungo 5 m e pesava 306 kg, rispetto ai normali 3 m e 200 kg. La vescica natatoria e le spine sono usate per la produzione di colla

EUROPEAN EEL

Scientific name:
Anguilla anguilla

Family: Anguillidae — freshwater eels
Typical size: 100 cm, 3.5 kg

Eels are catadromous, meaning that they spawn in the sea and return to freshwater streams to grow. European eels and the very similar American eels both breed in the Sargasso Sea. No-one knows how the stocks remain separate, or how the elvers find their way to rivers thousands of miles from the oceanic breeding grounds. Eels are increasingly being farmed; elvers are captured and raised in ponds.

DESCRIPTION
Because of their snake-like appearance, eels do not find favour with some consumers, who find it hard to believe that these animals are true fishes. The elvers entering European rivers are transparent, turning dark brown as they grow. Brown eels are caught during the years they are growing in fresh water. When fully grown, the sides turn bright silver. These silver eels are, of course, larger than brown ones but they are the same species, at different stages of their life cycle.

Eels can live out of water in a damp atmosphere for many days. They are captured in traps, then transported alive to market areas where they can be slaughtered and processed.

COMMON NAMES
D: Aal, Flussaal
DK: Ål, europæisk ferskvandsål
E: Anguila, anguila europea
F: Anguille
GR: Chéli
I: Anguilla, ragano
IS: All
J: Unagi
N: Ål
NL: Paling, aal
P: Enguia, enguia europeia
RU: Evropejskij rechnoj ugor
S: Aal
SF: Ankerias
TR: Yilan baligi
US: European eel

LOCAL NAMES
PO: Wegorz
CH: Sin

HIGHLIGHTS
A nutritious fish with a high oil content, eels can be hot smoked whole, in fillets or in chunks. They can be canned or pickled. They can be prepared from fresh or frozen state in numerous ways. Eels are one of the most versatile and tasty fish. Popular throughout Europe, they are being farmed in increasing quantities to meet the demand.

The eel's life in fresh water is fairly well understood, but its travels in the sea and its precise spawning grounds are still unknown, making successful management of eel resources virtually impossible.

Nutrition data:
(100 g edible weight)

Water	70.6 g
Calories	168 kcal
Protein	16.6 g
Total lipid (fat)	11.3 g
Omega-3	0.4 mg

EUROPEAN EEL *Anguilla anguilla*

FISHING METHODS

MOST IMPORTANT FISHING NATIONS

Italy, Denmark, France

PREPARATION

Live 

USED FOR

EATING QUALITIES

Eels are prepared by skinning and cleaning. It is better if they are bled, which is usually possible as they are frequently distributed alive. There are no pinbones and the meat is easily removed from the backbone, offering boneless, white and very tasty meat which is greatly enjoyed by many Europeans. In Germany, Holland and Scandinavia many eels are hot smoked. In the United Kingdom, a product boiled with vinegar, known as jellied eel, is a traditional and still popular food.

Skinned eel steaks or fillets are excellent fried, steamed or baked. the meat is white, with a small flake and a very sweet flavour. Because the meat is rather fat, it is best cooked in a way that permits some of the oil to drain away. Eels make an excellent soup.

IMPORTANCE

Total production of European eels is fairly stable at around 17,000 tons a year, of which an increasing proportion, already over 50 percent, is farmed. Italy, Holland, Denmark and France are the major producers of farmed eels.

In addition to markets in Europe, eels are exported to Japan, Taiwan and other parts of Asia, where the elvers, the larval young eels, are particularly prized and fetch astronomical prices. Elvers are also eaten in Spain and Italy, when consumers can compete with the prices offered by Asian buyers.

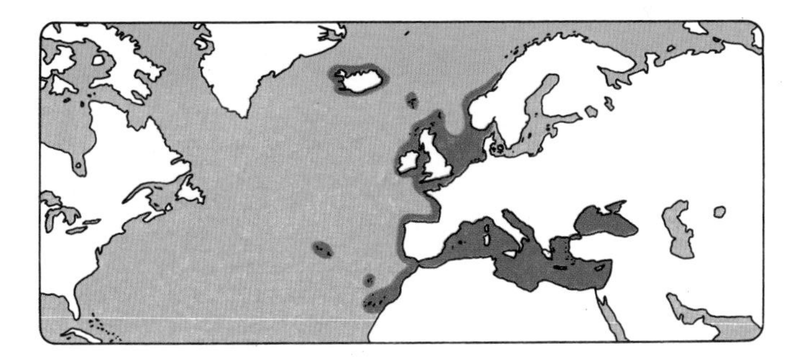

DK: Ålen finder stor og varieret anvendelse på det europæiske marked. Det hvide kød er delikat, og skind og ben er let at fjerne. Opdrættes i stigende grad. Distribueres ofte levende

D: Der Aal wird in grossen Mengen und sehr variabel in Europa verwendet. Sein weisses Fleisch ist delikat, Haut und Gräten lassen sich einfach entfernen. Zunehmende Aufzucht. Wird häufig lebend vertrieben

E: La anguila es muy demandada en el mercado europeo por sus usos variados. Su carne blanca es delicada, y su piel y sus espinas son fáciles de quitar. La cría está aumentando mucho, y se distribuye a menudo viva

F: Les modes d'utilisation de l'anguille sur les marchés européens sont très variés. La chair blanche est savoureuse et les arêtes sont faciles à ôter. De plus en plus, anguille est élevée en pisciculture. Est souvent distribuée vivante

I: L'anguilla è molto richiesta sul mercato europeo per svariati usi. La sua carne bianca è pregiata e la pelle e le spine sono facili da rimuovere. L'allevamento è in forte aumento e spesso viene distribuita viva

FLOUNDER

Scientific name:
Platichthys flesus

Synonyms: White fluke, European flounder, mud flounder, butt
Family: Pleuronectidae — right-eye flounders
Typical size: 50 cm

In some areas, as many as a third of these flounders are left-eyed, making the usual basic taxonomic distinction between flatfish of little value. However, it has small spiny prickles on the lateral line near the head and on the base of the anal and dorsal fins which help to make positive identification.

FISHING METHODS

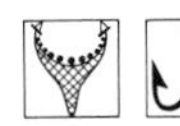

MOST IMPORTANT FISHING NATIONS

Denmark, Germany

PREPARATION

USED FOR

EATING QUALITIES

The meat of the European flounder is rather soft and watery, lacking a definite flavour. It is best cooked whole. The meat is white and is easily removed from the bones after the fish is cooked.

COMMON NAMES

D: Flunder, Butt, Struffbutt
DK: Skrubbe
E: Platija europea, acedia
F: Flet commun, flet
GR: Kalkáni
I: Passera pianuzza
IS: Flundra
J: Karei
N: Skrubbe
NL: Bot, ijbot
P: Solha das pedras
RU: Rechnaya kambala
S: Skrubba, skrubbskädda
SF: Kampela
TR: Derepsi baligi
US: Flounder

LOCAL NAMES

PO: Stornia

Nutrition data:
(100 g edible weight)

Water	81.2 g	Total lipid	
Calories	82 kcal	(fat)	1.8 g
Protein	16.4 g	Omega-3	0.4 mg

HIGHLIGHTS

Flounder is found in shallow water from the White Sea through the Mediterranean to the Black Sea. Although caught throughout this extensive range, it is not highly regarded except in the Baltic countries, where it is abundant and quite popular.

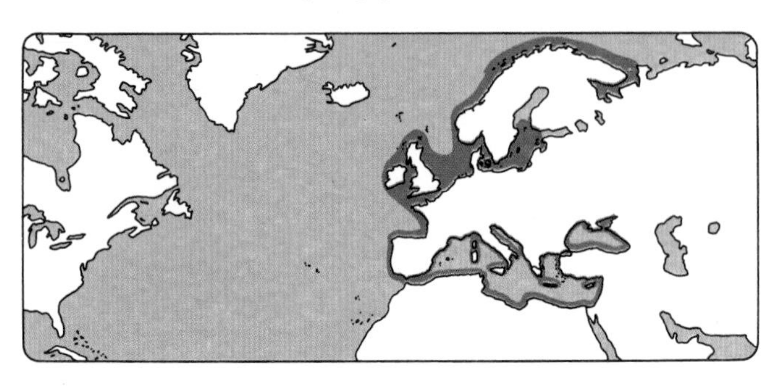

DK: Stor udbredelse fra Hvidehavet til Mid-delhavet og Sortehavet. Fiskes overalt, men er kun populær i landene omkring Østersøen, hvor bestanden er stor

D: Kommt vom Weissen Meer bis zum Mittelmeer und Schwarzen Meer vor. Wird überall gefischt, ist aber nur in den Ostseeländern populär, wo der Bestand gross ist

E: Especie abundante en el Mar Blanco, el Mediterráneo y el Mar Negro. Existe en todos los mares, pero sólo es popular y abundante en los países del Báltico

F: Très répandu de la mer Blanche à la Méditerranée et la mer Noire, le flet est pêché partout, mais il est surtout apprécié dans les pays situés sur la Baltique

I: Largamente diffuso dal Mar Bianco fino al Mediterraneo e al Mar Nero. Pescato dappertutto, ma popolare solo nei paesi baltici, dove il patrimonio è abbondante

FLUKE/SUMMER FLOUNDER

Scientific name:
Paralichthys dentatus

Synonyms: Gulf flounder, northern fluke
Family: Bothidae — left-eye flounders
Typical size: up to 1 m, 10 kg

Fluke or summer flounder is one of the larger North Atlantic flatfish species. American catches have fallen to about 6,000 tons a year and the fishery is heavily regulated to protect apparently declining resources for both commercial and expanding recreational fisheries.

FISHING METHODS

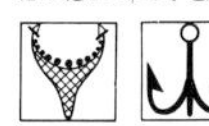

MOST IMPORTANT FISHING NATIONS
United States

PREPARATION

USED FOR

EATING QUALITIES
Fluke has firm, white meat with a small flake typical of flounders. It has a sweet flavour and is excellent battered and fried, or simply grilled. It has thick, meaty fillets from fish which can exceed 11 kg and are regularly as much as 4 to 5 kg.

COMMON NAMES
D: Sommerflunder
DK: Sommerhvarre
E: Falso halibut del Canadá
F: Cardeau d'été
GR: Chomatída tou kalokairioú
I: Rombo dentato
IS: Sumarflundra
J: Hirame
NL: Zomerbot
P: Carta de verao
RU: Letnyaya kambala
SF: Kesäkampela
US: Summer flounder, fluke

LOCAL NAMES
PO: Ptastugi

Nutrition data:
(100 g edible weight)

Water	78.0 g	Total lipid	
Calories	84 kcal	(fat)	0.5 g
Protein	20.0 g	Omega-3	0.1 mg

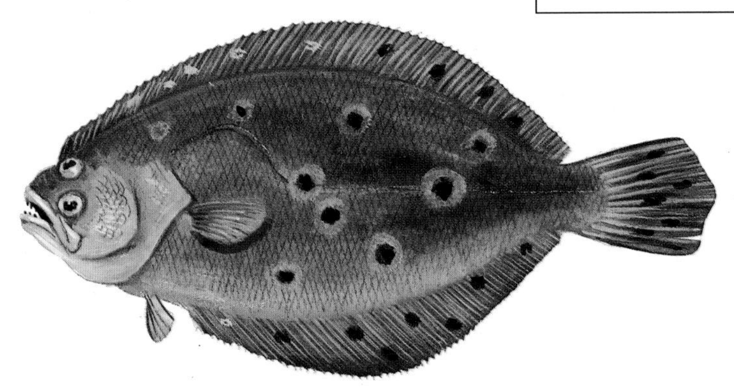

HIGHLIGHTS
Once one of the most important flounders of the American east coast, declining yields from the fishery have left fluke relatively rare and expensive. Japanese buyers, who airfreight the fish fresh and use it raw for sashimi, have added to the pressure on supplies.

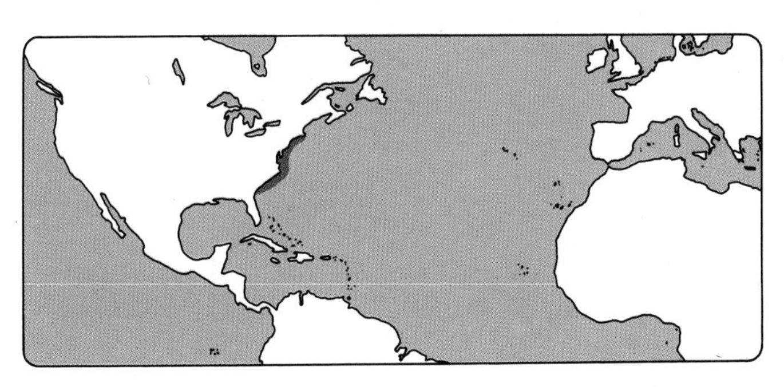

DK: Tidligere en af de vigtigste skrubber ved den amerikanske østkyst, men faldende fangster har idag gjort fisken til en dyr mangelvare

D: Früher eine der wichtigsten Flundern an der Ostküste Amerikas; rückgängige Fangmengen haben aber bewirkt, dass der Fisch heute teure Mangelware ist

E: Antes era una de las platijas más importantes de la costa este de América, pero han bajado mucho las capturas, y por eso es ahora un pescado escaso y caro

F: Autrefois, il était l'un des plus importants des flets de la côte Est américaine mais les pêches en déclin l'ont rendu à la fois relativement rare et cher

I: In precedenza una delle passere più importanti della costa orientale dell'America, ma con il calo delle risorse è diventato un prodottto scarseggiante e caro

GARFISH

Scientific name:
Belone belone

Synonyms: Garpike, billfish, greenbone, mackerel guide
Family: Belonidae - needlefishes
Typical size: 80 cm

The long, slender garfish is fairly common in the eastern Atlantic, from northern Norway as far south as the Canary Islands and throughout the Black and Mediterranean Seas. It is targeted by recreational fishermen as well as for commercial use.

FISHING METHODS

 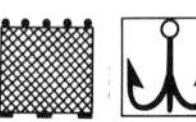

MOST IMPORTANT FISHING NATIONS
Denmark

PREPARATION

USED FOR

EATING QUALITIES
The meat is excellent, quite firm and white with a sweet flavour. The backbone, however, is greenish, which disconcerts consumers unaccustomed to the fish. Nevertheless, the garfish is prized in many areas, especially in Scandinavia and some Mediterranean countries.

COMMON NAMES
D: Hornhecht, Hornfisch
DK: Hornfisk
E: Aguja, agujeta, saltón
F: Aiguille de mer, orphie
GR: Zargána
I: Aguglia, agora
IS: Geirsíli
J: Datsu
N: Horngjel
NL: Geep, snip
P: Peixe agulha
RU: Obyknovenny sargan
S: Horngädda, näbbgädda
SF: Nokkakala
TR: Zargana
US: Garfish

LOCAL NAMES
EG: Khirm
IL: Belon
MO: Boulmarayet
TU: M'sella

Nutrition data:
(100 g edible weight)

Water	75.1 g	Total lipid	
Calories	106 kcal	(fat)	2.5 g
Protein	20.9 g	Omega-3	1.0 mg

HIGHLIGHTS
Unfortunately, the green bones of the garfish are its best-known feature, detracting from its appeal as an excellent food fish. Fast swimming and a good fighter, the garfish is known for its spectacular leaps out of the water when hooked.

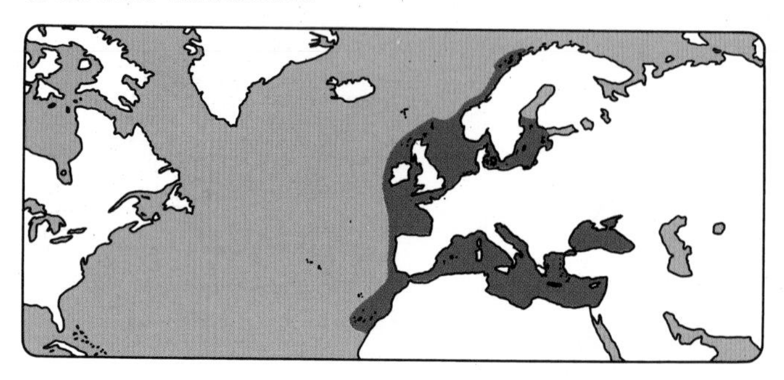

DK: Fortrinlig spisefisk med ret fast, hvidt kød og en sødlig smag. De grønne ben virker ofte afskrækkende på potentielle forbrugere

D: Vorzüglicher Speisefisch mit recht festem, weissem Fleisch und süsslichem Geschmack. Die grünen Gräten verschrecken häufig potentielle Verbraucher ab

E: Excelente pescado comestible con carne bastante firme y blanca, con sabor dulce. Las espinas verdes a menudo desaniman a consumidores potenciales

F: Excellent poisson de table de chair blanche, assez ferme et d'un goût légèrement sucré. Les arêtes verts tendent à décourager les consommateurs potentiels

I: Ottimo pesce commestibile con carni bianche e relativamente sode di sapore dolciastro. Le spine verdi spesso scoraggiano i potenziali consumatori

GIANT STARGAZER

Scientific name:
Kathetostoma giganteum

Family: Uranoscopidae — stargazers
Typical size: 40 cm, 2 kg; up to 90 cm

Found on the shelf around New Zealand, the giant stargazer is taken in small quantities in bottom trawls. Skinless, deboned fillets are sometimes available for export.

FISHING METHODS

MOST IMPORTANT FISHING NATIONS

New Zealand

PREPARATION

USED FOR

EATING QUALITIES

The stargazer has off-white meat with a very firm texture. It is similar in use to monkfish, but unfortunately contains pinbones, detracting from wider use as an alternative species. The meat does not flake, so can be used by penny-pinching or unscrupulous chefs as an extender in lobster or shrimp recipes.

COMMON NAMES

J: Omishima
RU: Gigantsky zvezdochet
US: Stargazer

LOCAL NAMES

MA: Kertatok
NZ: Bulldog

Nutrition data:
(100 g edible weight)

Water	79.6 g	Total lipid (fat)	2.5 g
Calories	90 kcal		
Protein	16.9 g	Omega-3	0.3 mg

HIGHLIGHTS

The giant stargazer lies buried in the sea bed until suitable prey passes by, when it propels itself out of the sand or mud and engulfs the prey with its upward facing mouth, which can hold a meal half the length of the stargazer itself.

DK: Ligger begravet på sandbunden, men hvirvler sig løs efter byttet og sluger det med den aparte opadvendte mund

D: Liegt im Sandboden vergraben, wirbelt sich aber frei nach der Beute und verschlingt sie mit seinem oberständigen Maul

E: Está enterrado en la arena del fondo del mar, pero al pasar una presa sale de su escondite para perseguirla y tragarla con su extraña boca, que da hacia arriba

F: Il s'enterre dans le sable du fond, et quand une proie s'approche, il se dégage en se tourbillonnant pour l'attraper dans sa bouche étrange orientée vers le haut

I: Si seppellisce nelle sabbie dei fondali, ma ne esce fuori come un vortice quando passa la preda, divorandola con la sua strana bocca rivolta verso l'alto

GRAYLING

Scientific name:
Thymallus arcticus

Family: Salmonidae — salmon and trout
Typical size: 35 cm, 750 g; up to 75 cm, 2.7 kg

Found mainly in lakes and rivers of the far north in Siberia and Canada, this Arctic grayling is closely related to the European grayling (*T. thymallus*). It is one of only a few fish in the far north that can be caught with fly-rods.

FISHING METHODS

MOST IMPORTANT FISHING NATIONS
Canada, Russia, United States

PREPARATION

USED FOR

EATING QUALITIES
Grayling have firm, white meat with a delicate flavour. They can be treated much as trout. Grayling are not available commercially, but sportsmen, especially fly-fishermen, spend large sums travelling to the far north to catch and eat them,

COMMON NAMES
D: Äsche
DK: Stalling
E: Timalo
F: Ombre
GR: Thymalos
I: Temolo
IS: Heimskautaharri
N: Harr
NL: Vlagzalm
P: Peixe sombra
RU: Kharius
S: Harr
SF: Pohjanharjus
US: Grayling

LOCAL NAMES
PO: Lipien

Nutrition data:
(100 g edible weight)

Water	69.2 g	Total lipid	
Calories	157 kcal	(fat)	7.9 g
Protein	21.4 g	Omega-3	1.6 mg

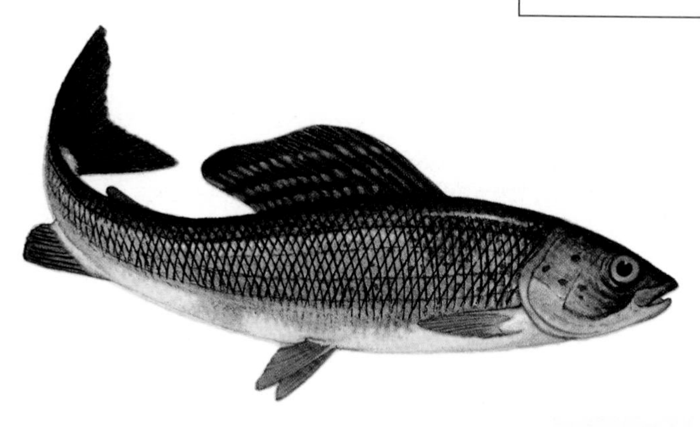

HIGHLIGHTS
Grayling are an attractive game fish: they school in moderate numbers, feed on surface insects and investigate almost everything that lands on the water, including artificial flies. They leap and fight when hooked and are excellent eating.

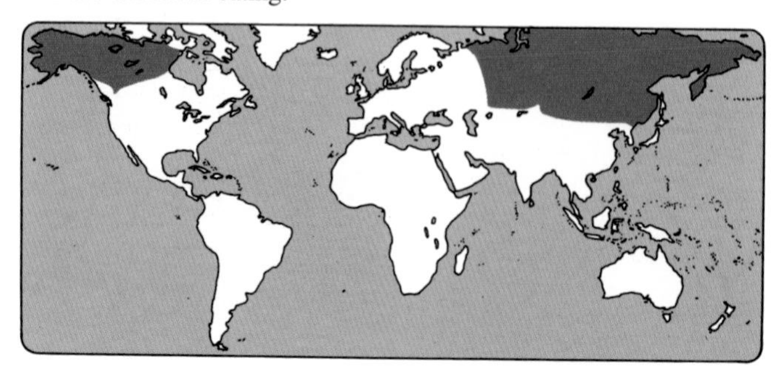

DK: En virkelig god spisefisk og populær sportsfisk: den har den rette kampånd og undersøger alt på vandoverfladen - også fiskerens flue

D: Ein sehr guter Speisefisch und Sportfisch: er besitzt den rechten Kampfgeist und untersucht alles an der Wasseroberfläche, auch die Fliegen der Fischer

E: Exquisito pescado deportivo y popular: tiene espíritu combativo y examina casi todo lo que se encuentre en la superficie del agua, también la mosca del pescador

F: Un excellent poisson sportif et comestible. Doté d'un esprit combatif et curieux de nature, il examine tout ce qui bouge à la surface - même la mouche du pêcheur

I: Un pesce apprezzato per le carni pregiate e, dai pescatori sportivi, per il suo spirito combattivo. Esamina tutto nella superficie - anche la mosca del pescatore

GREATER SANDEEL

Scientific name:
Hyperoplus lanceolatus

Family: Ammodytidae — sand lances
Typical size: 30 cm

The largest of the sandeels or sand lances, this species is found only from northern Spain to Norway and in the adjacent North and Baltic Seas. Large shoals hide by burrowing into sandy sea beds if frightened.

FISHING METHODS

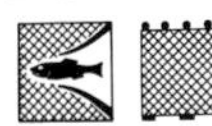

MOST IMPORTANT FISHING NATIONS
Not known

PREPARATION
Fishmeal and oil

USED FOR

EATING QUALITIES
Although mostly used for fishmeal and fish oil, the sandeel can be cleaned and fried like smelts. The meat is white and delicate, lifting easily from the many small bones. The species is mostly distributed fresh and is found mainly in coastal communities.

COMMON NAMES
D: Grosser Sandaal
DK: Tobiskonge
E: Lanzón, aguacioso, sula
F: Grand lançon
GR: Ammóchelo
I: Cicerello
IS: Trönusíli
N: Storsil, stortobis
NL: Smelt, zandspiering
P: Galeota maior, frachao
RU: Bolshaya peschanka
S: Tobiskung
SF: Isotuulenkala
TR: Kum baligi
US: Sand lance

Nutrition data:
(100 g edible weight)

Water	78.2 g	Total lipid (fat)	1.5 g
Calories	89 kcal		
Protein	18.8 g	Omega-3	0.3 mg

HIGHLIGHTS
Sandeels are important forage for major commercial species such as cod, flatfish and salmon. They are not important commercially, but they are sometimes targeted for fishmeal and oil, and occasionally used for bait.

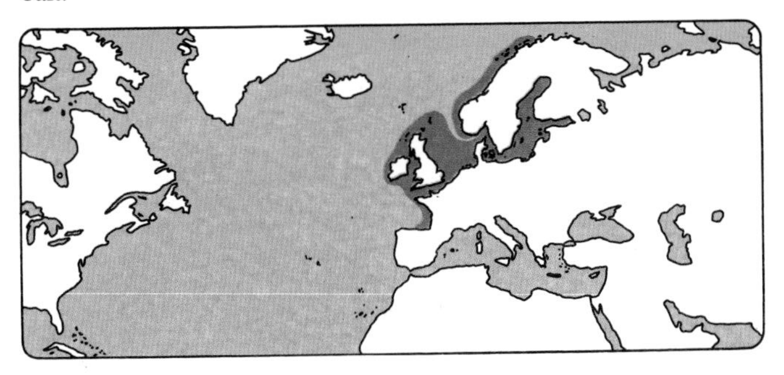

DK: Anvendes hovedsagelig til fiskemel og fiskeolie. Også en fin spisefisk med hvidt, delikat kød, som er let at udbene

D: Wird hauptsächlich zu Fischmehl und Fischöl verarbeitet. Ebenfalls ein schmackhafter Speisefisch mit weissem Fleisch, das sich leicht entgräten lässt

E: Se usa ante todo para harina y aceite de pescado. Es también un apreciado pescado comestible con carne blanca y delicada, cuyas espinas son fáciles de quitar

F: Est utilisé surtout pour farine et huile de poisson, mais c'est aussi un excellent poisson comestible de chair blanche très délicate, facile à désosser

I: Utilizzato prevalentemente per olio e farina di pesce. E' anche un ottimo pesce commestibile con carni bianche e gustose, facili da separare dalla lisca

GREATER WEEVER

Scientific name:
Trachinus draco

Synonym: Stingfish
Family: Trachinidae — weevers
Typical size: 25 to 30 cm, up to 41 cm

Weevers are found in deep water from Norway to Morocco. They are quite common in the Mediterranean, but not often caught because of their habit of burying themselves in the mud, where they are out of reach of trawls and other fishing gear.

FISHING METHODS

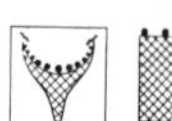

MOST IMPORTANT FISHING NATIONS

Data not available

PREPARATION

USED FOR

EATING QUALITIES

Weevers have good tasting, firm flesh, which can be used in bouillabaisse and other dishes which require lengthy cooking. The venomous spines should be removed by the fisherman or processor before the fish is sold to the consumer.

COMMON NAMES

D: Grosses Petermännchen
DK: Stor fjæsing
E: Escorpión, araña vera
F: Grande vive
GR: Drákaina
I: Tracina drago, ragno
IS: Fjörsungur
J: Toragisu
N: Fjesing
NL: Grote pieterman, arend
P: Peixe aranha maior
RU: Morskoj drakon, skorpion
S: Fjärsing
SF: Louhikala
TR: Trakonya
US: Weever

LOCAL NAMES

EG: Ballama
TU: Drachna

Nutrition data:
(100 g edible weight)

Water	78.4 g	Total lipid	
Calories	84 kcal	(fat)	0.7 g
Protein	19.4 g	Omega-3	0.1 mg

HIGHLIGHTS

Weevers are best known for the poisonous spines on their backs and gill covers. The venom can be very painful as well as disabling. Secondary infections from the wound can lead to permanent injury. A related species is found on beaches in shallow water.

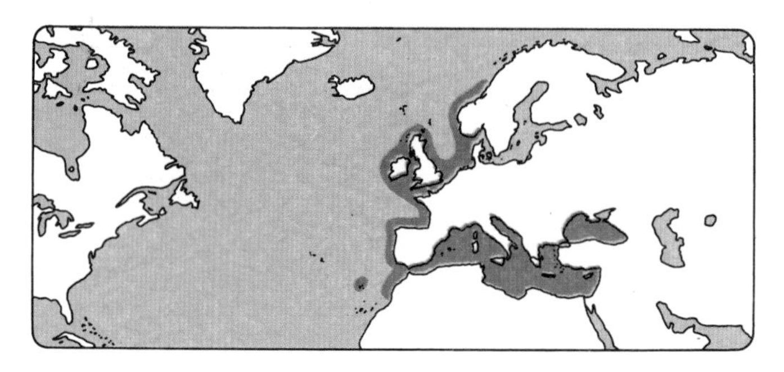

DK: Fjæsingen er mest kendt for giftpiggene på ryggen og gællelågene. Giften kan være meget smertefuld samt invaliderende

D: Grosse Petermännchen sind bekannt für ihre Giftstacheln auf Rücken und Kiemendeckel. Das Gift kann starke Schmerzen, Schwellungen und Entzündungen hervorrufen

E: El escorpión es mejor conocido por sus espinas venenosas de la espalda y opérculos branquiales. El veneno puede ser muy doloroso y aún dejarle a uno mutilado

F: La grande vive est connue pour ses épines venimeuses sur le dos et sur les opercules des branchies. Le venin peut être très douloureux et même paralysant

I: Il trachino drago è conosciuto soprattutto per le spine velenifere nella pinna dorsale e sugli opercoli. Il veleno può essere doloroso e menomante

GREENLAND HALIBUT/TURBOT

Scientific name:
Reinhardtius hippoglossoides

Synonyms: Black halibut, blue halibut, lesser halibut
Family: Pleuronectidae — right-eye flounders
Typical size: up to 1 m, 10 kg

Increasing world catches now exceed 125,000 tons yearly, with Norwegian production increasing substantially. This large flatfish is usually filleted and frozen. It is also made into blocks for the production of portions and coated products.

FISHING METHODS

 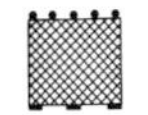 

MOST IMPORTANT FISHING NATIONS
Iceland, Norway, Greenland, Russia, Portugal, Canada

PREPARATION

USED FOR

EATING QUALITIES
Greenland turbot is comparatively fat for a flatfish, making the creamy coloured meat moist, sometimes watery, and rather soft. It is frequently sold as a substitute for flounder (in North America) and plaice (in Europe), although its eating quality is inferior to both.

COMMON NAMES
D: Schwarzer Heilbutt
DK: Hellefisk
E: Hipogloso negro
F: Flétan noir, halibut noir
GR: Hálibat tis Grilandías
I: Halibut di Groenlandia
IS: Grálúda
J: Karasugarei
N: Blåkveite
NL: Groenlandse heilbot
P: Alabote da Gronelândia
RU: Grenlandskij paltus
S: Lilla hälleflundran
SF: Grönlanninpallas
US: Greenland turbot

LOCAL NAMES
PO: Kulbak czarny

Nutrition data:
(100 g edible weight)

Water	70.3 g	Total lipid	
Calories	182 kcal	(fat)	13.8 g
Protein	14.4 g	Omega-3	1.0 mg

HIGHLIGHTS
Greenland halibut (greenland turbot) is found in deep cold boreal water on both sides of the North Pacific and North Atlantic Oceans. This is a major commercial species, sold mainly as a substitute for better, more expensive species than for its limited intrinsic qualities.

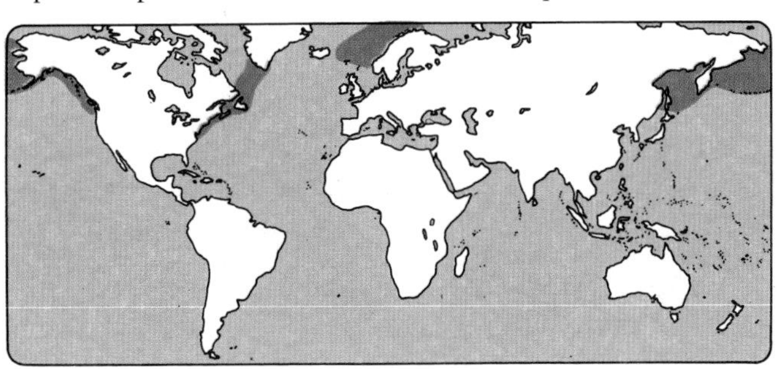

DK: En vigtig og rigt forekommende fladfisk, der for at dække efterspørgslen, sælges som erstatning for de bedre og dyrere arter

D: Ein wichtiger und zahlreich vorkommender Plattfisch, der um den Bedarf des Marktes zu decken als Ersatz für bessere und teurere Arten verkauft wird

E: Pleuronecto importante y abundante, que se vende en sustitución de especies mejores y más caras para satisfacer el mercado

F: Un poisson plat important et très abondant qui est souvent vendu en substitution d'espèces plus chères pour satisfaire à la demande du marché

I: Una specie dei pleuronettiformi di grande importanza commerciale, venduta in sostituzione delle specie migliori e più costose, per coprire il fabbisogno

GRENADIER

Scientific name:
Coryphaenoides rupestris

Synonym: Roundnose grenadier
Family: Macrouridae — grenadiers or rattails
Typical size: up to 100 cm

Grenadiers inhabit very deep water, sometimes down to 2,200 m. They are mostly caught with pelagic trawls by factory trawlers in late summer and fall when they form huge schools at 600 to 900 m depth. Catches of grenadiers have declined greatly, from over 80,000 tons to less than 10,000 tons. The resource is considered overutilized.

FISHING METHODS

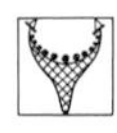

MOST IMPORTANT FISHING NATIONS
Russia, Denmark, Germany, Poland

PREPARATION
Fishmeal 

USED FOR

EATING QUALITIES
Grenadiers have good texture and taste, but the narrow, pointed fillet shape can sometimes be awkward to use. Block and surimi production have been attempted, so far with limited success.

COMMON NAMES
D: Randnasiger Grenadierfisch
DK: Skolæst, langhale
E: Granadero
F: Grenadier de roche
GR: Grenadiéros ton vráchon
I: Granatiere
IS: Slétthali, slétti langhali
J: Shiira
N: Skolest
NL: Grenadier
P: Lagartixa da rocha
RU: Tuporyly makrurus
S: Skoläst
SF: Lestikala
US: Rock grenadier

Nutrition data:
(100 g edible weight)

Water	79.0 g	Total lipid (fat)	0.3 g
Calories	79 kcal		
Protein	19.0 g	Omega-3	0.2 mg

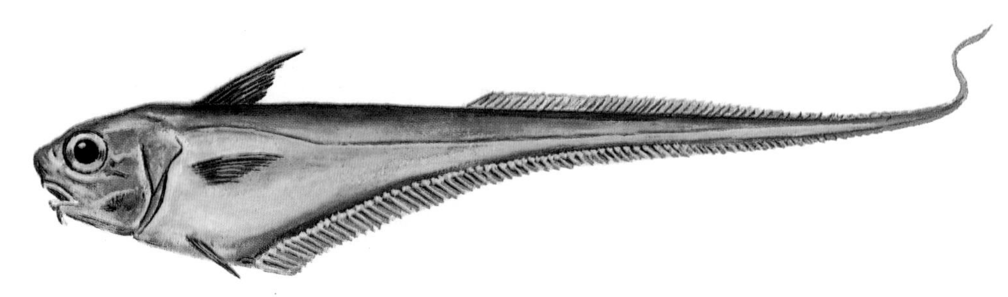

HIGHLIGHTS
This is the most important commercial species of grenadier of the many found in the world oceans. Larger than most other grenadiers, it is still difficult to harvest from its deep habitat. The species yields high quality fillets, making it worth while to hunt for them.

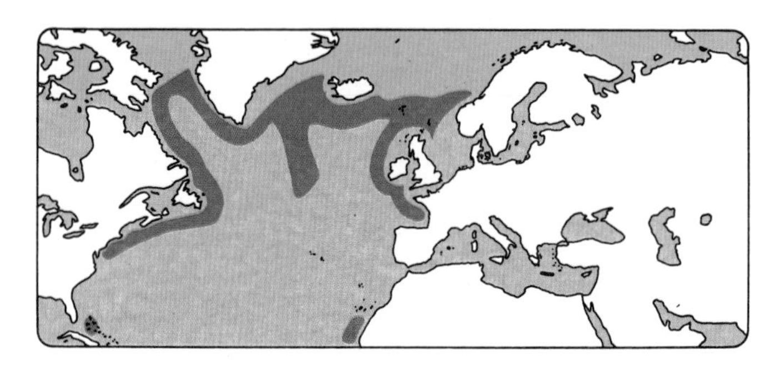

DK: Større end de fleste andre skolæste men alligevel en svært tilgængelig dybhavsbundfisk. Fin spisefisk med lange hvide filetter

D: Grösser als die meisten anderen Grenadierfische aber trotzdem ein schwer zugänglicher Tiefseebodenfisch. Schmackhafter Speisefisch mit langen weissen Filets

E: Más grande que la mayoría de los grenaderos, pero es un pescado de aguas profundas de acceso difícil. Exquisito pescado deportivo con filetes largos y blancos

F: Plus grand que la plupart des grenadiers, mais toujours un poisson des grands fonds, difficile à pêcher. Excellent poisson de table avec de longs filets blancs

I: Più grande degli altri pesci granatieri, ma tuttavia un pesce abissale difficilmente accessibile. Pesce commestibile squisito con filetti lunghi e bianchi

GROPER

Scientific name:
Polyprion oxygeneios

Synonyms: Hapuku, Juan Fernandez wreckfish
Family: Serranidae — sea basses
Typical size: 100 cm, 6 kg

Gropers grow as large as 20 kg. They are found throughout New Zealand and in southern Australia, where they are sometimes landed as a by-catch of the trawl fishery for blue eye (*Hyperoglyphe ant-arctica*). New Zealand has small quantities available for export, but the fish provides catches of only 1,000 to 2,000 tons yearly.

FISHING METHODS

 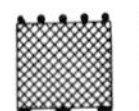 

MOST IMPORTANT FISHING NATIONS

New Zealand

PREPARATION

USED FOR

EATING QUALITIES

Groper meat is firm and off-white, with a mild flavour. It can be used in most cooking methods and recipes. The roes are large and can be smoked, then fried.

COMMON NAMES

D: Hapuku-Wrackbarsch
DK: Newzealandsk vragfisk
E: Cherna de Juan Fernández
F: Cernier de Juan Fernandez
GR: Vláchos tis N. Zilandías
I: Cernia Neozelandese
J: Nyuuziirando-ôhata
NL: Hapuku wrakbaars
P: Cherne da Nova Zelândia
RU: Gruper
S: Vrakfisk
SF: Hylkyahven
US: Groper

Nutrition data:
(100 g edible weight)

Water	77.0 g	Total lipid	
Calories	95 kcal	(fat)	1.8 g
Protein	19.8 g	Omega-3	0.3 mg

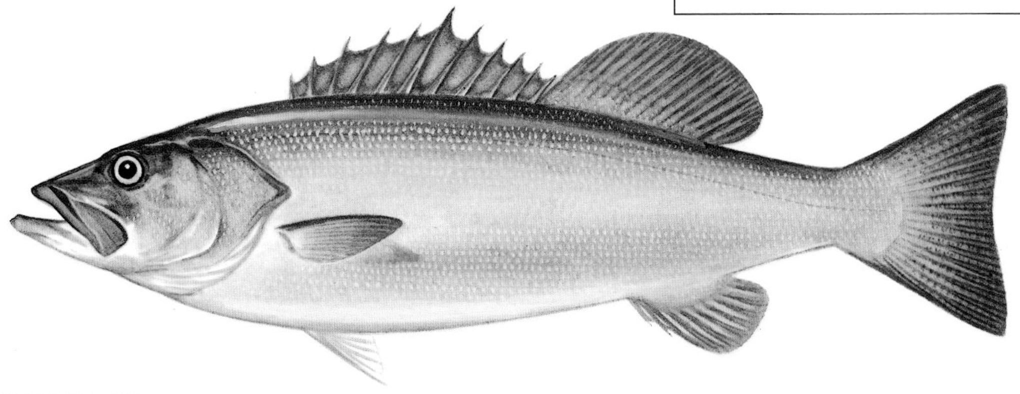

HIGHLIGHTS

Groper is caught and sold fresh throughout the year, but is mostly landed in summer, when the fish gather for spawning. Although the flesh is less good when the fish is ripe, the roes are desirable, adding to the value of the fishery at this time.

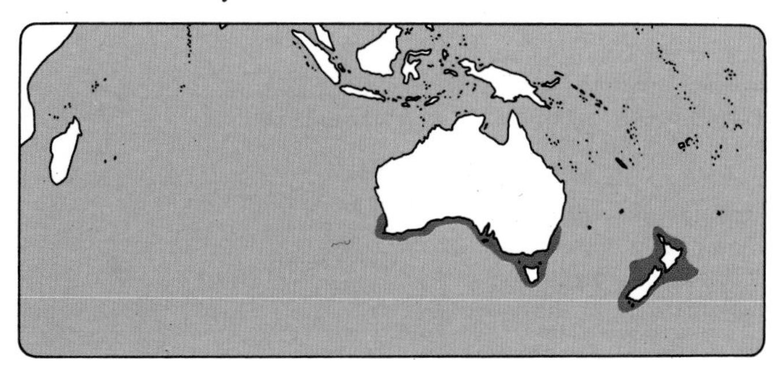

DK: En god spisefisk med en fortrinlig rogn. Markedet er året rundt forsynet med fersk kød, omend de største fangster tages i sommerens gydeperiode

D: Guter Speisefisch mit vorzüglichem Rogen. Ganzjährige Belieferung des Markts mit frischem Fisch, obwohl im Sommer zur Laichzeit am meisten gefangen wird

E: Buen pescado comestible, que se captura y vende fresco todo el año, pero las capturas son más importantes en verano durante el desove ofreciendo exquisitas huevas

F: Un bon poisson de table dont les oeufs sont très appréciés. Il est vendu frais sur les marchés toute l'année, mais les meilleures prises se font en été

I: Un buon pesce commestibile con uova eccellenti. Venduto fresco tutto l'anno, ma le maggiori quantità sono prese durante il periodo estivo della riproduzione

HADDOCK

Scientific name: *Melanogrammus aeglefinus*

Synonyms: Gibber, seed haddock (small haddock)
Family: Gadidae — cods
Typical size: 50 cm, 2 kg

Haddock is an important species for European and north American fishermen and consumers. Although its abundance is reduced, it is still greatly in demand, especially for smoking. The finnan haddock, a smoked product invented in Scotland in the last century, is popular on both sides of the Atlantic and is sold fresh, frozen or canned in a variety of forms and packaging.

DESCRIPTION
Similar to Atlantic cod in appearance, though smaller, the haddock is well regarded by consumers. It is found at depths down to 400 m, usually at temperatures between 4^0 and 10^0 C. Haddock migrate long distances to and from spawning grounds. They grow slowly and have low fecundity, so do not recover well from fishing pressure. The major fishing grounds are in the eastern Atlantic. The species is important locally in Canada and the USA, although stocks on the western side have greatly diminished.

The largest haddock recorded was from Iceland. It measured 112 cm and weighed 16.8 kg.

COMMON NAMES
D: Schellfisch
DK: Kuller
E: Eglefino, anon, liba
F: Eglefin, aiglefin, ânon
GR: Bakaliáros
I: Eglefino
IS: Ysa
J: Montsukidara
N: Kolje, hyse
NL: Schelvis
P: Arinca
RU: Piksha
S: Kolja
SF: Kolja
US: Haddock

LOCAL NAMES
PO: Lubacz

HIGHLIGHTS
Haddock catches are far below their historical levels. The resource throughout the North Atlantic has declined, in some places disastrously. Nevertheless, large quantities of haddock are caught, processed, sold and consumed.

Haddock is particularly valued in northern England and Scotland, for fish-and-chips and for smoking. The species is now mostly too expensive for salting and drying, but smoked haddock, sometimes made from cheaper white fish substitutes, remains a strong market.

Nutrition data:
(100 g edible weight)

Water	79.4 g
Calories	81 kcal
Protein	19.0 g
Total lipid (fat)	0.6 g
Omega-3	0.2 mg

HADDOCK *Melanogrammus aeglefinus*

FISHING METHODS

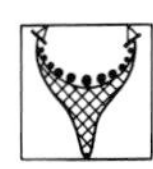

MOST IMPORTANT FISHING NATIONS

Iceland, United Kingdom, Norway, Canada, Faroe Islands

PREPARATION

USED FOR

EATING QUALITIES

Haddock is very similar to Atlantic cod, but smaller, with a smaller flake. Preferences for one over the other tend to be regional and traditional. Outsiders may not be able to tell the difference at all when sampling the two species. Overall, haddock is a little drier than cod, with a slightly more pronounced flavour. Like cod, it is a moist, white meat that adapts well to most recipes and cooking techniques.

Haddock smokes well. Finnan haddock is made as follows: first, split small haddock (under 1 kg), then brine it and hang it to dry. Cold smoke for up to six hours. The yellow colour develops in the smoke and subsequent cooling. Nowadays, finnans are often made from fillets, as consumers in Europe and north America increasingly eschew bones.

IMPORTANCE

Haddock is distinguished from cod by the dark colour of the lateral line and the large, dark blotch on each side just above the pectoral fin. These marks remain after the fish is processed. As a result, most haddock is sold with the skin on, so the buyer can determine that the species is indeed haddock, not cod.

Haddock catches dropped precipitously in the late 1980s from over 400,000 tons to only 200,000 tons annually. Iceland and the United Kingdom produce over half the catch. The decline has been throughout the range of the species, but has been more acute in the western Atlantic.

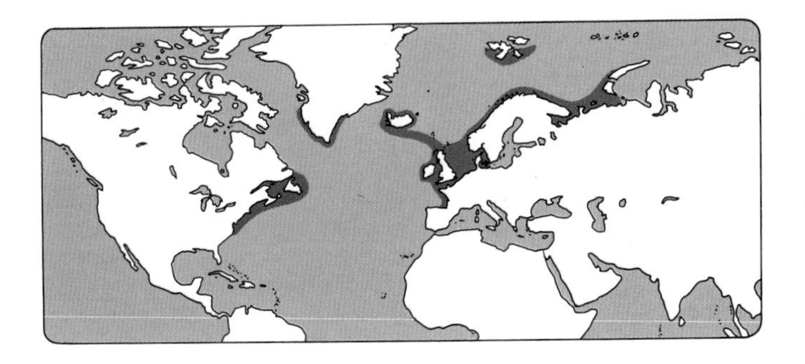

DK: Bestanden er stærkt reduceret. Stadig stor efterspørgsel efter denne fisk til især rygning. Anvendes i Skotland til 'finnan haddock' (koldrøget kullerfilet), en populær spise på begge sider af Atlanten

D: Der Bestand ist stark dezimiert bei anhaltender Nachfrage, besonders zum Räuchern. Wird in Schottland zu 'finnan haddock' (kaltgeräuchertes Schellfischfilet), einem beiderseits des Atlantiks beliebten Gericht, verarbeitet

E: Las existencias han bajado mucho, pero aún muy demandado, especialmente para ahumados. En Escocia se usa para 'finnan haddock' (filete de pescadilla ahumada en frío), popular a ambos lados del Atlántico

F: Bien que la population de l'espèce ait été fortement réduite, elle est toujours très demandée, surtout pour le fumage. Le 'finnan haddock' (filet d'aiglefin fumé), préparé en Ecosse, est un mets très prisé des deux côtés de l'Atlantique

I: La popolazione di questa specie è stata fortemente ridotta, data la grande domanda. Usato in Scozia per 'finnan haddock' (filetto di eglefino affumicato a freddo), un piatto apprezzato su ambedue i lati dell'Atlantico

HAKE

Scientific name:
Merluccius merluccius

Synonym: European hake
Family: Merlucciidae — merluccid hakes
Typical size: 30 to 60 cm, may reach 140 cm, 15 kg

Nomenclature of hakes and whitings is confused, especially in the United States where the two words are virtually interchangeable and whiting is generally regarded as a better marketing term. Although European hake, *Merluccius merluccius*, is not used in the USA, a number of other hakes and whitings are; they are mostly inexpensive, small fish. The European hake is never called whiting.

In Spain, where it is particularly well regarded, the name is "merluza." Names of the same derivation are used in many other European countries for this and similar fish.

DESCRIPTION
Hake is a fairly large relative of the cod, rather slender and long compared with other hakes and whitings. The inside of the mouth is almost black, accounting for the Danish name. It is found on the Atlantic coast of Europe and northwest Africa, from Iceland and Norway to Mauritania. It is also found in the Mediterranean and along the southern shores of the Black Sea. The Mediterranean and Black Sea stocks, which are generally smaller than the Atlantic fish, are considered by some experts to be a separate subspecies.

COMMON NAMES
D: Europäischer Seehecht
DK: Europæisk kulmule
E: Merluza europea
F: Merlu européen, colin
GR: Bakaliáros
I: Nasello
IS: Lysingur
N: Lysing
NL: Mooie meid, heek
P: Pescada, pescada branca
RU: Angolsky khek, merluzy
S: Kummel
SF: Kummeliturska
TR: Berlâm
US: Whiting

LOCAL NAMES
EG: Nazelli
MO: Lcola
TU: Nasalli
PO: Morszczuk

HIGHLIGHTS
Hake is found in moderately deep water around the coasts of the Eastern Atlantic and is mostly caught by trawling. The species is highly valued in Spain, France, Italy and other countries. Despite this high regard, catches have been falling; but some experts believe that the resources could provide more of this fine fish in the future.

Hake is most highly valued fresh, but it freezes well and in many instances may be better quality frozen than fresh.

Nutrition data:
(100 g edible weight)

Water	79.0 g
Calories	92 kcal
Protein	18.0 g
Total lipid (fat)	2.2 g
Omega-3	0.4 mg

HAKE *Merluccius merluccius*

FISHING METHODS

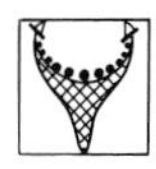  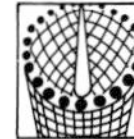 

MOST IMPORTANT FISHING NATIONS

Spain, Italy, France, Morocco, Portugal

PREPARATION

USED FOR

EATING QUALITIES

Hake has delicate, white meat which flakes easily from the bones once the fish is cooked. Although the texture is soft and the flakes tend to separate before and during cooking, the excellent flavour makes the fish well worth the extra care that is needed to prepare it. Some recipes suggest that dressed fish for baking should be tied to hold it together. Traditional cuisines in Italy, Spain, Portugal and the Adriatic have numerous recipes for hake.

Fresh hake tends to lose its flavour rather quickly, becoming what one writer calls "insipid." Freezing helps to delay this tendency.

IMPORTANCE

Hake has long been eaten fresh by Europeans. It has also been salted and dried for later use, though only the largest fish are used for curing. Recent landings have been around 110,000 tons annually, of which about 75 percent comes from the northeast Atlantic, the remainder from the Mediterranean, the Black Sea and West African waters. Scientists estimate that the northeast Atlantic resource alone could support annual harvests of 150,000 tons, or about twice current landings, indicating that catches might be increased in the future.

In Mediterranean countries, hake is often regarded as the best gadoid. Spanish freezer trawlers travel to distant waters to harvest the species. Other hakes are imported from around the world as substitutes.

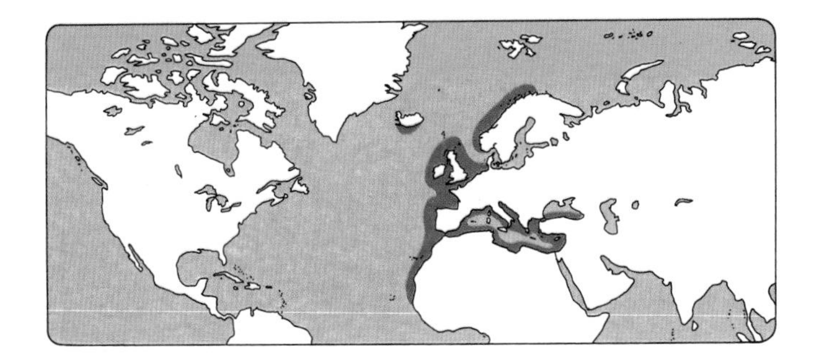

DK: Regnes for en virkelig god spisefisk i Spanien, Frankrig, Italien. Foretrækkes frisk, selvom kvaliteten faktisk forbedres ved frysning

D: Wird in Spanien, Frankreich und Italien als ausgesprochen guter Speisefisch geschätzt. Wird frisch bevorzugt, obwohl die Qualität eigentlich durch das Einfrosten verbessert wird

E: Es considerado un pescado comestible muy apreciado en España, Francia e Italia. Se prefiere fresco, aunque en muchos casos la calidad es mejor congelada que fresca

F: Est considéré comme un excellent poisson comestible en Espagne, France et Italie. On le préfère frais bien que la qualité soit effectivement améliorée par la congelation

I: Considerato un pesce commestibile molto pregiato in Spagna, Francia e Italia. E' preferito fresco, anche se la qualità migliora con il congelamento

HALIBUT

Scientific name:
Hippoglossus hippoglossus

Family: Pleuronectidae — right-eye flounders
Typical size: 120 cm, 22 kg.

Atlantic halibut are the largest of all flatfishes and one of the most valuable. Because they are large, live in deep water and are difficult to capture, recreational fisheries are not important, but commercial fishermen throughout the North Atlantic have no trouble selling every fish they can catch.

DESCRIPTION
Atlantic halibut grow as large as 2.4 m and 300 kg. They live in cool, but not the very coldest, waters on both sides of the North Atlantic. In the east, halibut are found from the Bay of Biscay to the Barents Sea, where the Gulf Stream pushes warmer water far to the north. In the west, where there is no warm current so far north, they range from the west coast of Greenland along the Labrador shelf to the Gulf of Maine, including the Grand Banks and Georges Bank. The most southerly specimen recorded came from Virginia in 1946, but the species is not normally encountered south of New Jersey.

Halibut grow fast and may live for 30 years or more. Individuals over 90 kg are now very rare, an indication that heavy exploitation has affected the resource.

COMMON NAMES
D: Atlantischer Heilbutt
DK: Helleflynder
E: Hipogloso, fletán, halibut
F: Flétan de l'Atlantique
GR: Hippóglossa, hálibat
I: Ippoglosso atlantico
IS: Flydra, lúda, heilagfiski
J: Ohyô
N: Kveite
NL: Heilbot
P: Alabote do Atlântico
RU: Paltus
S: Hälleflundra
SF: Ruijanpallas
US: Atlantic halibut

LOCAL NAMES
PO: Halibut, kulbak

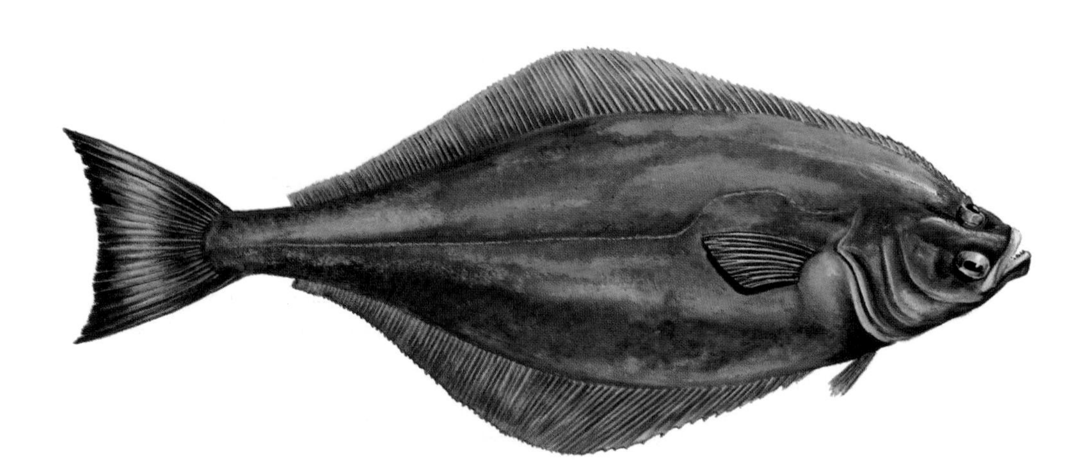

HIGHLIGHTS
Many experts consider that halibut is one of the best of all fish to eat. There are few which can provide such large pieces of boneless meat. The desirability and value of halibut has grown as the supply has decreased.

Halibut keeps well on ice and is excellent frozen, although the scarcity and high price of the Atlantic species means that most of the frozen halibut sold now is the more plentiful Pacific halibut, *Hippoglossus stenolepsis*.

Nutrition data:
(100 g edible weight)

Water	76.3 g
Calories	103 kcal
Protein	21.5 g
Total lipid (fat)	1.9 g
Omega-3	0.6 mg

HALIBUT

Hippoglossus hippoglossus

FISHING METHODS

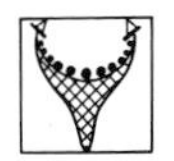 

MOST IMPORTANT FISHING NATIONS

Canada, Spain, Iceland

PREPARATION

USED FOR

EATING QUALITIES

Halibut is one of the finest of all white-fleshed fish. The meat is firm yet tender, standing up to every cooking method without toughening. The strand-like flakes hold together well, so the fish can be cooked without the skin. The large bones are easily seen and removed. The flavour is sweet and definite, without being at all fishy. Halibut makes excellent fish-and-chips, although often considered too expensive for such a mundane use. It also smokes well: hot smoked halibut is a fine, delicate food which deserves wider distribution.

Halibut makes a tasty soup. An additional benefit is that the bones provide a gel which helps to thicken the dish. Cooked meat retains its taste and texture well, so can be used successfully in salads.

IMPORTANCE

In 1852 Moses Perley, writing of Atlantic halibut in Nova Scotia, observed: "This fish is found in such abundance, and in so large size, that the localities are avoided by those engaged in cod-fishing, as a boat, or small vessel, becomes soon too heavily laden."

This once abundant species has almost disappeared from both sides of the North Atlantic. Total landings are stable at around 7,000 tons, of which Canada and Iceland take the major part. Large fish are mostly caught on longlines, smaller specimens are taken in trawls. Much of the production is now by-catch from other groundfish fisheries.

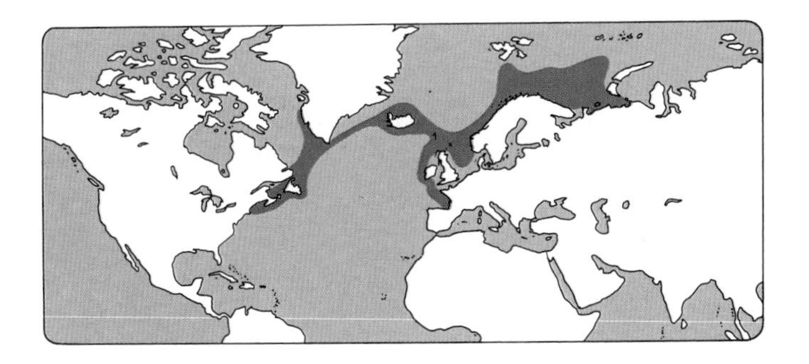

DK: Af mange experter regnet for en af verdens bedste spisefisk. Kun få andre arter giver så store stykker hvidt, velsmagende benløst kød. Priserne er steget proportionalt med faldende fangster

D: Viele Kenner halten ihn für einen der besten Speisefische der Welt. Nur wenige andere Arten ergeben so grosse weisse, schmackhafte, grätenlose Stücke. Die Preise sind proportional zu den rückgängigen Fängen gestiegen

E: Muchos expertos opinan que es uno de los mejores pescados comestibles del mundo. Pocas otras especies dan piezas tan grandes de carne blanca, sabrosa y sin espinas. Los precios han subido proporcionalmente a la caída de las capturas

F: Considéré par les experts comme l'un des meilleurs poisson de table du monde. Peu d'espèces fournissent d'aussi grands morceaux de chair blanche sans arêtes, d'un goût aussi savoureux. Les prix montent au rythme de la pénuerie d'arrivages

I: Da molti esperti considerato uno dei pesci commestibili migliori del mondo. Solo poche altre specie danno pezzi di carne bianca e senza spine cosi grandi e gustosi. I prezzi sono aumentati di pari passo con il calo delle risorse

HERRING

Scientific name:
Clupea harengus

Synonyms: Atlantic herring, digby, mattie, sild, yawling
Family: Clupeidae — herrings, sardines
Typical size: 20 to 25 cm, up to 40 cm
See also Pacific herring

Herrings have been salted in Europe for winter food for at least 1,000 years. Disputes about access to herring stocks started medieval wars and continue to be a problem for the European Union. Herring was a staple food on the east coast of North America and supported enormous industries in Maine and Newfoundland. Canada continues to produce large quantities of herring for European and Far East markets, including some roe for Japan, although the Pacific herring is more highly valued for this purpose.

Most herrings are cured or canned and the fat content is critical to producing a successful product. The many different herring stocks spawn at different times, so the raw material can be available with the "right" fat content over a lengthy period.

COMMON NAMES

D: Hering
DK: Sild
E: Arenque
F: Hareng de l'Atlantique
GR: Rénga
I: Aringa
IS: Síld
J: Nishin
N: Sild, strømming
NL: Haring
P: Arenque
RU: Mongopozvonkovaya seld
S: Sill, strömming
SF: Silli, silakka
TR: Ringa
US: Herring

LOCAL NAMES

PO: Sledz

DESCRIPTION

Herrings are small, silver fish which shoal in large numbers, usually near the surface. Their high oil content means that they turn rancid quickly, so must be handled with great speed to avoid spoilage. The same feature makes them amazingly versatile for many uses, including pickling, salting, smoking and canning.

HIGHLIGHTS

Although herring catches are far below their historical peaks, the species is still a very important one, especially in Europe. There are a number of subspecies such as the Baltic herring and some experts regard the Pacific herring also as a subspecies of the Atlantic fish. Whatever the precise taxonomy, all herring subspecies are valued and sought by fishermen throughout colder coastal seas of the North Atlantic. Stocks are now rigorously protected everywhere and the species shows signs of recovering some of its former abundance.

Nutrition data:
(100 g edible weight)

Water	68.0 g
Calories	190 kcal
Protein	17.8 g
Total lipid (fat)	13.2 g
Omega-3	1.8 mg

HERRING *Clupea harengus*

FISHING METHODS

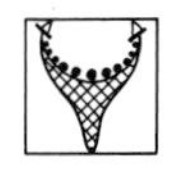 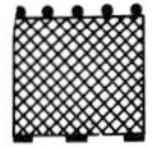 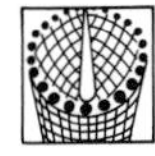

MOST IMPORTANT FISHING NATIONS

Canada, Norway, Denmark, Sweden, Russia, UK

PREPARATION

USED FOR

EATING QUALITIES

Fresh herring has mild, creamy coloured meat with a small flake. It is excellent fried, grilled or broiled and can also be marinated. Herring is mostly processed by smoking, curing or canning before it is eaten. There are multitudes of products, traditional and new. The best known smoked herrings include bloaters, buckling and kippers; pickled products include rollmops, bratherring, bismarck herring and matjes; canned herring come in every imaginable sauce; there are many salted herring preparations. Tastes are changing, moving away from saltier products, but herring processors have been able to develop saleable alternatives.

IMPORTANCE

While fishing pressure and natural cycles have reduced catches to under 1.5 million tons a year, herring remains a very important fish, especially in Europe where it is used in an enormous array of preparations and forms, including, of course, fresh and frozen. Herrings are even eaten raw and whole in Holland, although the practice is now discouraged because of the incidence of anisakis parasites in the gut of the fish, which is not normally eaten elsewhere.

Every coastal nation in the herring's range has valuable fisheries for the species; most of these people have been catching and eating herring for centuries. Although modern products are sometimes quite different from the traditional uses, the fish remains highly regarded, widely consumed and economically important, especially in northern Europe.

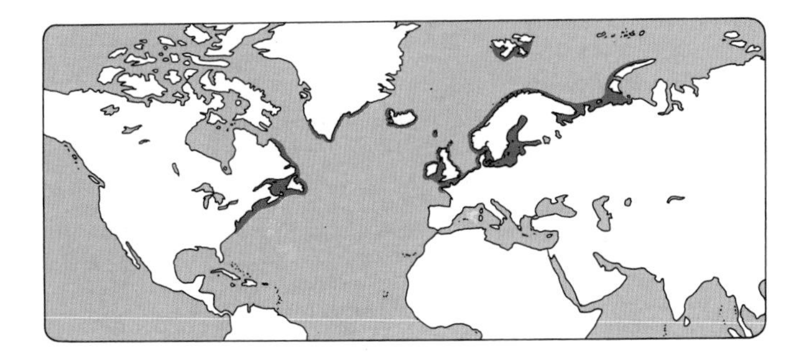

DK: Trods overfiskning og uheldige klimatiske ændringer, der har reduceret den årlige fangst til under 1,5 mill ton er silden, især i Europa, stadig en vigtig konsumfisk med utallige anvendelsesmuligheder

D: Trotz Überfischen und ungünstiger Klimaänderungen, die den jährlichen Fang auf weniger als 1,5 Mio. Tonnen beschränkt haben, ist der Hering weiter ein wichtiger Speisefisch, besonders in Europa, mit unzähligen Anwendungsmöglichkeiten

E: A pesar de la pesca excesiva y cambios climáticos que han reducido las capturas anuales a menos de 1,5 mill de tldas., el arenque sigue siendo un importante pescado de consumo, sobre todo en Europa donde tiene un sinfín de usos

F: Malgré une surpêche et des changements climatiques qui ont réduit la pêche à moins de 1,5 mill. de t/an, le hareng conserve son importance surtout en Europe, comme poisson alimentaire avec d'innombrables possibilités de préparation

I: Nonostante la pesca eccessiva ed i cambiamenti climatici sfavorevoli che hanno ridotto le catture annuali a meno di 1,5 milioni di tonn., l'aringa è ancora un pesce di consumo importante con tante possibilità di utilizzo

HOKI/BLUE GRENADIER

Scientific name:
Macruronus novaezelandiae

Synonym: New Zealand whip-tail
Family: Merlucciidae — merluccid hakes
Typical size: 60 to 100 cm, 1.5 kg
Similar species: Patagonian grenadier, *M. magellanicus*

The fishery for hoki or blue grenadier (note that the species is not a grenadier, although it is called by this name in Australia) is quite new. Until 1986, catches were under 50,000 tons yearly. Now, catches are as much as eight times greater, prompted by the development of markets for this lean, white fish. The fishery requires large vessels, capable of trawling in deep water and processing the catch on board. The similar Patagonian grenadier is exploited mainly by Chile, with landings well over 100,000 tons.

DESCRIPTION
This long, thin fish from deep, cold waters of the southern hemisphere is sometimes called whiptail, a descriptive name which indicates its appearance. It is blue-green above, silvery below and has large scales which are easily removed. Found in large schools, the species may grow as long as 120 cm.

Hoki are mostly caught with bottom and midwater trawls at depths of 500 to 900 m. The major fishery is during the southern winter, from June to September, but it is fished year round in different parts.

COMMON NAMES
D: Neuseeländischer Grenadier
DK: Newzealandsk langhale
E: Cola de rata azul
F: Grenadier bleu de N. Zél.
GR: Grenadiéros tis N. Ziland.
I: Merluzzo granatiere
IS: Hokinhali
J: Hoki
NL: Blauwe grenadier, hoki
P: Granadeiro azul
RU: Novozelandsky makruronus
US: New Zealand whiting

LOCAL NAMES
NZ: Whip-tail, hoki, blue hake
AU: Blue grenadier

HIGHLIGHTS
Hoki provide long, thin fillets with excellent eating characteristics, although the shape can present some difficulties for some users. Putting the fillets into blocks solves that problem; blocks are widely acceptable for most portion and coated products. Hoki can be made into surimi, when surimi markets are strong enough to attract the raw material into this use.

The large resource is concentrated around New Zealand, with smaller schools in Australian waters.

Nutrition data:
(100 g edible weight)

Water	80.4 g
Calories	85 kcal
Protein	16.9 g
Total lipid (fat)	1.9 g
Omega-3	0.3 mg

HOKI/BLUE GRENADIER *Macruronus novaezelandiae*

FISHING METHODS

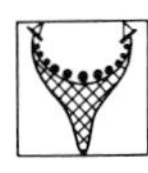

MOST IMPORTANT FISHING NATIONS

New Zealand, Japan, Russia, Australia

PREPARATION

Surimi  

USED FOR

EATING QUALITIES

Hoki meat is delicate, succulent, and white when cooked. It has a medium flake and is suitable for most cooking methods used for white-meated species. Fresh fillets are available in Australia and sometimes in New Zealand. Frozen fillets are sold in numerous world markets.

The fillets have a strip of fat under the lateral line which should be removed to extend shelf life and improve the flavour of fillets and fillet blocks. Defatted blocks are excellent raw material for coated portions and other processed fish products.

IMPORTANCE

There is a very large resource of hoki, currently producing in the region of 300,000 to 400,000 tons of fish yearly. Formerly known as New Zealand whiptail in the United States, the species has earned a solid market following under the name of hoki, selling as an alternative to cod, whiting and other groundfish species.

New Zealand produces over half of the catch, while much of the rest is taken in New Zealand fishing zones under licence. Japanese, Russian and Korean vessels participate in the fishery. Almost all of this catch is taken by large factory trawlers and processed at sea. Hoki is also caught off southern Australia, where it is a significant part of the South-East Fishery. Some of this catch is sold fresh into Melbourne and other local markets.

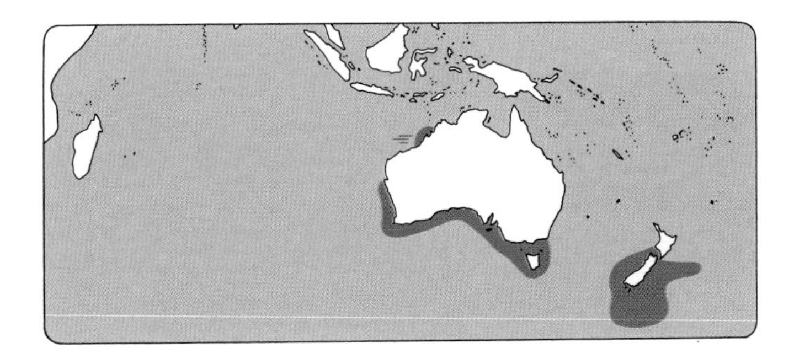

DK: Kvalitetsspisefisk, som giver lange, tynde, lyse filetter. Filetternes form kan ved videreforarbejdningen skabe problemer, som dog løses ved sammenpresning i blokke. Bruges også til 'surimi'

D: Qualitätsspeisefisch, der lange, dünne, weisse Filets liefert. Diese Form der Filets kann bei der Weiterverarbeitung Probleme bereiten, die sich durch das Zusammenpressen zu Blöcken beheben lassen. Wird auch zu 'Surimi' verarbeitet

E: Pescado comestible de calidad, que da filetes largos, delgados y blancos. La forma de los filetes puede crear problemas para su preparado posterior, pero se soluciona comprimiéndolos en bloques. Sirve también para 'surimi'

F: Poisson comestible de qualité. Lors de la transformation, la forme des longs filets minces de chair blanche peut créer des problèmes, qui sont cependant résolus en les comprimant en blocs. Est aussi utilisé pour préparer du 'surimi'

I: Pesce commestibile di alta qualità che dà filetti bianchi, lunghi e sottili. Questa forma può rendere difficile la lavorazione dei filetti, ma il problema si risolve, comprimendoli in blocchi. Usato anche per 'surimi'

HORSE MACKEREL

Scientific name:
Trachurus novaezelandiae

Family: Carangidae — jacks
Typical size: 30 to 40 cm, 850 g

Horse mackerel is a large and apparently underutilized resource in the southern hemisphere. The meat is of moderate quality but the fish is often used, either alive or dead, as bait for tuna fishing by both commercial and sport fishermen.

COMMON NAMES
J: Nyujirando maaji
RU: Novozelandskaya stavrida
US: Jack mackerel, scad

LOCAL NAMES
AU: Yellowtail scad

FISHING METHODS

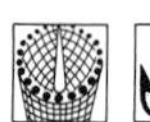 

MOST IMPORTANT FISHING NATIONS
New Zealand, Australia, Japan

PREPARATION

USED FOR

EATING QUALITIES
The dark meat of the horse mackerel lightens when cooked. It has a strong flavour, but it is still milder than the oilier horse mackerels of the northern hemisphere. In Australia, the fish is held in ice slurries for best quality.

Nutrition data:
(100 g edible weight)

Water	72.0 g	Total lipid	
Calories	139 kcal	(fat)	6.3 g
Protein	20.5 g	Omega-3	1.5 mg

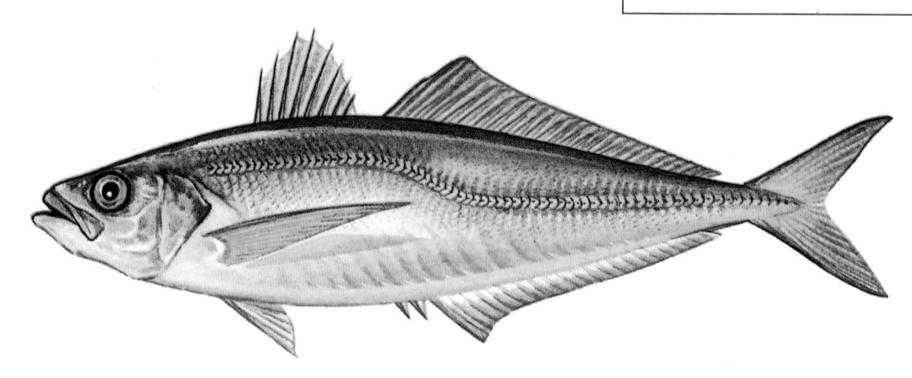

HIGHLIGHTS
Catches of the very similar greenback horse mackerel (*T. declivis*) are mainly used for fishmeal, but increasing quantities of both these species are canned, either for pet food or human consumption. Horse mackerel, if properly handled, can be utilized in frozen form.

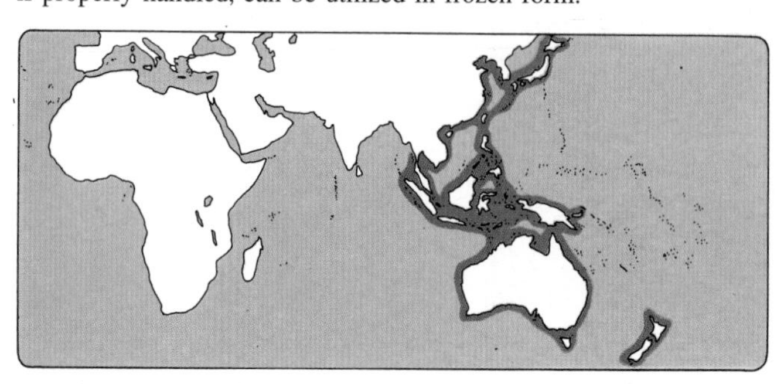

DK: Store, uudnyttede forekomster på den sydlige halvkugle. Fisken er af middelmådig kvalitet og anvendes ofte som madding i tunfiskeriet

D: Grosse, ungenutzte Bestände auf der südlichen Halbkugel. Der Fisch ist von mässiger Qualität und wird häufig als Köder beim Thunfischfang eingesetzt

E: Grandes existencias sin aprovechar en el hemisferio austral. La carne es de mediana calidad y se usa a menudo para cebo de la pesca de atunes

F: Dans l'hémisphère austral, les énormes ressources sont encore peu exploitées. De qualité assez médiocre, il est surtout utilisé comme appât pour la pêche au thon

I: Ci sono grandi risorse non sfruttate sull'emisfero australe. Il pesce è di qualità mediocre e viene spesso adoperato come esca nella pesca dei tonni

HORSE MACKEREL/SCAD

Scientific name:
Trachurus trachurus

Family: Carangidae — jacks
Typical size: 25 cm; up to 50 cm, 1.5 kg

Mostly processed as steaks or fillets canned in oil, the scad is a low value fish available in very large quantities. It is considered to be an excellent bait for swordfish and tuna; part of the catch is frozen for that purpose.

FISHING METHODS

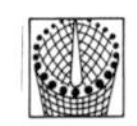 

MOST IMPORTANT FISHING NATIONS
Netherlands, Denmark, Norway, Ireland, Spain, Portugal, Germany

PREPARATION

USED FOR

EATING QUALITIES
Horse mackerel is a moderate quality fish for eating, with dark coloured meat which has a strong flavour and can be stringy. It is best hot smoked and can also be used for making well flavoured fish stock and fumet.

COMMON NAMES
D: Bastardmakrele, Stöcker
DK: Hestemakrel
E: Jurel, saurel, chincharro
F: Chinchard d'Europe, saurel
GR: Savrídi
I: Suro, sugarello
IS: Brynstirtla
J: Muroaji, maaji, aji
N: Taggmakrell
NL: Horsmakreel, maasbanker
P: Carapau, chicharro
RU: Yuzhnoafrikansk. stavrida
S: Taggmakrill
SF: Piikkimakrilli
TR: Istavrit
US: Scad

LOCAL NAMES
EG: Seif
IL: Trakhon gedol-magen
MO: Chrene
TU: Shourou

Nutrition data:
(100 g edible weight)

Water	76.9 g	Total lipid	
Calories	102 kcal	(fat)	2.5 g
Protein	19.9 g	Omega-3	0.5 mg

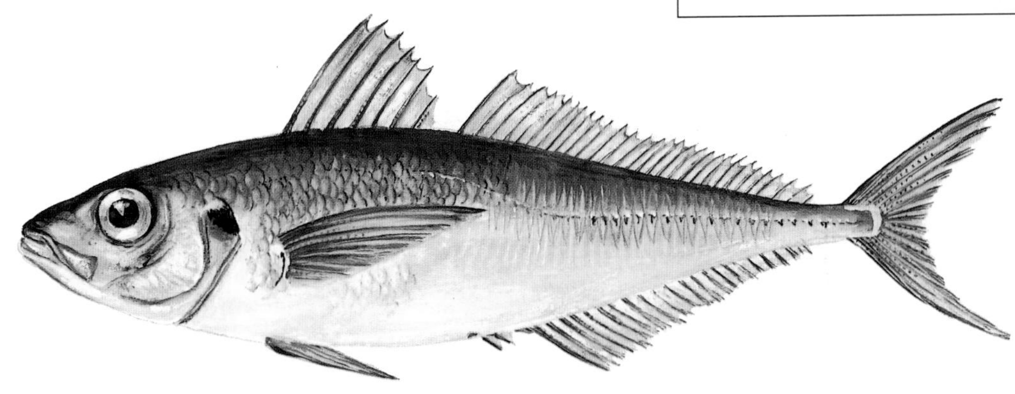

HIGHLIGHTS
European countries now land over 400,000 tons of horse mackerel a year from the eastern Atlantic, with increasing quantities being targeted. In South Africa, the species, known there as mossbunker, is the basis for a large industry canning, smoking and selling fresh.

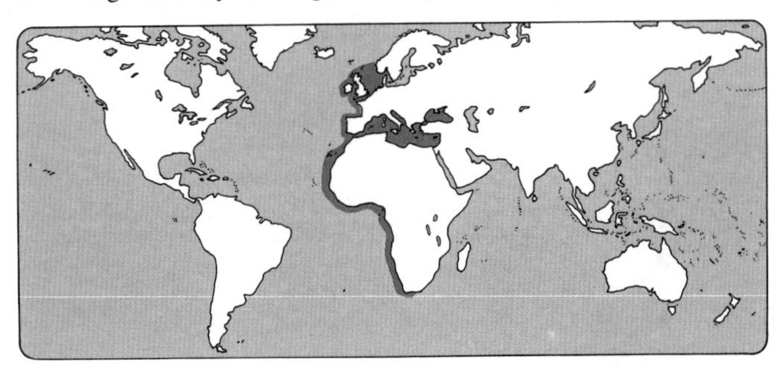

DK: De europæiske lande fisker årligt store mængder af hestemakrel i det østlige Atlanterhav, og fiskeriet er stadig stærkt stigende

D: Die europäischen Länder fangen jährlich grosse Mengen Bastardmakrele im Ostatlantik. Die Fänge sind weiterhin stark im Anstieg begriffen

E: Los países europeos capturan grandes cantidades de jurel cada año en el este del Atlántico, y su pesca aumenta constantemente

F: Les pays européens pêchent annuellement des quantités énormes de chinchards dans l'Atlantique Est, et la pêche est en progression constante

I: I paesi europei pescano annualmente grossi quantitativi di sugarello nell'est Atlantico e la pesca va fortemente crescendo

JACK MACKEREL

Scientific name:
Trachurus symmetricus

Synonym: Pacific horse mackerel
Family: Carangidae — jacks
Typical size: up to 76 cm, 2.3 kg

Jack mackerel is excellent smoked, but most commercial catches are canned and sold into markets for canned mackerel. There is a sub-stantial sport fishery, especially in California and Mexico where the species is most abundant and found close to shore.

FISHING METHODS

 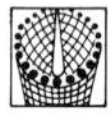

MOST IMPORTANT FISHING NATIONS
United States

PREPARATION

USED FOR

EATING QUALITIES
Jack mackerel should be bled when caught. This lightens the meat and improves the flavour, as well as extending the shelf life. Like other jacks, this one has fairly thick skin, which is best removed with the red subcutaneous fat, to make the flavour milder.

COMMON NAMES
D: Pazifische Bastardmakrele
DK: Stillehavshestemakrel
E: Jurel del norte, chicharro
F: Chinchard du Pacifique
GR: Fengarópsaro
I: Suro del Pacifico, suro
IS: Kyrrahafsbrynstirtla
NL: Pacifische horsmakreel
P: Carapau do Pacífico
RU: Peruanskaya stavrida
US: Jack mackerel

Nutrition data: (100 g edible weight)			
Water	69.2 g	Total lipid	
Calories	150 kcal	(fat)	6.3 g
Protein	23.2 g	Omega-3	1.3 mg

HIGHLIGHTS
Jack mackerel resources and catches fluctuate wildly from year to year. The species is reported to be one of the more abundant fishes off California, but lack of markets and regulatory controls inhibit greater exploitation.

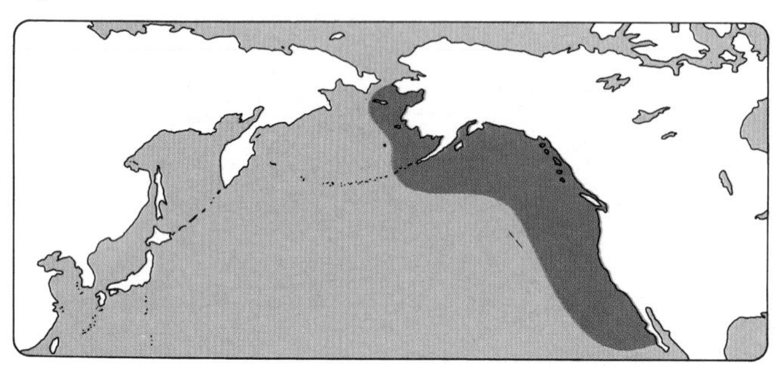

DK: En af de mere talrige arter ved Californien, men manglende markeder og lave fangstkvoter forhindrer et mere intensiveret fiskeri

D: Eine der zahlreicheren Arten vor Kalifornien, die aber wegen fehlender Märkte und niedriger Fangquoten nicht intensiv genutzt wird

E: Una de las especies más abundantes de California, pero la falta de mercados y bajas cuotas impide una pesca más intensiva

F: Une des espèces abondantes rencontrées au large de la Californie. La pénurie de marchés et les faibles quotas de capture interdisent une pêche plus intensive

I: Questa specie è una delle più abbondanti della California, ma l'assenza di mercati e i bassi contingenti costituiscono un ostacolo alla pesca intensiva

JAPANESE ATKA MACKEREL

Scientific name:
Pleurogrammus azonus

Family: Hexagrammidae — greenlings
Typical size: up to 45 cm

Although Japan and Korea exploit Atka mackerel intensively, there is a very large resource of an almost identical species, *A. monopterygius*, available in American waters which has hardly been fished since it was used as a cheap, salted staple for feeding miners during the 1849 California gold rush.

COMMON NAMES

D: Terpug
E: Lorcha de Atka
F: Maquereau de Atka
J: Hokke
NL: Atka makreel
P: Lorcha de Atka
RU: Japonsky morskoj lenok
US: Atka mackerel

FISHING METHODS

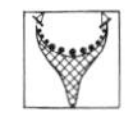 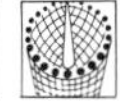

MOST IMPORTANT FISHING NATIONS

Japan, Russia, USA, Korea

PREPARATION

USED FOR

EATING QUALITIES

An oily fish, atka mackerel has dark meat and a strong flavour. It is popular in Japan, where it is used in a wide variety of dishes and is fairly inexpensive.

Nutrition data:
(100 g edible weight)

Water	78.0 g	Total lipid	
Calories	100 kcal	(fat)	3.9 g
Protein	16.3 g	Omega-3	0.8 mg

HIGHLIGHTS

Japanese atka mackerel is a greenling, related to lingcod. It is not a mackerel, although it has similarly dark and oily meat. Like other greenlings, it occasionally has a greenish tint to the bones, although this disappears when the fish is cooked or salted.

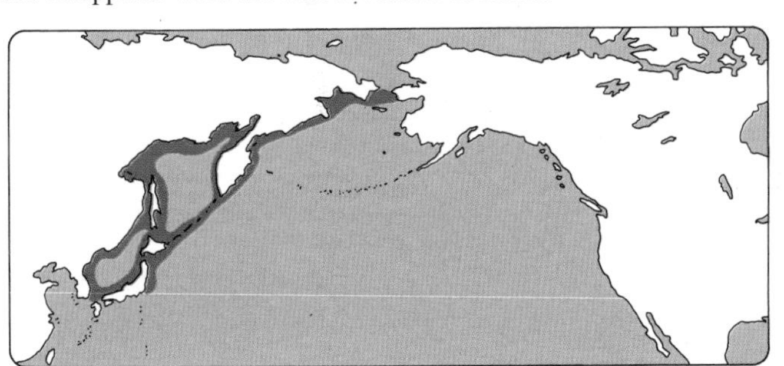

DK: En talrig og næsten identisk art fra de Amerikanske farvande blev under Guldfeberen i 1849 saltet til et billigt næringsmiddel

D: Eine zahlreiche und sehr ähnliche Art in den amerikanischen Gewässern wurde während des Goldrausches von 1849 zu einem billigen Lebensmittel eingesalzen

E: Especie abundante y casi idéntica a la de las aguas americanas. Durante la Fiebre del Oro en 1849 fue salado y sirvió de comida barata

F: Espèce abondante et presque identique à celle des eaux américaines qui, à l'époque de la Ruée vers l'or en 1849, se vendait bon marché à l'état salé

I: Una specie numerosa e quasi identica delle acque americane fu conservata sotto sale durante la febbre dell'oro nel 1849, costituendo un alimento a buon mercato

JAPANESE CHAR

Scientific name:
Salvelinus leucomaenis

Family: Salmonidae — salmonids
Typical size: up to 30 cm, 3 kg

The Japanese or East Asian char is similar to the Arctic char. It ranges from Japan north to the Sea of Okhotsk. Like other chars, it has greater value as a game fish than a commercial fish for human consumption. The species is not farmed.

FISHING METHODS

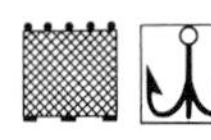

MOST IMPORTANT FISHING NATIONS
Japan, Russia

PREPARATION

USED FOR

EATING QUALITIES
Japanese char have white, amber or red flesh, well-flavoured, moist and firm, somewhat similar to brook trout. They are particularly good to eat if taken in the sea or soon after they migrate into fresh water. Red-meated populations are preferred.

COMMON NAMES
D: Fernöstlicher Saibling
DK: Japansk fjeldørred
J: Iwana
NL: Japanse ridder
RU: Kundzja

Nutrition data:
(100 g edible weight)

Water	71.5 g	Total lipid	
Calories	153 kcal	(fat)	9.0 g
Protein	18.0 g	Omega-3	1.4 mg

HIGHLIGHTS
Japanese char are not normally seen commercially, although the fish is important as a subsistence species in some parts of its range on the mainland of East Asia. Some experts dispute whether this is really a separate species, or simply a distinct race of the Arctic char.

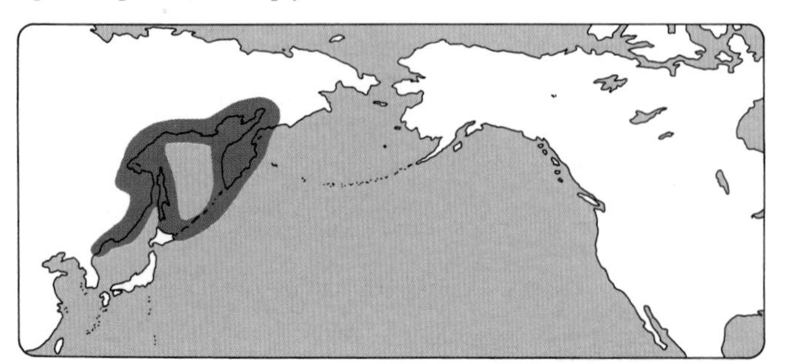

DK: Minder meget om almindelig fjeldørred, og som de øvrige ørreder er den primært en sportsfisk. Fiskes fra Japan og nordpå til Det ochotske Hav

D: Ähnelt sehr dem Seesaibling und wird wie die übrigen Saiblinge hauptsächlich von Sportfischern gefangen; von Japan bis nördlich zum Ochotskischen Meer

E: Similar al salvelino corriente y como los demás de esta especie es ante todo un pescado deportivo. Se pesca desde Japón y hacia el norte hasta el Mar de Ojotsk

F: Ressemble à l'omble chevalier. Comme les autres truites, c'est surtout un poisson de pêche sportive. Il est pêché du Japon jusqu'à la mer d'Okhotsk

I: Molto simile al salmerino artico e, come le altre trote, è prevalentemente un pesce sportivo. Si pesca dal Giappone e verso nord fino al Mare di Ohotsk

JAPANESE EEL

Scientific name:
Anguilla japonica

Family: Anguillidae — freshwater eels
Typical size: 100 cm, 3.5 kg

Ranging from Korea along the Asian coast to Taiwan, the Japanese eel is the most important species of eel, by both volume and value, in world fisheries. Total production is about 60,000 tons, almost all of which is produced by aquaculture in Taiwan.

FISHING METHODS

MOST IMPORTANT FISHING NATIONS

Taiwan, Japan, Korea

PREPARATION

Live, kabayaki    

USED FOR

EATING QUALITIES

White meat with firm flesh characterise the Japanese eel, which is very similar to its European cousin. It is used in a multitude of traditional dishes in Japan and other Asian countries, where the elvers are also a considerable and valuable delicacy.

COMMON NAMES

D: Japanischer Aal
DK: Japansk ål
E: Anguila japonesa
F: Anguille du Japon
GR: Chéli tis Iaponías
I: Anguilla giapponese
IS: Japans-áll
J: Unagi
N: Japansk ål
NL: Japanse paling
P: Enguia japonesa
RU: Japonskij rechnoj ugor
S: Japansk ål
SF: Japaninankerias
TR: Japon yilan baligi
US: Japanese eel

Nutrition data:
(100 g edible weight)

Water	63.9 g	Total lipid	
Calories	217 kcal	(fat)	15.8 g
Protein	18.8 g	Omega-3	3.2 mg

HIGHLIGHTS

Eel is an extremely valuable farmed species; Taiwan currently dominates production, though farmers in Japan and Korea are also developing the necessary techniques. Elvers are captured and grown: farmers cannot yet breed eels for production.

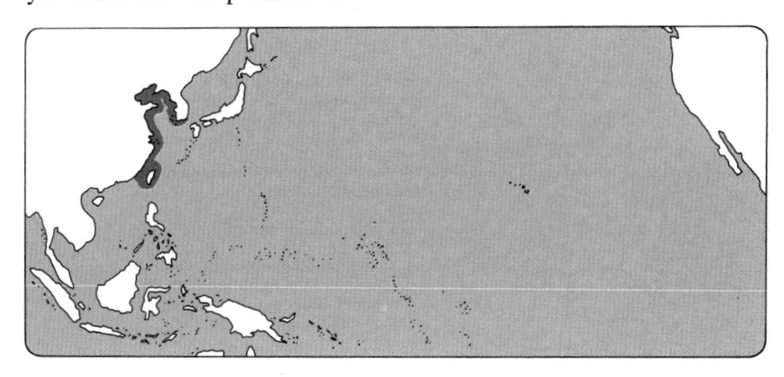

DK: En utrolig vigtig, opdrættet art. Taiwan er hovedproducenten efterfulgt af Japan og Korea, som dog stadig er på udviklingsstadiet

D: Eine sehr wichtige, gezüchtete Art. Taiwan ist Hauptproduzent, gefolgt von Japan, während Korea sich noch im Versuchsstadium befindet

E: Especie de cría extremadamente importante. Taiwan es el principal productor, seguido por Japón y Corea, cuyas técnicas todavía están en estado de desarrollo

F: Espèce d'élevage excessivement importante. Le Taiwan est le producteur principal suivi du Japon et de la Corée, ces derniers étant au stade de développement

I: Una specie di allevamento estremamente importante. La Taiwan è il produttore principale seguita, allo stadio di sviluppo, dal Giappone e dalla Corea

JAPANESE PILCHARD

Scientific name:
Sardinops melanostictus

Family: Clupeidae — herrings, sardines
Typical size: 15 to 20 cm

The Japanese pilchard or sardine is one of the most important species in world fisheries, supplying up to 5 million tons a year, mainly from Japanese waters but also along the coasts of China, Korea and Russia. The species is used in many traditional preparations in Japan, including salting. Canned, it is an inexpensive staple.

FISHING METHODS

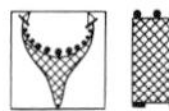  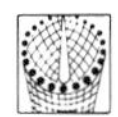

MOST IMPORTANT FISHING NATIONS
Japan, Russia, South Korea

PREPARATION
Fishmeal   

USED FOR

EATING QUALITIES
Canning softens the small bones and makes them edible. The meat is similar to that of other sardines, sweet and easily flaked from the bones of cooked fish.

COMMON NAMES
D: Japanische Sardine
DK: Japansk sardin
E: Sardina japonesa
F: Pilchard du Japon
GR: Sardéla tis Iaponías
I: Sardina giapponese
J: Ma-iwashi
N: Japansk sardin
NL: Japanse pelser
P: Sardinopa japonesa
RU: Japonomorskaya sardina
S: Japansk sardin
SF: Japaninsardiini
TR: Japon sardalyasi
US: Japanese sardine

LOCAL NAMES
KO: Chong-o-ri

Nutrition data:
(100 g edible weight)

Water	58.5 g	Total lipid	
Calories	260 kcal	(fat)	20.0 g
Protein	20.0 g	Omega-3	4.0 mg

HIGHLIGHTS
This abundant coastal pelagic species is the basis for a major fishing industry. The fish is very similar to other pilchards, including the Californian and Chilean species. In fact, these fish are so alike that some taxonomists regard them as subspecies.

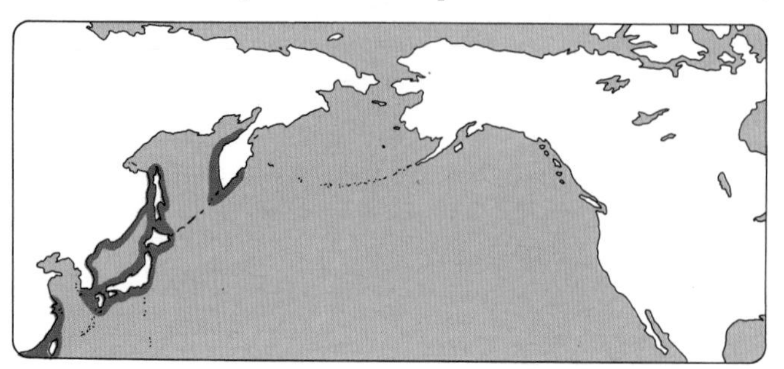

DK: En af verdens vigtigste arter med årlige fangster på indtil 5 millioner tons. En fisk med mange anvendelses-muligheder

D: Eine der wichtigsten Arten in der Welt, deren jährliche Fänge bis zu 5 Mio. Tonnen ausmachen. Ein Fisch mit vielen Anwendungsmöglichkeiten

E: Una de las especies más importantes del mundo, llegándose a capturar hasta 5 millones de toneladas al año. Pescado que tiene muchos usos

F: L'une des espèces les plus importantes de la pêche mondiale se chiffrant à jusqu'à 5 millions de t/an. Un poisson utilisé dans de nombreuses préparations

I: Una delle specie più importanti del mondo, di cui si pescano ben 5 milioni di tonnellate all'anno. Un pesce che offre tante possibilità di utilizzo

JAPANESE SPANISH MACKEREL

Scientific name:
Scomberomorus niphonius

Family: Scombridae — tunas and mackerels
Typical size: up to 100 cm, 4.8 kg

A slender Spanish mackerel, the meat is considered to be especially good in the winter, though it is caught and eaten throughout the year as it migrates inshore and through the Yellow Sea and the Sea of Japan on its spawning cycle.

FISHING METHODS

MOST IMPORTANT FISHING NATIONS
China, Korea, Japan

PREPARATION

USED FOR

EATING QUALITIES
Spanish mackerels are among the best eating fish. The flavour is delicate and the finely-textured meat has a small flake. The brownish colour of the raw meat turns white when cooked. Excellent for smoking; very versatile in most fish recipes.

COMMON NAMES
D: Japanische Makrele
DK: Japansk kongemakrel
E: Carite oriental
F: Thazard oriental
GR: Skoumbrí tis Iaponías
I: Maccarello reale giappone.
J: Sawara
NL: Japanse makreel
P: Serra oriental
RU: Japonskaya korolev. makrel
US: Mackerel

LOCAL NAMES
VI: Cá thu áu cham xanh

Nutrition data:
(100 g edible weight)

Water	76.3 g	Total lipid	
Calories	101 kcal	(fat)	2.5 g
Protein	19.7 g	Omega-3	0.5 mg

HIGHLIGHTS
About 150,000 to 200,000 tons a year of Japanese Spanish mackerel are caught. Although most of the catch is now taken by China, this high quality seafood is especially highly valued by consumers in Japan, where it is the best known Spanish mackerel.

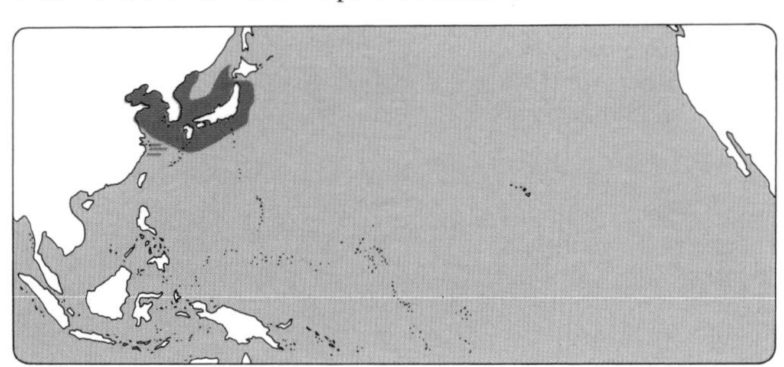

DK: Fiskes især af Kina, men er en eftersprugt kvalitetsfisk på det japanske marked, hvor den bedst kendes som 'spansk makrel'

D: Wird besonders von China gefischt, ist aber ein auf dem japanischen Markt nachgefragter Qualitätsfisch, am bekanntesten unter dem Namen 'Spanische Makrele'

E: Sobre todo se pesca en China, pero es un pescado de calidad solicitado en el mercado japonés, donde se lo conoce mejor bajo el nombre de 'caballa española'

F: Pêché surtout en Chine, c'est un poisson de haute qualité très demandé sur le marché japonais, où il est connu sous le nom de 'thazard atlantique'

I: Pescato soprattutto dalla Cina, ma è un pesce di qualità richiesto dal mercato giapponese dove è conosciuto meglio con il nome di 'sgombro atlantico'

JOHN DORY

Scientific name:
Zeus faber

Synonyms: Dory, Peter-fish
Family: Zeidae — dories
Typical size: 35 cm, 1400 g

John Dory or St. Peter's fish is found widely in many parts of the world. It is absent from the western Atlantic, where the very similar American john dory, *Zenopsis ocellata*, is found. Comments on eating quality and characteristics apply equally to both species.

The mirror dory, *Zenopsis nebulosus*, is also similar. This is widely distributed in the warm temperate Indo-Pacific. In Australia, where the two species are sometimes caught together, the john dory fetches a higher price as a superior fish for the table.

DESCRIPTION
This is a strange looking fish, with a compressed body and large jaw, angled upwards. It grows as large as 66 cm and 3 kg.

In the eastern Atlantic, the species is found, usually inshore in depths less than 200 m, from the north of Scotland and Norway to South Africa, including the Mediterranean and Black Seas. It is quite common around New Zealand, especially the northern parts of North Island; it is taken in the trawl fishery of eastern Australia; its range extends to Japan and Korea.

COMMON NAMES
D: Petersfisch, Heringskönig
DK: Sanktpetersfisk
E: Pez de San Pedro, gallo
F: Saint Pierre, Jean Doré
GR: Christópsaro
I: Pesce San Pietro
IS: Pétursfiskur
J: Matôdai
N: St. Petersfisk
NL: Sint pietervis, zonnevis
P: Peixe galo, galo negro
RU: Obyknovennyj solnechnik
S: St. Persfisk
SF: Pietarinkala
TR: Dülger baligi
US: John Dory

LOCAL NAMES
TU: Hout sidi sliman
EG: Afreet
IL: Morig
MO: Boukhatem-chatra

HIGHLIGHTS
The name of St. Peter's fish derives from the characteristic black thumbprint on each side, said to be the legacy of St. Peter, who took the tribute from the mouth of the fish, leaving the mark of his thumb on the species.

John Dory is a solitary fish, never found in large numbers, generally taken as a by-catch with other, more abundant, species. Nevertheless, it is retained and eaten wherever it is caught, because of its value and high quality meat.

Nutrition data:
(100 g edible weight)

Water	78.1 g
Calories	89 kcal
Protein	19.0 g
Total lipid (fat)	1.4 g
Omega-3	0.5 mg

JOHN DORY *Zeus faber*

FISHING METHODS

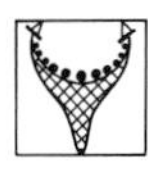 

MOST IMPORTANT FISHING NATIONS
Australia, New Zealand, South Africa, Japan, Morocco, France

PREPARATION

USED FOR

EATING QUALITIES
One of the finest of all fish to eat, the John Dory is not well known to consumers because of its limited availability. The meat is white when cooked and very sweet, with a firm texture. The bones are easily removed from cooked fish, another advantage to this species. Fillets can easily be cut without bones.

John Dory and related dories can be processed by almost any cooking method and are versatile in many recipes designed primarily for other fish species, including many of the classic French preparations for sole. The meat yield is low, but the carcass makes excellent fish stock, soup or fumet.

As a comment on the quality of the meat, commercial fishers are reported to take these fish home to eat themselves, rather than selling them.

IMPORTANCE
John Dory is landed from many of the world's oceans. FAO records catches from all sectors of the eastern Atlantic, from the Mediterranean and from the southwest and southeast Pacific. It is likely that total catches are greater than the approximately 5,000 tons a year reported to the international agency.

John Dory is a favourite catch of recreational fishermen, for eating rather than for its fighting abilities.

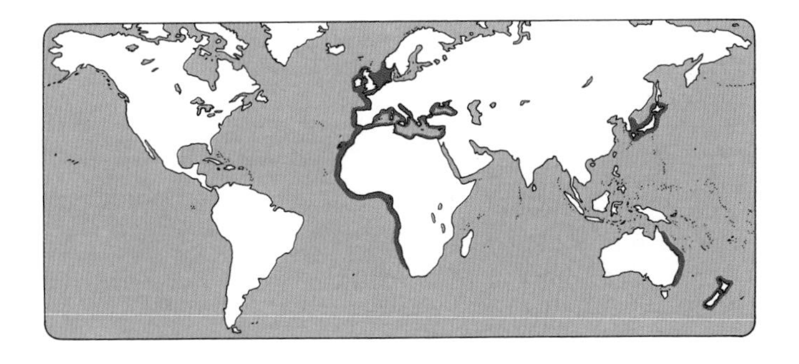

DK: Fisken har fået navn efter apostlen Skt. Peter, som ifølge legenden trak en mønt ud af dens mund og derved efterlod sit tommelfingeraftryk på siden

D: Der Fisch ist nach dem Apostel St. Peter benannt, der eine Münze aus dem Fischmaul zog, und dabei den Abdruck seines Daumens an dessen Seite hinterliess

E: El pescado tiene su nombre del apóstolo San Pedro, que sacó una moneda de su boca, dejando así la marca de su pulgar en el lado del pescado

F: Le poisson tient son nom de Pierre l'Apôtre, qui tira une pièce de monnaie de sa bouche en laissant ainsi son empreinte sur le flanc

I: Il pesce ha avuto il suo nome dall'apostolo San Pietro che secondo la leggenda estrasse una moneta dalla sua bocca, lasciando l'impronta del pollice sul suo lato

JUREL/S. JACK MACKEREL

Synonym: Chilean jack mackerel
Family: Carangidae — jacks
Typical size: 25 cm

Southern jack mackerel is one of the most important species by volume in world fisheries, contributing in some years well over three million tons. Much of this is still made into fishmeal.

FISHING METHODS

MOST IMPORTANT FISHING NATIONS
Chile, Peru, Russia

PREPARATION
Fishmeal 

USED FOR

Scientific name:
Trachurus murphyi

COMMON NAMES
D: Chilenische Bastardmakrele
DK: Chilensk hestemakrel
E: Jurel del Pacífico Sur
F: Chinchard jurel
GR: Savrídi tis Chilís
I: Sugarello cileno
J: Chiri-maaji
NL: Chileense horsmakreel
P: Carapau chileno
RU: Peruanskaya stavrida
US: Jack mackerel

EATING QUALITIES
Southern jack mackerel has moist, oily meat similar to mackerel. Mostly, it is canned if for human use. Fresh, it can be grilled or baked like mackerel. Generally, it is caught in such large quantities that it is damaged in the nets, making it unsuitable for processing and sale fresh or frozen.

Nutrition data:
(100 g edible weight)

Water	73.8 g	Total lipid	
Calories	128 kcal	(fat)	5.6 g
Protein	19.4 g	Omega-3	1.1 mg

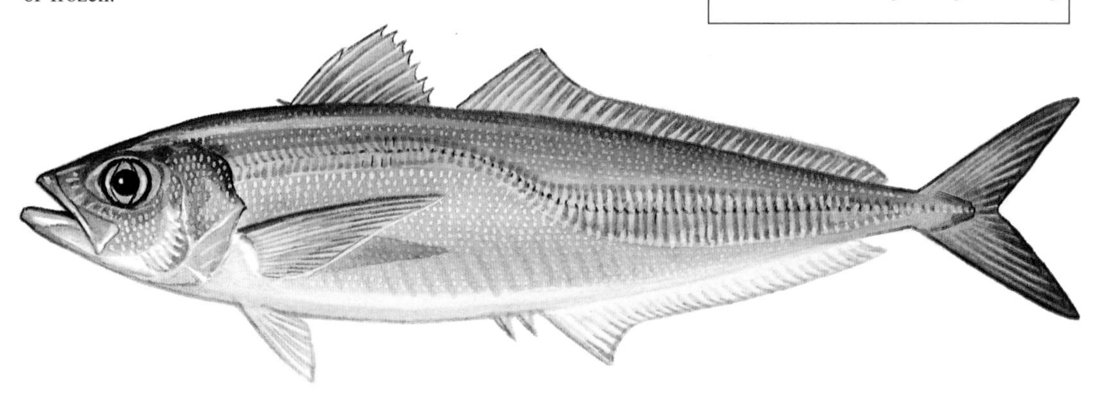

HIGHLIGHTS
Efforts to develop more valuable uses for this species continue. It can be made into canned products that are acceptable on world markets. If catching rates are slowed, it can possibly be used in fresh or frozen form to generate added value for the producers.

DK: Benyttes i øjeblikket kun til fiskemel og konserves, men der forskes i at bruge fisken til dyrere, forædlede produkter

D: Wird gegenwärtig nur zu Fischmehl und Konserven verarbeitet, obwohl die Verarbeitung zu teureren, veredelten Produkten Forschungsgegenstand ist

E: Por el momento sólo se usa para harina y productos enlatados, pero se está investigando para obtener con el pescado productos más caros y perfeccionados

F: Exploité jusqu'à présent uniquement pour la farine de poisson et pour les conserves, mais des recherches sont en cours en vue de mieux valoriser ce poisson

I: Attualmente viene usato solo per farina di pesce e conserve alimentari, ma sono in atto tentativi di utilizzare questo pesce per altri prodotti più redditizi

KAHAWAI

Scientific name:
Arripis trutta

Synonym: Ruff
Family: Arripidae — Australian salmon
Typical size: 40 to 60 cm, 3 kg

Commonly known as Australian salmon, although quite unrelated to salmonids, the kahawai is found in temperate waters of Australia and throughout New Zealand. It supports commercial food fisheries in both countries. In Australia, smaller specimens are often used for bait for rock lobsters. Canned Kahawai is used for both humans and pets.

FISHING METHODS

MOST IMPORTANT FISHING NATIONS

New Zealand, Australia

PREPARATION

USED FOR

EATING QUALITIES

Kahawai has dark meat, which turns pink when it is canned, accounting for the salmon name used in Australia. It is strongly flavoured and quite firm, best if smoked or pickled.

COMMON NAMES

D: Australischer Lachs
DK: Australsk laks
E: Salmón australiano
F: Saumon australien, kahawai
GR: Solomós tis Australías
I: Salmone australiano
NL: Australische zalm
P: Peixe grosa australiano
RU: Avstralijskij losos
SF: Arripi
US: Kahawai

LOCAL NAMES

AU: Australian salmon

Nutrition data:
(100 g edible weight)

Water	69.3 g	Total lipid	
Calories	162 kcal	(fat)	8.6 g
Protein	21.2 g	Omega-3	2.7 mg

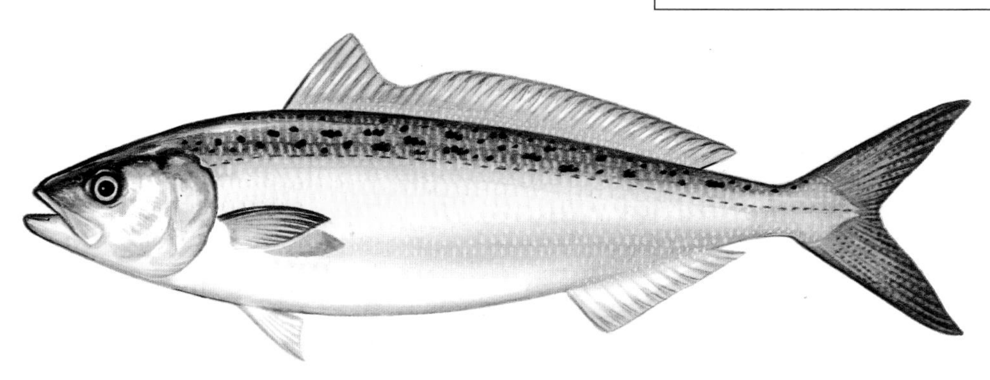

HIGHLIGHTS

Kahawai contributes up to 10,000 tons yearly to the fisheries and is taken mostly close inshore, sometimes from the beach. A low value commercial species, it is also targeted by recreational fishermen. The largest Australian salmon recorded weighed 9.4 kg.

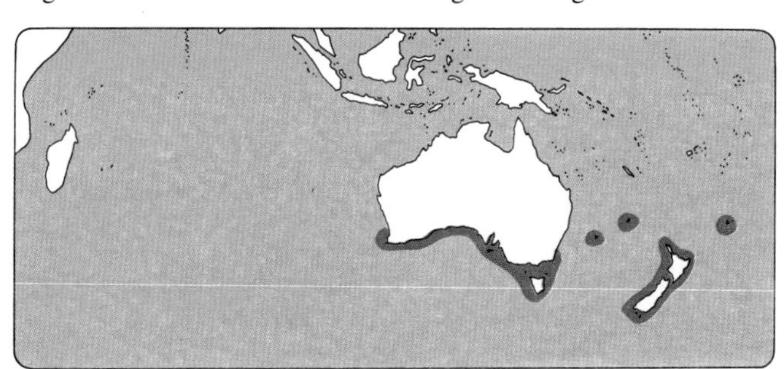

DK: De årlige fangster af denne konsum- og sportsfisk er på omkring 10.000 tons. Fanges tæt ved kysten, undertiden også fra stranden

D: Die jährlichen Fänge dieses Speise- und Sportfisches betragen rund 10.000 Tonnen. Wird dicht an der Küste, manchmal sogar vom Strand aus gefangen

E: De este pescado deportivo y de consumo se capturan cerca de la costa hasta 10 mil toneladas al año, a veces también desde la playa

F: La pêche de ce poisson sportif comestible est de l'ordre de 10.000 tonnes par an. Il est pêché près des côtes, même souvent directement depuis la plage

I: La produzione annuale di questo pesce sportivo e di consumo arriva a circa 10.000 tonnellate. E' pescato vicino alla costa e a volte anche dalla spiaggia

KINGCLIP

Scientific name:
Genypterus capensis

Family: Ophidiidae — cusk-eels
Typical size: up to 1.5 m, 15 kg

Found only off South Africa and Namibia from Walvis Bay to Algoa Bay, the kingclip fishery has dropped from 16,000 tons to 3,000 as South African and other vessels have been excluded from Namibian waters. The resource remains available for renewed exploitation.

FISHING METHODS

MOST IMPORTANT FISHING NATIONS

South Africa

PREPARATION

USED FOR

EATING QUALITIES

Kingclip yields a fine, off-white fillet which when cooked is attractively white. The texture is firm and the flavour mild but sweet. Generally inexpensive, it is good value for most culinary purposes, blending well with stuffings and sauces.

COMMON NAMES

D: Südafrikanischer Kingklip
DK: Sydafrikansk kingklip
E: Rosada del Cabo
F: Abadèche royale du Cap
GR: Ofídio tou Akrotiríou
I: Abadeco del Sudafrica
NL: Kaapse koningklip
P: Abadejo do Cabo
RU: Karpskij oshiben
SF: Kapinrihmanilkka
US: Kingklip

Nutrition data:
(100 g edible weight)

Water	78.7 g	Total lipid	
Calories	82 kcal	(fat)	0.7 g
Protein	19.0 g	Omega-3	0.1 mg

HIGHLIGHTS

South African kingclip fillets, under numerous names, have been a popular item in Europe for many years. The species can support a significant trawl fishery. The liver is said to be excellent to eat, similar in flavour and texture to chicken liver.

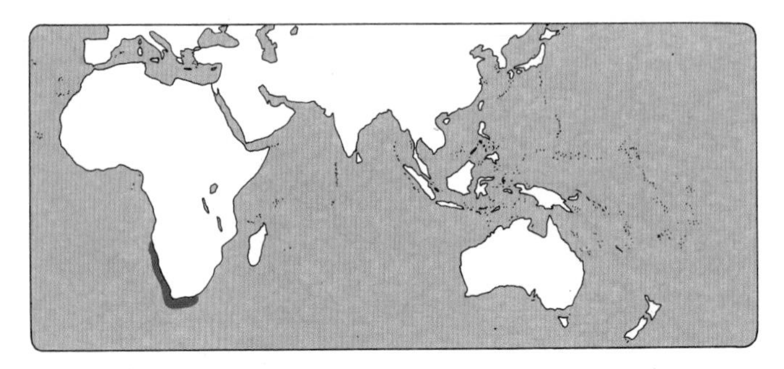

DK: Filetterne har gennem mange år med succes været markedsført under flere navne i Europa. Leveren er fortrinlig, meget lig kyllingelever

D: Die Filets werden seit vielen Jahren erfolgreich unter verschiedenen Namen auf europäischen Märkten angeboten. Die Leber ist vorzüglich und ähnelt Geflügelleber

E: Durante muchos años sus populares filetes se han vendido en Europa bajo varios nombres. El hígado es exquisito, muy similar al hígado de pollo

F: Les filets sont commercialisés en Europe avec succès sous des noms différents depuis de nombreuses années. Le foie excellent ressemble au foie de volaille

I: Da tanti anni i filetti vengono commercializzati con successo in Europa sotto nomi diversi. Il fegato, molto simile a quello del pollo, è ottimo

KINGCLIP/PINK CUSK EEL

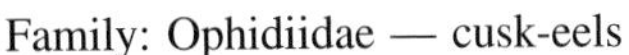

Scientific name:
Genypterus blacodes

Family: Ophidiidae — cusk-eels
Typical size: 90 cm, 7 kg; up to 1.6 m, 20 kg

Found in colder southern waters, the kingclip is abundant around the Falkland Islands, New Zealand and southern Australia. It is believed that catches can be greatly increased as markets are developed. Stocks in Australasia in particular appear to be under-exploited.

FISHING METHODS

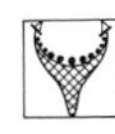

MOST IMPORTANT FISHING NATIONS
Argentina, Chile, New Zealand

PREPARATION

USED FOR

EATING QUALITIES
Kingclip has rather dense, firm white meat with a mild flavour not unlike orange roughy. It can withstand most cooking methods and is excellent smoked. In Japan, the meat of small specimens is sometimes eaten raw as sashimi.

COMMON NAMES
D: Goldener Schlangenfisch
DK: Rosa kingklip
E: Rosada
F: Abadèche rose
GR: Kokkinofídio
I: Abadeco
J: Kingu, ringu
NL: Roze kingklip
P: Maruca da Argentina
RU: Chernyj kongrio
US: Kingklip

LOCAL NAMES
AU: Ling, pink ling
NZ: Ling

Nutrition data:
(100 g edible weight)

Water	78.3 g	Total lipid	
Calories	87 kcal	(fat)	0.8 g
Protein	19.9 g	Omega-3	0.2 mg

HIGHLIGHTS
Kingclip fillets, usually skinless and boneless, are an attractive inexpensive product gaining increasing acceptance on world markets. Marketing efforts led by New Zealand have been exploited by Argentina and Chile, where the catches are considerably greater.

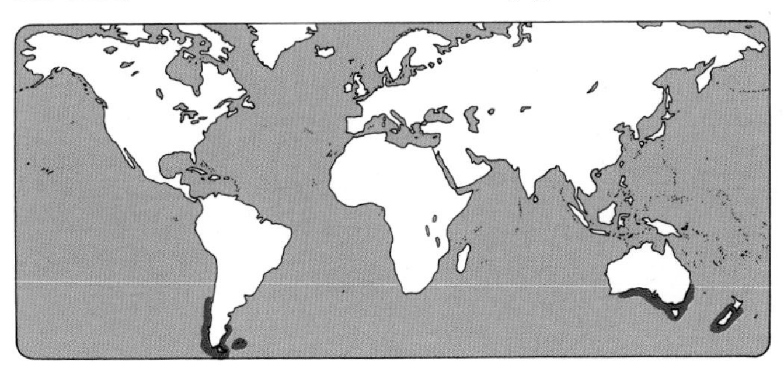

DK: De flåede, benløse filetter, et efterspurgt og billigt produkt, vinder gradvist større indpas på verdensmarkedet

D: Die enthäuteten und grätenlosen Filets, ein gefragtes und preiswertes Produkt, halten allmählich Einzug auf dem Weltmarkt

E: Los filetes despellejados y sin espinas son un producto demandado y barato, que va ganando aceptación en los mercados mundiales

F: Les filets sans peau et sans arêtes constituent un produit recherché et bon marché qui commence à pénétrer les marchés mondiaux

I: I filetti senza pelle e spine costituiscono un prodotto economico e richiesto che si afferma sempre di più sul mercato mondiale

LAKE HERRING/TULLIBEE

Scientific name:
Coregonus artedii

Synonyms: Lakefish, chub
Family: Salmonidae — salmonids
Typical size: 50 cm

There are a number of species of lake herring, also called cisco and tullibee, which are very similar and often confused. All are found in northern lakes of North America. There are significant commercial fisheries as well as important recreational use of these resources.

FISHING METHODS

MOST IMPORTANT FISHING NATIONS
Canada, United States

PREPARATION

USED FOR

EATING QUALITIES
Lake herrings have delicate, white flesh and excellent flavour. They are best broiled or grilled, but can also be used very much like trout. Hot smoked dressed fish, generally called chubs, are popular in many parts of North America.

COMMON NAMES
D: Felchen, Maräne
DK: Amerikansk helt
E: Arenque de lago
F: Cisco de lac
GR: Limnórenga
I: Coregone americano
IS: Vatnasíld
NL: Ciscomarene, witvis
P: Arenque de lago
RU: Ozerny seld
SF: Amerikanmuikku
TR: Göl ringasi
US: Cisco

Nutrition data:
(100 g edible weight)

Water	78.9 g	Total lipid	
Calories	93 kcal	(fat)	1.9 g
Protein	19.0 g	Omega-3	0.4 mg

HIGHLIGHTS
Tullibees vary in size, appearance and taste from lake to lake, but all are desirable. Catches, mainly in Canada, are less than 2000 tons yearly. The species was extremely abundant in the last century, but has been heavily fished, especially from the Great Lakes.

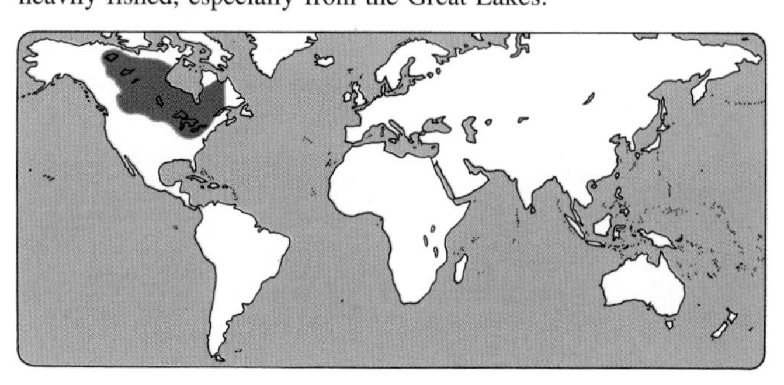

DK: Forveksles ofte med flere lignende arter, der også lever i de nordlige søer i Nordamerika og har betydning for såvel lyst- som erhvervsfiskeriet

D: Wird oft mit vielen anderen Arten verwechselt, die auch in den nördlichen Seen Nordamerikas leben und für die Sport- und Berufsfischerei von Bedeutung sind

E: Se confunde a menudo con varias especies similares que viven en los lagos del norte de Norteamérica, y tienen importancia para la pesca deportiva y profesional

F: Souvent confondu avec plusieurs espèces similaires qui vivent aussi dans les lacs du nord des Etats-Unis. Poisson important pour la pêche sportive et commerciale

I: Confondibile con altre specie affini che vivono nei laghi del nord dell'America settentrionale e importante per la pesca sportiva e professionale

Scientific name:

LAKE WHITEFISH

Coregonus clupeaformis

Synonym: Common whitefish
Family: Salmonidae — salmonids
Typical size: 1 m, 10 kg

An important food, recreational and subsistence species in the far north of Canada and Alaska, the whitefish is still the most valuable commercial freshwater fish in Canada, although landings are greatly reduced from historical levels to about 8,000 tons a year.

FISHING METHODS

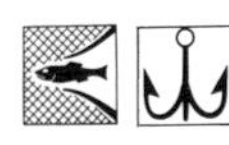

MOST IMPORTANT FISHING NATIONS

Canada, United States

PREPARATION

Caviar

USED FOR

EATING QUALITIES

Whitefish has delicate, white meat which tastes more like salmon than trout. It can be used with distinction in most trout and salmon recipes. Much whitefish is dyed yellow and hot smoked. Smaller fish are smoked dressed, larger ones as fillets.

COMMON NAMES

D: Nordamerikanisches Felchen
DK: Søhelt
E: Coregono de lago
F: Corégone de lac
GR: Korégonos, limnokorégonos
I: Coregone dei grandi laghi
IS: Násíld
N: Sik
NL: Amerikaanse marene
P: Coregono de lago
RU: Lapsharyba
S: Sik
SF: Sillisiika
US: Whitefish

Nutrition data:
(100 g edible weight)

Water	72.3 g	Total lipid (fat)	5.9 g
Calories	130 kcal		
Protein	19.1 g	Omega-3	1.2 mg

HIGHLIGHTS

Prized for its meat as well as for its roe, which is made into an excellent caviar, the whitefish remains an important commercial species. It is caught through thick ice in the winter and supports recreational fisheries throughout the year.

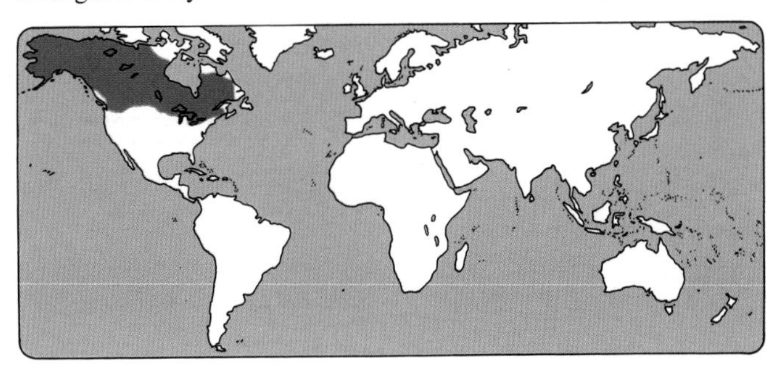

DK: En vigtig konsum- og sportsfisk, værdsat for såvel kødet som rognen, hvoraf der fremstilles en fortrinlig kaviar

D: Ein wichtiger Speise- und Sportfisch, der wegen seines Fleisches und Rogens beliebt ist, aus dem ein vorzüglicher Kaviar hergestellt wird

E: Pescado comestible y deportivo, apreciado por su carne y sus huevas, de las que se obtiene un caviar excelente

F: Poisson de consommation et de pêche sportive important, apprécié tant pour sa chair que pour les oeufs dont on prépare un excellent succédané de caviar

I: Un importante pesce sportivo e di consumo, apprezzato sia per le carni che per le uova da cui si produce un ottimo caviale

LAMPERN/RIVER LAMPREY

Scientific name:
Lampetra fluviatilis

Synonym: Stone eel
Family: Petromyzontidae — lampreys
Typical size: 30 to 35 cm
Also known as *Petromyzon fluviatilis*

One of a group of related fishes, the river lamprey is found in small numbers throughout western Europe. It is still important in the Baltic, where it is sometimes smoked or salted. There are many ancient recipes for lamprey, but few are still used.

COMMON NAMES

D: Flussneunauge, Pricke, Uhl
DK: Flodlampret, flodniøje
E: Lamprea de río, lamprea
F: Lamproie de rivière
GR: Lámprena
I: Lampreda di fiume
IS: Fisksuga
J: Yatsumeunagi
N: Elveniøje
NL: Rivierprik, negenoog
P: Lampreia do rio
RU: Minoga
S: Flodnejonöga
SF: Nahkiainen

FISHING METHODS

MOST IMPORTANT FISHING NATIONS

Not available

PREPARATION

USED FOR

EATING QUALITIES

River lampreys were once a common part of European cuisine, from Portugal to Finland. Few people now have the opportunity to sample them, but they are said to resemble eel in taste and texture, as they do also in general appearance.

Nutrition data:
(100 g edible weight)

Water	70.6 g	Total lipid	
Calories	168 kcal	(fat)	11.3 g
Protein	16.6 g	Omega-3	0.7 mg

HIGHLIGHTS

An anadromous fish which spends most of its adult life close to fresh water, the lampern is parasitic, sucking blood from sea trout, salmon, shad and other species. Once an important food source, stocks have largely disappeared as pollution in rivers destroyed spawning areas.

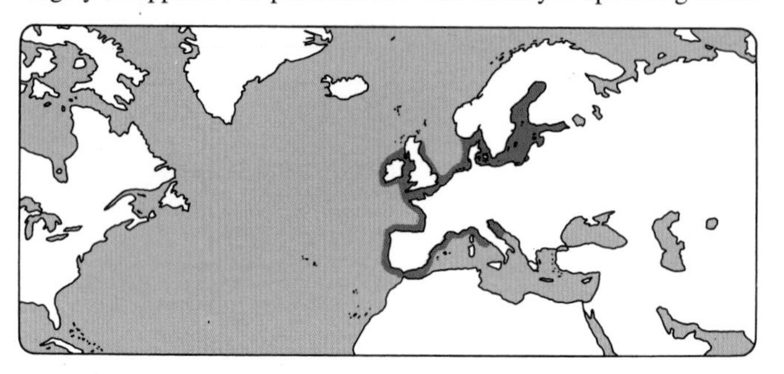

DK: Denne anadrome fisk tilbringer størsteparten af sit voksenliv ved flodmundinger. Den er parasit og suger blod af bl.a. laksefisk

D: Dieser anadrome Fisch verbringt den grössten Teil seines Erwachsenendaseins an Flussmündungen. Er lebt parasitär und saugt z.B. das Blut aus Lachsen

E: Este pescado anádromo pasa la mayor parte de su vida adulta en las embocaduras de los ríos. Es parásito y succiona sangre del salmón y otras especies afines

F: Ce poisson anadrome passe le plus clair de sa vie d'adulte aux embouchures des rivières. C'est un parasite qui vit du sang du saumon et autres espèces

I: Questo pesce anadromo passa la maggior parte della sua vita adulta nelle foci. E' un parassita che succhia il sangue tra l'altro dai salmoni

LEMON SOLE

Scientific name:
Microstomus kitt

Synonyms: Lemon dab, lemon fish, Mary sole, smear dab
Family: Pleuronectidae — right-eye flounders
Typical size: 45 cm, up to 65 cm

The United Kingdom catches about half of Europe's 12,000 ton production of lemon sole, but the fish is valued throughout northern Europe. Although mostly too small to fillet, the fish is cooked on the bone and enjoyed for its excellent flavour and texture.

FISHING METHODS

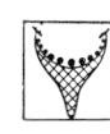

MOST IMPORTANT FISHING NATIONS

United Kingdom, Denmark, Iceland, France

PREPARATION

USED FOR

EATING QUALITIES

Lemon sole is a very good flatfish, though perhaps not quite as good as the best soles. The meat is white and delicate, with a sweet flavour. In Belgium, with other soles and flounders, it is sometimes salted and dried, then eaten without further preparation.

COMMON NAMES

D: Echte Rotzunge, Limande
DK: Rødtunge
E: Falsa limanda, mendo limón
F: Limande sole
GR: Lemonóglóssa
I: Limanda
IS: Thykkvalúra
N: Lomre
NL: Tongschar, steenschol
P: Solha limao
RU: Malorotaya kambala
S: Bergtunga, bergskädda
SF: Pikkupääkampela
US: Lemon sole

Nutrition data:
(100 g edible weight)

Water	81.2 g	Total lipid	
Calories	83 kcal	(fat)	1.5 g
Protein	17.4 g	Omega-3	0.1 mg

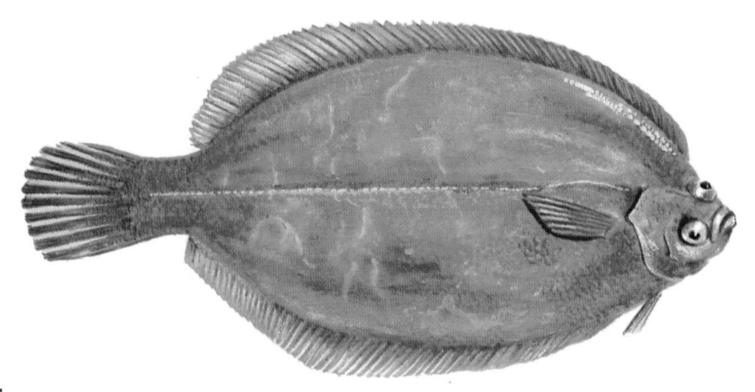

HIGHLIGHTS

The lemon sole is a right-eyed flatfish of the eastern Atlantic, found mainly on stony bottoms from the Bay of Biscay northwards to the White Sea. It is most abundant in the Irish Sea. This species is mostly sold fresh, though small quantities are frozen.

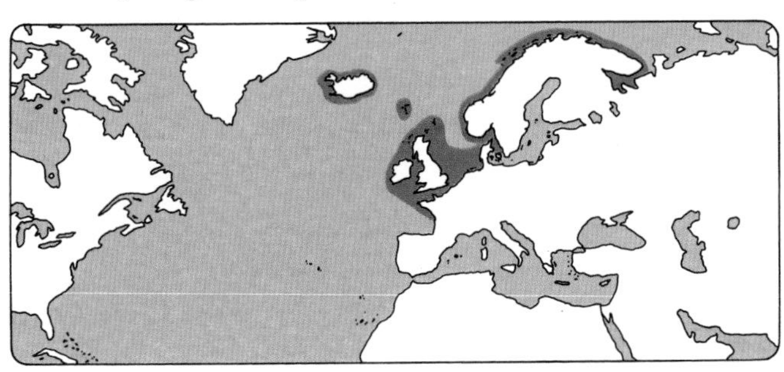

DK: En virkelig velsmagende fladfisk. Det hvide, sødlige kød er imidlertid ikke lige så delikat som søtungens

D: Ein sehr wohlschmeckender Plattfisch. Das weisse, süssliche Fleisch ist jedoch nicht ganz so delikat wie das der Seezunge

E: Pleuronecto muy sabroso. Sin embargo, su carne blanca y dulce no es tan delicada como la carne del lenguado

F: Poisson plat de goût vraiment excellent, mais la chair de la limande, blanche et d'un goût très doux, n'est pas aussi délicieuse que la chair de la sole

I: Un pesce appartenente ai pleuronettiformi. Le sue carni bianche e dolciastre sono eccellenti, ma meno apprezzate di quelle della sogliola

LING

Scientific name:
Molva molva

Synonym: Drizzie
Family: Gadidae — cods
Typical size: 100 to 150 cm, up to 200 cm

Ling is found mainly on rocky bottoms in fairly deep water, where they are not easy to catch with nets. Line-caught fish are often better handled than netted fish. Ling, often taken on lines, benefits from the fishing method as well as the intrinsic quality of the meat.

FISHING METHODS

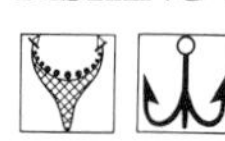

MOST IMPORTANT FISHING NATIONS
Norway, France, Iceland, Spain

PREPARATION

USED FOR

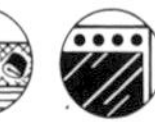

EATING QUALITIES
Ling has firm, white, sweet meat similar to cod, but perhaps a little firmer. It can be used in any recipe calling for cod. In the past, when the species was more abundant, it was widely used for salt-dried fish, for which it is excellent raw material.

COMMON NAMES
D: Leng, Lengfisch
DK: Lange
E: Maruca, barruenda
F: Lingue, grande lingue
GR: Pontíki
I: Molva
IS: Langa
J: Kurojimanagadara, ringu
N: Lange
NL: Leng
P: Maruca, donzela
RU: Morskaya schuka, molva
S: Långa
SF: Molva
TR: Gelincik
US: Ling

LOCAL NAMES
PO: Molwa

Nutrition data:
(100 g edible weight)

Water	79.3 g	Total lipid	
Calories	82 kcal	(fat)	0.7 g
Protein	18.8 g	Omega-3	0.1 mg

HIGHLIGHTS
Landings of ling average close to 60,000 tons a year, mostly from Norwegian waters. The species is highly regarded for its excellent quality meat and is a particularly favoured white-meated fish in Spain and France.

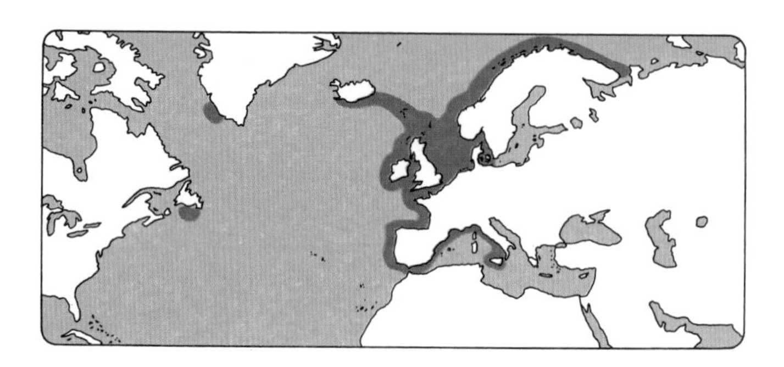

DK: En fin spisefisk og en specielt foretrukken fisk med lyst kød. Fint råmateriale til tørsaltede produkter

D: Ein wohlschmeckender und beliebter Speisefisch mit besonders hellem Fleisch. Guter Rohstoff für trockengesalzene Erzeugnisse

E: Exquisito pescado comestible de carne blanca muy apreciada. Materia prima excelente para productos secados con sal

F: Poisson comestible très délicat, particulièrement apprécié pour sa chair blanche. Une excellente matière première pour des produits salés à sec

I: Pesce commestibile di alta qualità con carne bianca molto apprezzata. Eccellente materia prima per prodotti essiccati al sale

Scientific name:
Ophiodon elongatus

LINGCOD

Synonyms: Blue cod, buffalo cod, green cod, greenling
Family: Hexagrammidae — greenlings
Typical size: up to 150 cm and 35 kg

Neither a cod nor a ling but a greenling, this species is favoured by recreational and commercial fishermen throughout its range. Frozen steaks are sometimes substituted for halibut, although the skin differences are quite significant.

COMMON NAMES

D: Langer Grünling
E: Bacalao largo, lorcha
F: Rascasse verte
I: Ofiodonte
IS: Græningi
NL: Lingcod
P: Lorcha
RU: Zubasty terpug, zmeezub
US: Lingcod

FISHING METHODS

MOST IMPORTANT FISHING NATIONS

Canada, United States

PREPARATION

USED FOR

EATING QUALITIES

Good quality, firm, mild meat characterize the lingcod; occasionally, the meat has a naturally greenish tint before cooking, which is something of a deterrent to consumers unfamiliar with the fish. Lingcod makes excellent fish and chips.

Nutrition data:
(100 g edible weight)

Water	81.0 g	Total lipid (fat)	1.0 g
Calories	80 kcal		
Protein	17.7 g	Omega-3	0.2 mg

HIGHLIGHTS

Lingcod is an important species, providing about 8,000 tons a year of fresh and frozen fillets and steaks. It is sometimes available headless dressed, mainly from Alaska. Line-caught and trapped fish is considered better than trawled specimens.

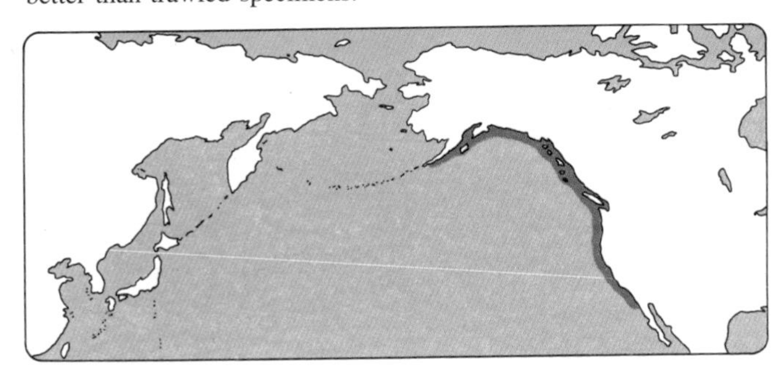

DK: Det virker afskrækkende på nye forbrugere, at det rå kød har en grønlig nuance, som dog forsvinder under tilberedningen

D: Das rohe Fleisch des Fisches hat manchmal eine grünliche Färbung, was neue Verbraucher abschreckt

E: Los consumidores que no lo conocen, pueden llegar a perder el interés por este pescado, cuya carne cruda tiene un color verdoso que desaparece al cocinarlo

F: Les nouveaux consommateurs sont parfois effrayés à la vue de la couleur verdâtre de la chair crue. Cette couleur disparaît cependant à la cuisson

I: Il colore verdognolo della carne fresca, che però sparisce una volta preparato, scoraggia i consumatori non a conoscenza di questo fatto

LONG ROUGH DAB

Scientific name:
Hippoglossoides platessoides

Family: Pleuronectidae — right-eye flounders
Typical size: 60 cm, 2.5 kg

Found on both sides of the North Atlantic, but far more abundant in the west, the long rough dab is frequently called American plaice in trade. Most production is filleted and sold as flounder; it is indistinguishable in use from the other major flounder species.

FISHING METHODS

MOST IMPORTANT FISHING NATIONS

Canada, United States, Iceland

PREPARATION

USED FOR

EATING QUALITIES

Long rough dab is a good quality flounder with tender, white, sweet-tasting meat. It is the most important flounder species in many areas of the Canadian Maritimes. Occasional instances of jellied meat are caused by the sexual ripening of the fish, not by parasites.

COMMON NAMES

D: Kliesche, Doggerscharbe
DK: Håising
E: Platija americana
F: Faux flétan, balai
GR: Kalkáni tou Kanadá
I: Passera canadese
IS: Skrápflúra
J: Hirame
N: Gapeflyndre
NL: Lange schar
P: Solha americana
RU: Kambala-jorsh
S: Lerskädda, lerflundra
SF: Liejukampela
US: American plaice

LOCAL NAMES

PO: Ptastugi

Nutrition data:
(100 g edible weight)

Water	80.4 g	Total lipid	
Calories	86 kcal	(fat)	2.7 g
Protein	15.5 g	Omega-3	0.5 mg

HIGHLIGHTS

Catches of around 40,000 tons a year have fallen steeply in recent years, as stocks of this and other groundfish resources of the western North Atlantic have come under increasing pressure. Canada is the major harvester. The species remains economically important .

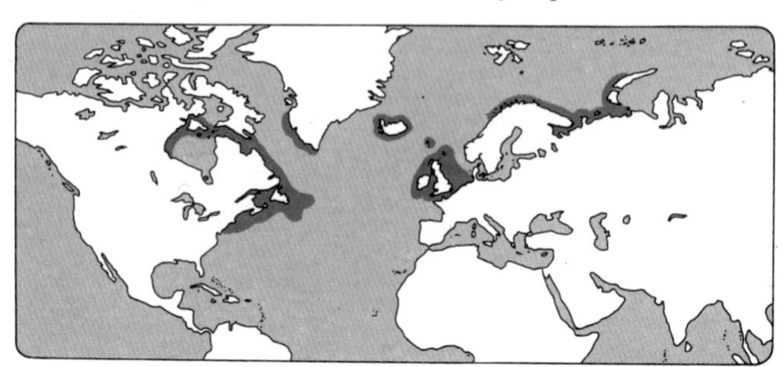

DK: Fangsterne er faldet voldsomt i de seneste år som følge af generelt overfiskeri af bundfisk i den vestlige del af Nordatlanten

D: Wegen der allgemeinen Überfischung von Bodenfischen im westlichen Nordatlantik sind die Fänge stark zurückgegangen

E: Las capturas han bajado mucho durante los últimos años como consecuencia de la pesca excesiva de pescados de fondo en el oeste del Atlántico Norte

F: Ces dernières années, les pêches sont fortement en déclin dû à une surpêche générale de poissons des fonds dans la partie ouest de l'Atlantique Nord

I: Le catture sono diminuite notevolmente in questi ultimi anni in seguito allo sfruttamento generale delle risorse dei pesci di fondo nell'ovest del Nordatlant.

LONGFINNED EEL

Scientific name:
Anguilla dieffenbachii

Family: Anguillidae — freshwater eels
Typical size: 100 cm

This large eel is native to both main islands of New Zealand and is not known elsewhere. The fishery produces several hundred tons yearly. Most of this is smoked. Eels are caught in fyke nets and other traps on their migration to the sea, when they are largest and fattest. Some experimental work on farming the species has been done.

FISHING METHODS

MOST IMPORTANT FISHING NATIONS
New Zealand

PREPARATION

USED FOR

EATING QUALITIES
The meat is white and firm, with a small flake and a sweet flavour. The fat content tends to be lower than that of the European eel (*A. anguilla*), but the longfinned eel is excellent smoked.

COMMON NAMES
D: Neuseeland-Aal
DK: Newzealandsk ål
E: Anguila de Nueva Zelanda
F: Anguille de N.-Zélande
GR: Chéli tis Néas Zilandías
I: Anguilla neozelandese
J: Ounagi
NL: Nieuwzeelandse paling
P: Enguia da Nova Zelândia
RU: Dlinnoplavnikovyj ugor
US: New Zealand eel

LOCAL NAMES
MA: Belut

Nutrition data:
(100 g edible weight)

Water	68.5 g	Total lipid	
Calories	179 kcal	(fat)	11.7 g
Protein	18.3 g	Omega-3	0.2 mg

HIGHLIGHTS
The longfin eel is common in lakes and rivers throughout New Zealand. They turn almost black when mature, but the flesh remains white and the smoked meat is an attractive golden brown colour. Females are approximately twice the size of males.

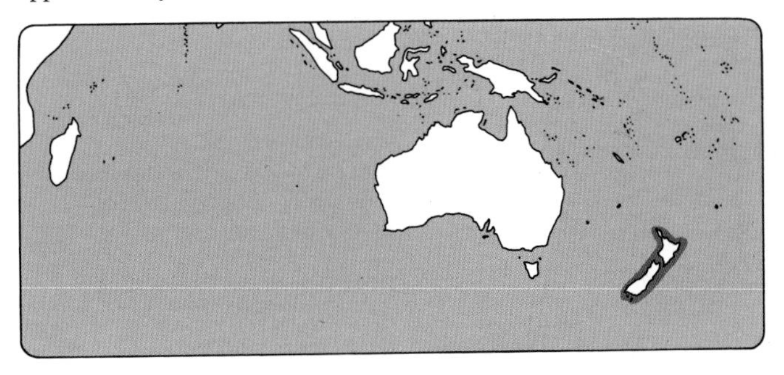

DK: Almindelig i søer og floder overalt i New Zealand. Det hvide kød får en flot gyldenbrun farve ved rygning

D: In ganz Neuseeland üblich in Seen und Flüssen. Das weisse Fleisch schimmert nach dem Räuchern goldbraun

E: Especie corriente en los lagos y ríos de toda Nueva Zelanda. Ahumado, su carne blanca obtiene un atractivo color marrón dorado

F: Poisson commun des lacs et des rivières de toute la Nouvelle-Zélande. Fumée, la chair blanche prend une magnifique couleur mordorée

I: Comune nei laghi e nei fiumi di tutta la Nuova Zelanda. Con l'affumicatura le carni bianche acquistano un bel colore marrone dorato

LUMPFISH

Scientific name:
Cyclopterus lumpus

Synonyms: Lumpsucker, sea-hen, paddle-cock, henfish
Family: Cyclopteridae — lumpfishes and snailfishes
Typical size: 35 to 60 cm, 2 to 7 kg.

Lumpfish eggs are salted, then shipped to market areas where they are desalted, dyed, flavoured and vacuum packed. Some producers are beginning to complete all phases of production immediately after the roes are landed. This results in a superior quality product.

FISHING METHODS

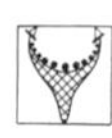

MOST IMPORTANT FISHING NATIONS

Norway, Iceland, Denmark, Canada

PREPARATION

Caviar

USED FOR

EATING QUALITIES

The meat is not normally eaten, although it can be salted and dried for use as human food. Lumpfish are valuable for their eggs, which make an inexpensive caviar. The eggs vary greatly in colour from green to red, but are dyed black for sale as caviar.

COMMON NAMES

D: Seehase, Lump, Lumpfisch
DK: Stenbider, kvabso (hun)
E: Libre de mar, ciclóptero
F: Lompe, mollet
GR: Kotópsaro
I: Ciclottero, lompo
IS: Hrognkelsi
J: Dango-uo
N: Rognkjeks, rognkall (hun)
NL: Snotdolf, strontvreter
P: Peixe lapa, galinha do mar
RU: Pinagor
S: Sjurygg, stenbit, kvabbso
SF: Rasvakala
US: Lumpfish

Nutrition data:
(100 g edible weight)

Water	81.8 g	Total lipid	
Calories	118 kcal	(fat)	9.5 g
Protein	8.1 g	Omega-3	1.9 mg

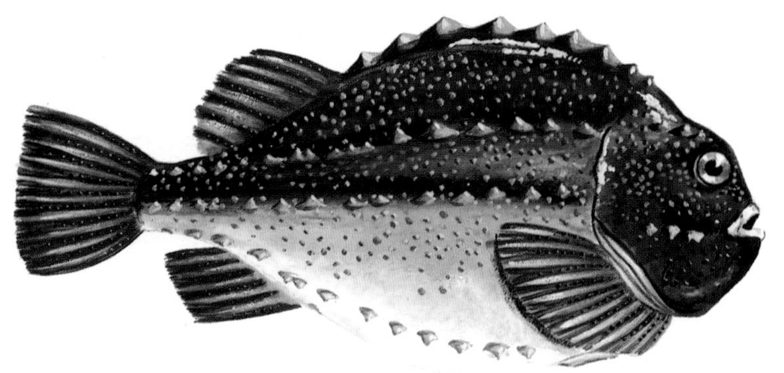

HIGHLIGHTS

Lumpfish roe production now exceeds 6,000 tons yearly and it is estimated that the resources could support substantially greater harvesting. Roe constitutes between 15 and 30 percent of the weight of the female fish. Males are small and usually escape the net.

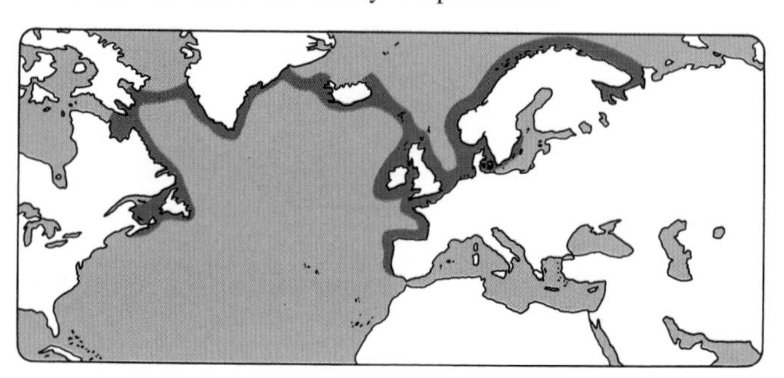

DK: Det anslås, at med den nuværende bestand vil den årlige produktion på over 6.000 tons stenbiderrogn kunne forøges væsentligt

D: Es wird geschätzt, dass bei dem jetzigen Bestand die Jahresproduktion von mehr als 6000 Tonnen Seehasenrogen erheblich erweitert werden kann

E: La producción de huevas del ciclóptero supera las 6 mil toneladas al año, y se estima que existe la base para aumentar las capturas notablemente

F: La production d'oeufs de lompe dépasse actuellement 6.000 t/an, et il est estimé qu'elle pourra être augmentée sensiblement sans risque pour l'espèce

I: Con le risorse attuali, si prevede che la produzione annuale di oltre 6.000 tonnellate di uova di lompo, potrà essere aumentata notevolmente

MENHADEN

Scientific name:
Brevoortia tyrannus

Synonyms: Atlantic menhaden, shad, bunker, pogy, mossbunker
Family: Clupeidae — herrings
Typical size: 18 to 28 cm

The USA catches about 400,000 tons a year of Atlantic menhaden and even more of the related Gulf species, *B. patronus*. All of the catch is used for fishmeal and oil. Menhaden are oily, bony and small, all features which militate against their use for human food.

COMMON NAMES

D: Nordatlantischer Menhaden
DK: Atlantisk menhaden
E: Lacha tirana
F: Menhaden tyran, menhaden
I: Alaccia americana
N: Menhaden
NL: Menhaden
P: Menhadem
S: Menhaden
SF: Menhaden
US: Menhaden

FISHING METHODS

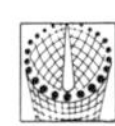

MOST IMPORTANT FISHING NATIONS

United States

PREPARATION

Fishmeal, animal feed

USED FOR

EATING QUALITIES

This species is not eaten by humans, although development work on using menhaden for surimi is proceeding. So far, the dark colour of the meat has proved to be a problem in surimi production, restricting potential uses of the paste.

Nutrition data:
(100 g edible weight)

Water	67.4 g	Total lipid (fat)	13.0 g
Calories	189 kcal		
Protein	17.9 g	Omega-3	2.6 mg

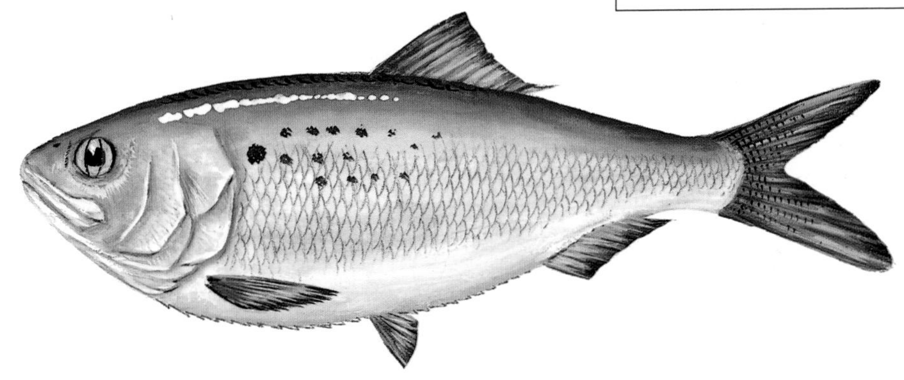

HIGHLIGHTS

Menhaden supports one of the largest fisheries in North America and is the most important industrial species. It is not yet clear whether menhaden can be used for surimi; currently poor markets for surimi have caused development work to be postponed.

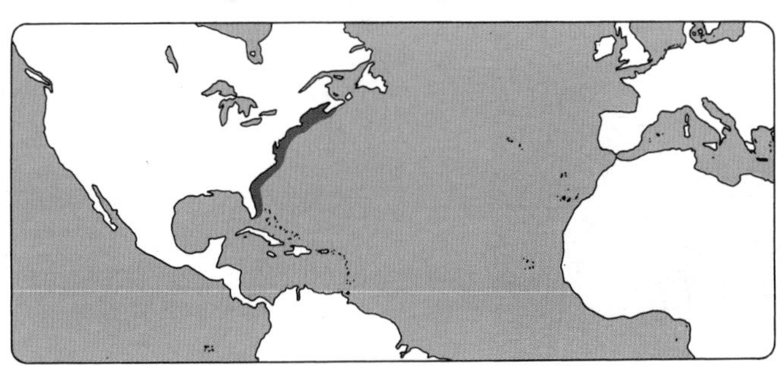

DK: Danner basis for store fangster i Nordamerika, hvor den er den vigtigste industrifisk. Potentiel art til surimi-fremstilling

D: Grosse Fangmengen in Nordamerika, wo er der wichtigste Wirtschaftsfisch ist. Potentielle Art zur Herstellung von 'Surimi'

E: Constituye la base de grandes capturas en Norteamérica, donde es el pescado industrial más importante. Especie potencial para la producción de 'surimi'

F: Ce poisson est à la base de grandes pêches en Amérique du Nord où c'est le plus important poisson industriel. Espèce potentielle pour la production de surimi

I: Crea le basi per grosse catture nell'America del Nord, dove è il pesce più importante dell'industria ittica. Specie potenziale per la produzione di surimi

MOLA/OCEAN SUNFISH

Scientific name:
Mola mola

Family: Molidae — molas
Typical size: 1 to 2 m, 80 to 225 kg

Molas are often seen floating on the sea. Slow moving, they are easily captured. They may grow to great sizes. One specimen 2.4 m long weighed over 800 kg. There is an unconfirmed report from Argentina in the 19th Century of a mola weighing 1,197 kg.

FISHING METHODS

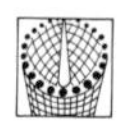 

MOST IMPORTANT FISHING NATIONS
Not available

PREPARATION

USED FOR

EATING QUALITIES
Sunfish is not normally eaten, though occasionally sampled as a curiosity. It is said to be oily, soft and insipid, but palatable. Molas are often heavily and visibly parasitized, both internally and externally, which deters sampling.

COMMON NAMES
D: Mondfisch, Klumpfisch
DK: Klumpfisk
E: Pez luna, rodador, atalo
F: Poisson-lune, môle commune
GR: Fengarópsaro
I: Pesca luna
IS: Tunglfiskur
J: Manbô
N: Månefisk
NL: Maanvis
P: Peixe lua, lua
RU: Ryba-solntse
S: Klumpfisk
SF: Möhkäkala
TR: Pervane
US: Ocean sunfish

Nutrition data:
(100 g edible weight)

Water	n.a.	Total lipid	
Calories	n.a.	(fat)	n.a.
Protein	n.a.	Omega-3	n.a.

HIGHLIGHTS
Found occasionally in all temperate and tropical seas, molas drift with ocean currents, living on jellyfish and similar organisms. They are a curiosity to fishermen, more use for attracting tourists than for food. Little is known of their life history or abundance.

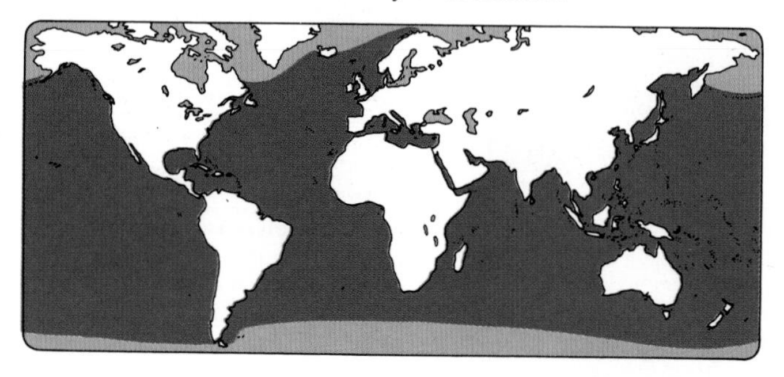

DK: Klumpfisken flyder med strømmene og lever af bløddyr og lignende organismer. Ringe viden om adfærd og bestand

D: Der Mondfisch schwimmt mit dem Strom und ernährt sich von Weichtieren und ähnlichen Organismen. Verhalten und Bestand sind wenig erforscht

E: El pez luna flota con las corrientes y vive de moluscos y organismos similares. Se sabe muy poco de su vida y abundancia

F: Le poisson-lune se laisse aller au fil de l'eau vivant de mollusques et autres organismes similaires. On ne sait que peu sur son comportement et sur la population

I: Il pesce luna si lascia trascinare dalle correnti e vive di molluschi e simili organismi. Se ne conosce poco del suo comportamento e dell'abbondanza

MORWONG/BLACK PERCH

Scientific name:
Nemadactylus morwong

Family: Cheilodactylidae — morwongs
Typical size: 60 cm

This is one of several related species which support significant fisheries off southeast and eastern Australia. However, the name "morwong" is also given to a number of quite different fish, especially various sweetlips (*Plectorhynchus spp.*), emperors and other reef species.

COMMON NAMES

NL: Morwong
RU: Morwong
US: Morwong

FISHING METHODS

MOST IMPORTANT FISHING NATIONS

Australia

PREPARATION

USED FOR

EATING QUALITIES

Morwong has greyish meat, with firm texture and mild flavour. It can be used in many different fish recipes. It is mostly sold fresh, though small quantities are being frozen for export markets.

Nutrition data:
(100 g edible weight)

Water	75.4 g	Total lipid	
Calories	108 kcal	(fat)	2.9 g
Protein	20.5 g	Omega-3	0.5 mg

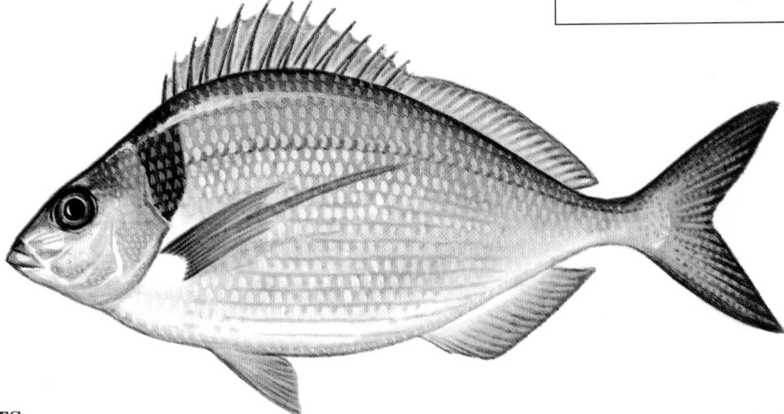

HIGHLIGHTS

This is one of several related species that together supply about 2,000 tons yearly to Australian markets. A good quality fish, morwong are only now beginning to be known outside their range, although frequently confused with the rather different sweetlips.

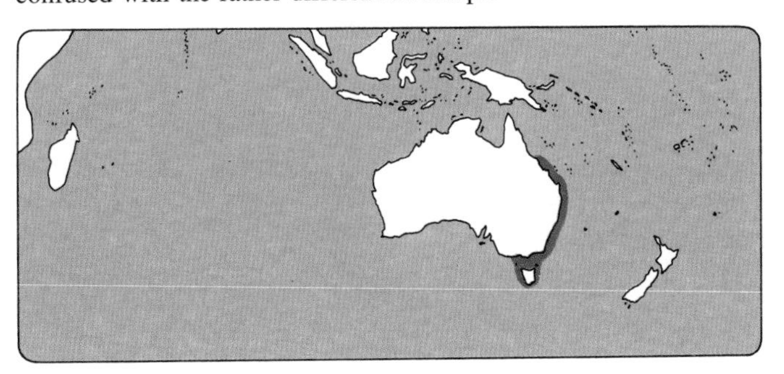

DK: En fin spisefisk, som ofte forveksles med den helt anderledes gryntefisk. Først nu finder den indpas på markeder uden for Australien

D: Ein guter Speisefisch, der häufig mit dem ganz anders gearteten Grunzer verwechselt wird. Beginnt erst jetzt Märkte ausserhalb Australiens zu erobern

E: Exquisito pescado comestible, que se confunde a menudo con el roncador, que es muy diferente. Empieza a ser aceptado en los mercados de fuera de Australia

F: Un excellent poisson de table, souvent confondu avec le grondeur pourtant très différent. Ce n'est que maintenant qu'il pénètre des marchés hors de l'Australie

I: Un ottimo pesce commestibile che spesso si confonde con il pesce burro, anche se è del tutto diverso. Solo adesso sta entrando in uso al di fuori dell'Australia

NORTHERN ANCHOVY

Scientific name:
Engraulis mordax

Synonym: North Pacific anchovy
Family: Engraulididae — anchovies
Typical size: 15 cm

The northern or California anchovy is very similar to the Peruvian anchovy or anchoveta, *E. ringens*. The species goes through cycles of great abundance followed by scarcity. In periods of abundance it is found from southern Baja California, Mexico, to the Queen Charlotte Islands of British Columbia, Canada. In times of scarcity, the range shrinks to a less extensive area centred on southern California and northern Mexico.

The anchovy co-exists with the California sardine, but prefers slightly different ocean conditions. The two species tend to move in opposite trends of abundance. At the moment, the California sardine is staging a comeback and the anchovy appears to be on a declining trend.

COMMON NAMES
D: Amerikanische Sardelle
DK: Nordpacifisk ansjos
E: Anchoveta del Pacífico
F: Anchois du nord
GR: Gávros tou Irinikoú
I: Acciuga del Nord Pacifico
IS: Kyrrahafsansjósa
N: Amerikansk ansjos
NL: Noordpacifische ansjovis
P: Biqueirao do Pacífico
RU: Kalifornijskij anchous
S: Amerikansk ansjovis
SF: Kaliforniansardelli
US: California anchovy

DESCRIPTION
Anchovy is used mainly for fishmeal and bait, although in recent times some has been canned for human consumption. Mexican catches crashed in recent years following heavy fishing. The resource off California is thought to be adequate, but managers have not permitted much fishing and are likely to allow even less in future.

HIGHLIGHTS
Mostly found within 30 km of the shore, the northern anchovy forms large, dense schools which are easily caught with purse seines and pumped into vessel holds.

Historically, both California and British Columbia used to restrict the use of anchovies so they could be used only for bait. Mexican fishermen have long targeted the species for fishmeal and oil as well as bait. The very oily meat is not considered very palatable.

Nutrition data:
(100 g edible weight)

Water	72.7 g
Calories	142 kcal
Protein	18.0 g
Total lipid (fat)	7.8 g
Omega-3	1.6 mg

NORTHERN ANCHOVY *Engraulis mordax*

FISHING METHODS

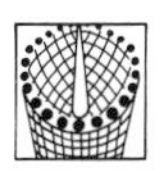

MOST IMPORTANT FISHING NATIONS
Mexico, USA

PREPARATION
Fishmeal 

USED FOR

EATING QUALITIES
Anchovies have an exceptionally high oil content and are considered generally unappetizing, although canned anchovies, especially if highly salted, can be palatable, if not as good as the more delicate European anchovy.

Fishmeal markets are volatile, although the rapid development of salmon, shrimp and other aquaculture industries has greatly increased demand for top quality fishmeal as a constituent of feed for farmed fish and shellfish.

IMPORTANCE
Mexican landings of northern anchovy have slumped from over 300,000 tons in 1982 to almost nothing in recent years. American landings, mainly from California, have remained at a few thousand tons (although there is evidence that more fish is available to be caught if management restrictions were removed). Catches from further north have not been significant since the 1940s, when the anchovy was abundant even in Canadian waters.

The abundance of the northern anchovy resource appears to be heavily dependent on hydrological conditions in the ocean. Recent occurrences of El Niño, the warm current in the South Pacific which affects ocean temperatures and weather patterns in many parts of the Pacific, may have disrupted the breeding and survival patterns of this species.

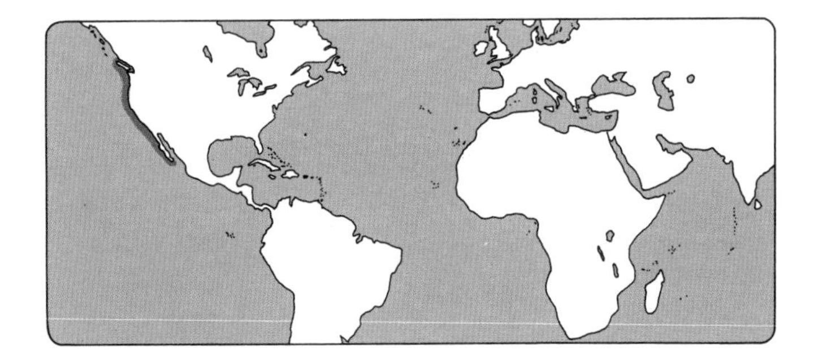

DK: Findes ofte indtil 30 km fra kysten, hvor fisken lever i tætte stimer, som nemt fanges med not og pumpes ned i lastrummet. En meget fedtrig art, som anvendes til fiskemel og madding

D: Kommt häufig bis zu 30 km vor der Küste vor, wo der Fisch in dichten Schwärmen lebt und sich einfach durch Ringwaden fangen und in den Lastraum pumpen lässt. Eine sehr fetthaltige Art, die zu Fischmehl verarbeitet wird und als Köder dient

E: Se encuentra a menudo hasta 30 km de la costa, donde vive el pescado en densos bancos, fáciles de capturar con red de cerco y de bombear a la bodega. Especie muy rica en grasa. Sirve para harina y cebo

F: Rencontré généralement jusqu'à 30 km des côtes, où il vit en bancs denses, qui sont pêchés en sennes coulissantes et mis en cale au moyen d'un pompage. Riche en graisses, il est transformé en farine de poisson et en appât

I: Si trova maggiormente fino a 30 km dalla costa, dove vive in branchi fitti che sono facili da pescare con il clanciolo e pompare nella stiva. Una specie ricca di grasso, usata per farina di pesce e come esca

NORTHERN BLUEFIN TUNA

Scientific name:
Thunnus thynnus

Synonyms: Bluefin tuna, tunny, tuna
Family: Scombridae — tunas and mackerels
Typical size: 200 cm, 400 to 500 kg

Northern bluefin tolerate cooler temperatures than most other tunas. They are found in warm and temperate waters of the North Atlantic and North Pacific (the populations in each ocean are each considered a subspecies). Small numbers also occur in widely scattered areas of the southern hemisphere, including New Zealand and South Africa.

The species supports a traditional trap fishery in the Mediterranean, which targets the fish as they follow their annual migration routes to the eastern Mediterranean and the Bosphorus.

Pressure on the resource has led to increasingly severe catch restrictions for both commercial and recreational purposes.

DESCRIPTION
The largest of the tunas, and one of the largest of all fishes, the bluefin can grow over 300 cm long and weigh as much as 680 kg. The largest fish are mostly caught in the western Atlantic, particularly off the northeastern United States and Canada. The eastern fish, although very large, seldom achieve giant sizes.

COMMON NAMES
D: Thunfisch, roter Thun
DK: Atlantisk tun
E: Atún rojo, cimarrón, atún
F: Thon rouge commun
GR: Tónos
I: Tonno rosso, tonno
IS: Túnfiskur
J: Kuromaguro
N: Makrellstørje
NL: Rode tonijn, gewone tonijn
P: Atum rabilho, atum
RU: Obyknovennyj golub. tunets
S: Tonfisk
SF: Tonnikala
TR: Orkinoz
US: Bluefin tuna

LOCAL NAMES
AR: Atún aleta azul
IL: Tunna kehula
PO: Tunczuk
TU: Toun ahmar
ME: Atún de aleta azul
VE: Atún aleta azul

HIGHLIGHTS
The largest of all the tunas is now a declining resource, with severe harvest restrictions. The giants may be as much as 20 years old, so stocks may need a considerable time to re-establish.

Demand for bluefin tuna remains very strong; the best quality northern bluefin can fetch prices unequalled by any similar fish. Experimental work on farming or ranching the species is being undertaken to increase production, but is unlikely to result in significant production for some years.

Nutrition data:
(100 g edible weight)

Water	68.1 g
Calories	137 kcal
Protein	23.3 g
Total lipid (fat)	4.9 g
Omega-3	1.2 mg

FISHING METHODS

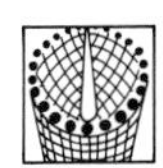 

MOST IMPORTANT FISHING NATIONS
France, Japan, Italy, United States, Spain

PREPARATION
Sashimi

USED FOR

EATING QUALITIES
Bluefin tuna are among the oiliest fish when in prime condition. The meat is firm and dark, almost red. The meat along the lateral line is particularly dark and strongly flavoured, while the belly meat, also dark, has the most oil. The freshest and oiliest fish are sold for exceptionally high prices for Japanese sashimi. Specially handled and prepared fish are flown to Tokyo for auction; the price depends on the freshness, colour and oiliness of the fish. Japanese vessels sometimes have special equipment to super-freeze the huge carcasses for sashimi.

Smaller fish (which may still be 100 to 200 kg) are less valuable, though in both North America and Europe there are good markets for fresh bluefin of any size. The meat is best brined before it is cooked, to reduce the pronounced fishy flavour. Small quantities used to be canned, as part of the light meat pack, but the high value of the species has largely ended the practice.

IMPORTANCE
Catches of northern bluefin are falling, as more restrictions are placed on fishing to protect the resource. Nevertheless, the extremely high value of the largest fish ensures that harvesters will catch every fish permitted.

In North America, the giant bluefin fishery attracts recreational and commercial interests, both because of its value and because of the thrill of catching these large and powerful fish.

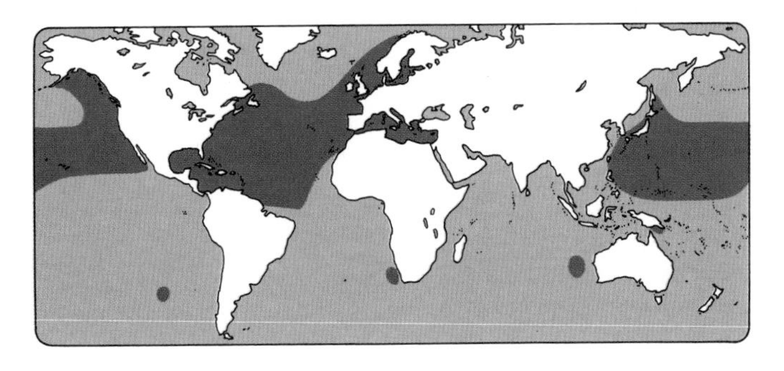

DK: Overfiskeri af denne, den største tunfisk, har ført til meget lave fangstkvoter. Er stærkt efterspurgt, og for de fineste eksemplarer er markedsprisen langt højere end for tilsvarende fisk

D: Das Überfischen dieser grössten Thunfischart hat zu äusserst niedrigen Fangquoten geführt. Ist sehr nachgefragt. Für die besten Exemplare liegt der Marktpreis weit höher als bei ähnlichen Fischarten

E: Las capturas son muy pequeñas por la pesca excesiva de esta especie, la más grande de los atunes. Es muy demandado y para los ejemplares más finos el precio de mercado es mucho más elevado que para cualquier otro pescado similar

F: Le plus grand de tous les thons. La surpêche a conduit à des quotas de pêche très bas. Fortement demandé, le prix du marché pour les meilleurs exemplaires est bien plus élevé que pour d'autres poissons de qualité équivalente

I: Il più grande dei tonni. La pesca eccessiva ha diminuito molto le quantità esistenti di questa specie. E' molto richiesto e per gli esemplari più belli il prezzo di mercato supera di molto quello degli altri pesci simili

OARFISH/KING OF THE HERRINGS

Scientific name:
Regalecus glesne

Family: Regalecidae — oarfishes
Typical size: up to 7 m

The king of the herrings is found throughout the world's oceans. Or so scientists believe, mainly on the basis that dead or dying specimens are occasionally found on widely scattered beaches across the globe.

In fact, very little is known of this extraordinary looking fish. It appears to inhabit deep water, around 1,000 m, where it lives on euphausiids (krill), which are shrimp-like organisms. Most examples of this fish known to science are either sick or dead, found floating on or near the surface or lying on the shore. These fish often have the ends of their long tails bitten off. It is thought that healthy fish may be able to outswim trawl nets.

COMMON NAMES
D: Riemenfisch, Heringskönig
DK: Sildekonge
E: Pez remo
F: Roi des harengs, régalec
GR: Vasiliás tis réngas
I: Re di aringhe
IS: Síldakóngur
N: Sildekonge
NL: Riemvis, haringkoning
P: Relangueiro
RU: Selyanoj korol
SF: Airokala
US: Oarfish

DESCRIPTION
The king of the herring is a brightly coloured, eel-shaped fish with a very long, mane-like dorsal fin running the length of its body, which may reach as much as 7 metres.

It is brilliantly silver in colour, with blue and purplish iridescence. The crest, which resembles that of a cockatoo, is crimson or bright pink. This colour is usually continued on all the fins.

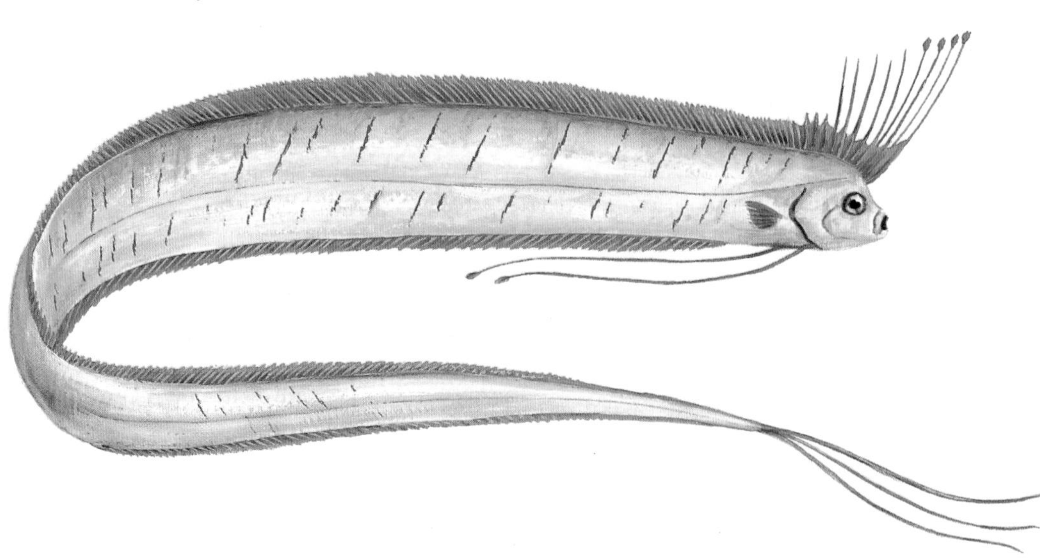

HIGHLIGHTS
The king of the herrings is named for a Norwegian belief that the fish accompanies herring shoals and that harming it would result in a disastrous season for herring fishing. Many Norwegians once dep-ended for their livelihood on herring fishing, so this would have been a frightening prospect.

Since herring are surface dwellers and the king of the herring appears to live in extremely deep water, it is not clear how these fishermen made such a connection between the two species.

Nutrition data:
(100 g edible weight)

Water	n.a.
Calories	n.a.
Protein	n.a.
Total lipid (fat)	n.a.
Omega-3	n.a.

FISHING METHODS

MOST IMPORTANT FISHING NATIONS
None

PREPARATION
Not eaten

USED FOR
Not eaten

EATING QUALITIES
The oarfish is not eaten, because it is not caught either commercially or by anglers. The quality of its meat is not known.

IMPORTANCE
The king of the herrings has no significance commercially or economically, but is an object of curiosity among scientists and lay people alike.

The fish appears to be fragile: the ends of the fins and tail are often missing and the silvery colour on the skin can be wiped off with the tip of a finger. It has no teeth and the flesh appears to be soft.

Fish harvesting technology is rapidly reaching a point where exploitation is becoming feasible of some species living in deep ocean water but far from the bottom. Lanternfishes, oarfishes and other mesopelagic species may offer large resources of usable protein. This might be used for direct consumption by humans if the fish is acceptable and can be marketed at a satisfactory price.

Alternatively, if the meat is of poor quality or the yield too small to permit machine processing of the species, it can be used for food for farmed fish and other animals. The need for feed for fish farming is increasing rapidly as aquaculturists continue to expand their existing operations and add new species.

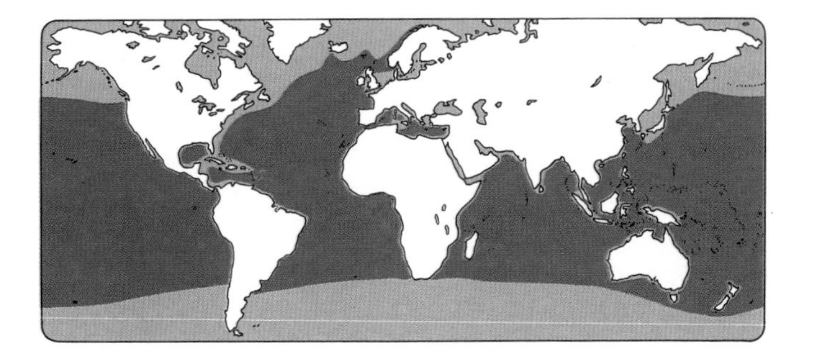

DK: Sildekongen har fået navn efter norsk overtro: fisken følger sildestimer, og at påføre den skade vil resultere i en katastrofal sæson for sildefiskeriet og industrien

D: Der Riemenfisch folgt den Heringsschwärmen. Nach norwegischem Aberglauben würde die Verwundung dieses Fisches für die Heringssaison katastrophale Folgen haben

E: El rey de los arenques debe su nombre a la superstición noruega según la cual el pescado sigue a los bancos de arenques, y dañar a este pescado significaría una catástrofe para la pesca y la industria del arenque

F: Le roi des harengs doit son nom à la superstition norvégienne: le poisson suit les bandes de harengs, et lui faire de mal équivaut à une saison catastrophique pour la pêche aux harengs et l'industrie de transformation

I: Il re delle aringhe, nominato cosi secondo una credenza norvegese, segue i branchi delle aringhe e ad arrecarlo danno, comporterà una stagione catastrofica per la pesca delle aringhe e per l'industria ittica

ORANGE ROUGHY

Synonym: Slimehead
Family: Trachichthyidae — slimeheads
Typical size: 30 to 40 cm, 1.5 kg

Scientific name:
Hoplostethus atlanticus

COMMON NAMES
D: Atlantischer Sägebauch
DK: Orange savbug, soldatfisk
E: Reloj anaranjado
F: Hoplostète orange
GR: Kathreptópsaro
I: Pesce specchio atlantico
IS: Búrfiskur, búri
J: Orenzi-rafii
NL: Kaizersbaars, valse beryx
P: Olho de vidro laranja
RU: Atlantichesky bolshegolov
S: Lyktfiskar
US: Orange roughy

Known to scientists as slimehead, this species got its commercial start when United States authorities were persuaded to allow it to be sold as orange roughy. Clearly, the name slimehead had no commercial future, but with its new name and mild taste the species took American buyers by storm. Unusually, the breakthrough was made in supermarkets, looking for a white, bland and inexpensive fish. Generally, new species are first tried and popularized through restaurants, where consumers traditionally tend to be more adventurous about trying previously unknown seafoods.

DESCRIPTION
Orange roughy has a massive head with bony ridges and cavities. The flesh yield is comparatively small for the size of the fish. Caught in very deep water, most are dead by the time they reach the lighter pressure of the surface.

Because of the rapid rise in popularity of the species and because it is only sold in fillet form, there have been a number of cases of substitution of less costly species for roughy. Oreo dory and even hoki have been used in such malpractice.

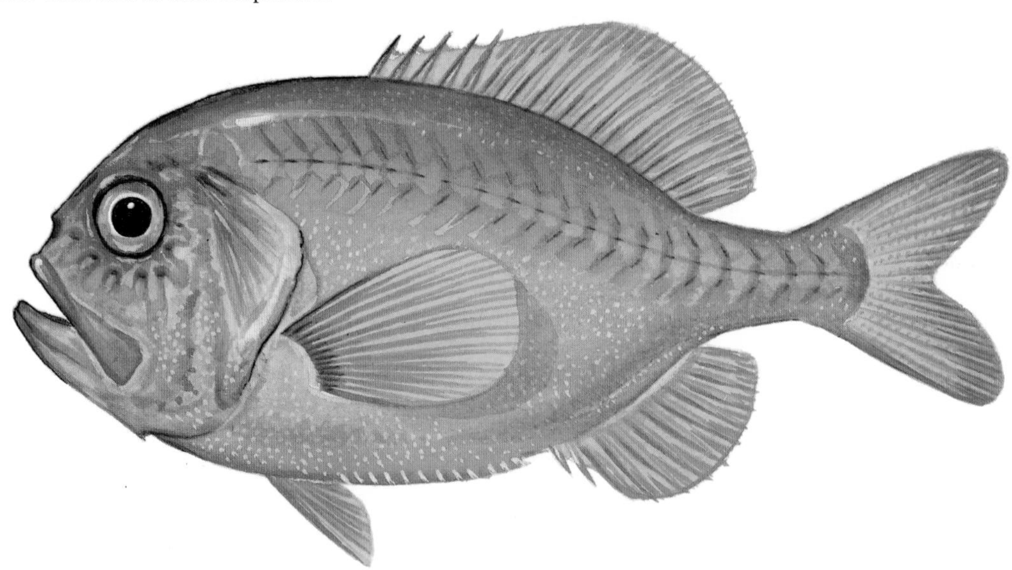

HIGHLIGHTS
Originally identified as an Atlantic species (as its scientific name suggests), large quantities of this species, which lives in water as deep as 1,100 m, were found off New Zealand in the late 1970s. Modern fishing techniques made trawling in such deep water possible.

Imaginative marketing coupled with high quality handling and processing rapidly built export markets in the United States and Europe, especially among people who preferred fish without a strong flavour.

Nutrition data:
(100 g edible weight)

Water	75.3 g
Calories	135 kcal
Protein	14.7 g
Total lipid (fat)	8.5 g
Omega-3	0.1 mg

FISHING METHODS

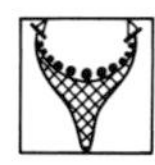

MOST IMPORTANT FISHING NATIONS
New Zealand, Australia

PREPARATION

USED FOR

EATING QUALITIES
Orange roughy has firm, pearly white meat which holds together well when it is cooked without the skin. The flavour is delicate; some authorities say it has a shellfish flavour. It is suitable for most cooking methods and has a good shelf life if properly handled.

Orange roughy is always skinned when processed, to remove the subcutaneous fat which consists of undigestible waxy esters, unlike the fat on other fishes. Some product is dressed and frozen at sea, then thawed, filleted and skinned later on land.

IMPORTANCE
Southern hemisphere orange roughy resources in deep water off Australia and New Zealand support carefully regulated fisheries of about 70,000 tons a year. New stocks have been discovered quite recently, leading to hopes that catches can be increased. Further large stocks have now been located in the North Atlantic, off Iceland and Norway, although commercial exploitation of these resources had not begun at the time of writing.

Orange roughy are irregular in distribution. Researchers have found a number of densely populated patches. It is not clear whether these are entirely separate populations or whether there is some mixing. The populations are cosmopolitan in deep, temperate waters, from the North Atlantic to the South Pacific and Indian Oceans.

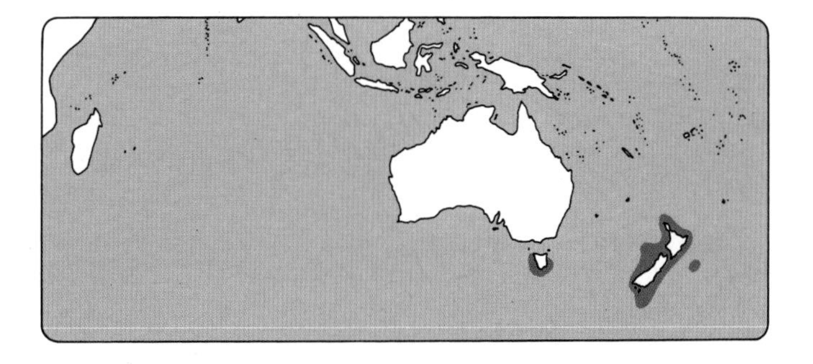

DK: Smart markedsføring kombineret med en effektiv rensnings- og forarbejdningsproces har skabt et stort marked i USA og Europa. Fisken appellerer til forbrugere, der ønsker et mildtsmagende produkt

D: Geschicktes Marketing verbunden mit einem effektiven Säuberungs- und Verarbeitungsprozess haben für grosse Nachfrage in den USA und Europa gesorgt. Der Fisch spricht die Verbraucher an, die ein mildes Produkt bevorzugen

E: Un eficaz proceso de limpieza, elaboración y comercialización han creado un gran mercado en EE.UU. y Europa. El pescado atrae a los consumidores que prefieren un producto suave

F: La commercialisation imaginative, combinée avec un processus de nettoyage et de transformation efficace, a créé un grand marché aux E.-U. et en Europe pour ce poisson qui attire les consommateurs désirant un produit de goût très doux

I: Un'intensa commercializzazione, combinata con un efficace processo di lavorazione, ha creato un importante mercato negli USA e in Europa, specialmente fra i consumatori che preferiscono un pesce di sapore non troppo caratteristico

PACIFIC BARRACUDA

Scientific name:
Sphyraena argentea

Family: Sphyraenidae — barracudas
Typical size: up to 1.2 m

The Pacific barracuda is an important game fish in California, Mexico and Central America. It is caught commercially in Mexico and Central America, but Californian catches have been reduced by bans on gillnetting in coastal waters.

FISHING METHODS

MOST IMPORTANT FISHING NATIONS
Mexico, United States

PREPARATION

USED FOR

EATING QUALITIES
Firm, rather dark, meat and a good flavour characterize the Pacific barracuda, which is excellent barbecued. It is important to bleed the fish when it is caught; without bleeding, the meat is dark and turns sour and rancid very quickly, even if frozen.

COMMON NAMES
D: Kalifornischer Barrakuda
DK: Stillehavsbarracuda
E: Picuda, barracuda
F: Barracuda du Pacifique
GR: Barakoúda tis Kalifórnias
I: Barracuda della California
NL: Californische barracuda
P: Bicuda da Califórnia
RU: Tikhookeanskaya barrakula
SF: Hopeabarrakuda
US: California barracuda

Nutrition data:
(100 g edible weight)

Water	76.5 g	Total lipid	
Calories	98 kcal	(fat)	2.0 g
Protein	20.0 g	Omega-3	0.4 mg

HIGHLIGHTS
A voracious feeder, the Pacific barracuda takes bait or strikes lures eagerly and fights hard, making it popular with anglers. Because of its large, sharp teeth and aggressive disposition, it is regarded as potentially dangerous, though there are no reports of actual attacks.

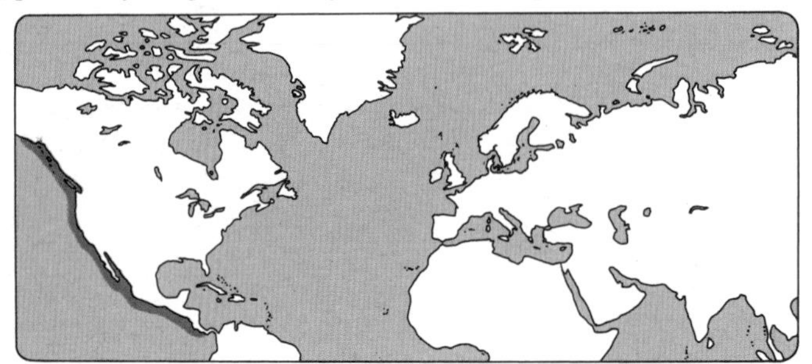

DK: Trods manglende konkrete beviser, anses barracudaen for farlig på grund af dens lange skarpe tænder og aggressive adfærd

D: Trotz fehlender Beweise wird der Barrakuda wegen seiner langen scharfen Zähne und seines aggressiven Verhaltens als gefährlich angesehen

E: La barracuda es considerada peligrosa debido a sus largos dientes afilados y su temperamento agresivo, aunque no existen pruebas concretas

F: En raison de ses longues dents tranchantes et son comportement agressif, il est considéré comme dangereux, malgré le manque de preuves des méfaits prétendus

I: Nonostante la mancanza di prove concrete, il barracuda è considerato un pesce pericoloso a causa dei denti acuti e il suo comportamento agressivo

PACIFIC BONITO

Synonym: Eastern Pacific bonito
Family: Scombridae — tunas and mackerels
Typical size: 60 cm, 4 kg

Scientific name:
Sarda chilensis

The Pacific bonito is a good game fish and a significant commercial species, especially in Peru, which accounts for about three quarters of the world catch of 50-55,000 tons. Twenty years ago, the catch in Peru and Chile was many times larger.

FISHING METHODS

MOST IMPORTANT FISHING NATIONS

Peru, Mexico, United States

PREPARATION

USED FOR

EATING QUALITIES

The Pacific bonito has rather dark flesh, which cans well (but may not be sold as tuna in the United States). Commercial catches are mostly canned. Broiled or grilled, the flavour can be good if the fish is properly bled and handled by the harvester.

COMMON NAMES

D: Chilenische Pelamide
DK: Chilensk bonit
E: Bonito del Pacífico
F: Bonite du Pacifique
GR: Palamída tou Irinikoú
I: Tonnetto cileno
IS: Síletúnfiskur
NL: Pacifische boniet
P: Bonito do Pacífico
RU: Tikhookeanskaya pelamida
S: Chilensk bonit
SF: Chilensarda
US: Bonito

LOCAL NAMES

CL: Bonito
ME: Bonito
PE: Aguadito

Nutrition data:
(100 g edible weight)

Water	68.6 g	Total lipid	
Calories	155 kcal	(fat)	7.0 g
Protein	22.9 g	Omega-3	1.4 mg

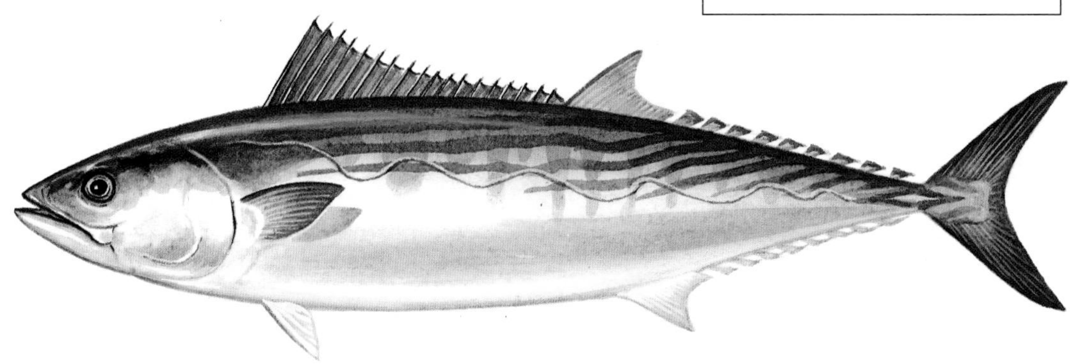

HIGHLIGHTS

The bonito, which may reach 100 cm and 11 kg, shoals close to shore, making it an available and popular game fish, especially in California and Mexico. Canned, it is inexpensive, with dark meat. Steaks, fresh or frozen, are well regarded in Latin America.

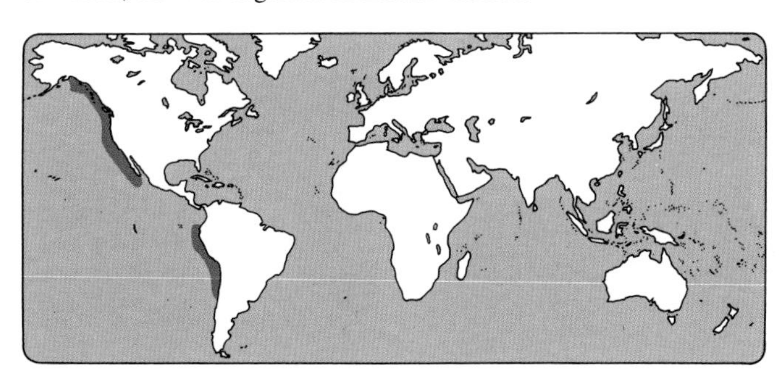

DK: Populær sportsfisk og en vigtig kommerciel art. Billigt, mørkt kød til konserves, men spises også frisk/frossen i skiver

D: Beliebt bei Sportfischern und wichtiger Industriefisch. Billig, dunkles Fleisch für Konserven, wird aber auch frisch/gefroren in Scheiben gegessen

E: Un popular pescado deportivo y una importante especie comercial. Carne barata y oscura para conservas, pero se consume también fresco o congelado en rodajas

F: Espèce commerciale importante et très cotée pour la pêche sportive. Chair foncée à bas prix pour les conserves; se mange aussi frais ou surgelée en tranches

I: Un pesce sportivo popolare e una specie commerciale importante. La carne scura a basso prezzo si trova conservata, ma anche fresco o congelato in fette

PACIFIC COD

Scientific name:
Gadus macrocephalus

Family: Gadidae — cods
Typical size: 85 to 100 cm up to 120 cm

Pacific cod are found mainly within 200 miles of the shore, within fishery zones claimed by the coastal states, that is Russia, Canada and the USA. Comparatively small and declining quantities are taken by vessels from other countries, fishing in the "doughnut hole" area which separates the Russian and American waters in the center of the North Pacific.

DESCRIPTION
Caught over the continental shelf and upper slope of the North Pacific, especially in the Bering Sea and adjacent waters of Alaska and Siberia, the species supports large fleets of factory trawlers as well as numerous shore-based processing facilities in these remote areas.

The fish is brownish, sometimes grey, with spots on the upper two thirds of the body. It is generally found at depths between 100 and 400 m. Size and location of schools vary considerably from year to year with natural conditions; the fish is seldom caught in the southern parts of its range on either side of the Pacific. Most landings are made from the colder waters.

COMMON NAMES
D: Pazifischer Kabeljau
DK: Stillehavstorsk
E: Bacalao del Pacífico
F: Morue du Pacifique
GR: Bakaliáros tou Irinikoú
I: Merluzzo del Pacifico
IS: Kyrrahafs-thorskur
J: Madara
N: Stillehavstorsk
NL: Pacifische kabeljauw
P: Bacalhau do Pacífico
RU: Tikhookeanskaya treska
S: Stillahavstorsk
SF: Tyynenmerenturska
TR: Pàsifik morinasi
US: Pacific cod

HIGHLIGHTS
Very similar to Atlantic cod in appearance, taste and texture, Pacific cod is an important contributor to world demand for white-meated groundfish. It is processed into numerous forms of fillets, portions and other products and is seldom distinguished from its Atlantic cousin in markets for these items.

The substantial and important fishery for Pacific cod tends to be overshadowed by the much larger fishery for Alaska pollock (*Theragra chalcogramma*) and the longer history of Atlantic cod.

Nutrition data:
(100 g edible weight)

Water	81.3 g
Calories	77 kcal
Protein	17.9 g
Total lipid (fat)	0.6 g
Omega-3	0.1 mg

PACIFIC COD *Gadus macrocephalus*

FISHING METHODS

  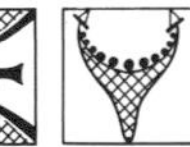 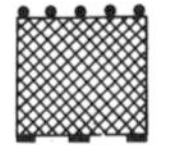 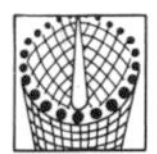 

MOST IMPORTANT FISHING NATIONS

United States, Russia, Japan, Canada, Korea

PREPARATION

USED FOR

  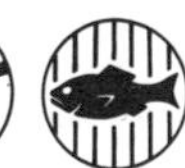   

EATING QUALITIES

Pacific cod has moist, mild-tasting white meat with a medium size flake. The fish and the flake are generally smaller than Atlantic cod. The texture of the meat is very similar; Pacific cod is perhaps a little more moist than its Atlantic cousin.

Pacific cod is used for salting and drying as well as for fish blocks, which are a major raw material for processed products. Much of the catch is processed at sea. Some cod is landed in China and other low-wage countries for thawing, filleting, re-freezing and other further processing. This product is often a little cheaper than the standard, single-frozen fillets produced by factory trawlers.

IMPORTANCE

Landings of Pacific cod average around 400,000 tons annually. The USA produces about 60 percent of the catch, Russia another 30 percent. Both nations have progressively excluded others from their fishing zones, keeping more of the resource for their own fishermen.

Pacific cod are fast growing and fecund fish, able to recover quickly from fishing pressure. This means that the fish can be exploited intensively. Political and economic factors rather than biological ones have resulted in a complex network of controls that probably keep catches well below the level at which the resource and fishing yield can be maintained.

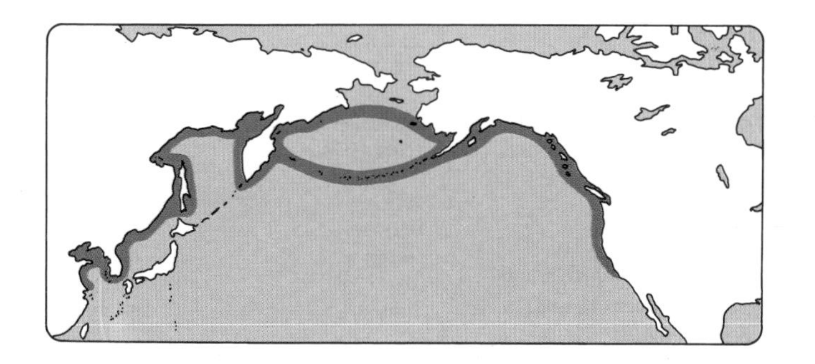

DK: Da smagen, konsistensen og det lyse kød er meget lig atlantisk torsk, er denne art med til at dække det store globale behov. Talrige forarbejdningsmuligheder

D: Da sie vom Geschmack, der Konsistenz und dem weissen Fleisch her dem atlantischen Dorsch recht ähnlich ist, trägt diese Art dazu bei, den grossen weltweiten Bedarf zu decken. Zahlreiche Verarbeitungsmöglichkeiten

E: Como su sabor, consistencia y carne blanca se parece mucho a la del bacalao atlántico, esta especie contribuye a cubrir la gran demanda global. Se puede consumir de muchas formas

F: Le goût, la consistance et la chair blanche ressemblant beaucoup au cabillaud atlantique, cette espèce contribue à couvrir le grand besoin global. De nombreuses possibilités de transformation

I: Dato che il sapore, la consistenza e le carni bianche sono molto simili al merluzzo atlantico, questa specie contribuisce a coprire il grande fabbisogno globale. Offre tante possibilità di impiego

PACIFIC HALIBUT

Scientific name:
Hippoglossus stenolepis

Family: Pleuronectidae — right-eye flounders
Typical size: 35 kg

Pacific halibut was important to the economy of indigenous people in northern North America. When railroads opened eastern markets, salted halibut was shipped in large quantities. Fishing techniques improved and resources closer to shore were fished out. International controls, imposed in 1932 to protect and renew the stocks, coupled with favourable ocean conditions in the ocean, have restored resources to a level which can supply current demand at reasonable prices.

DESCRIPTION
Some experts consider that Atlantic halibut (*H. hippoglossus*) and Pacific halibut are the same species. They are certainly very similar. Pacific halibut does not grow as large as its Atlantic cousins, but still reaches as much as to 260 cm and 220 kg.

These huge flatfish are found in the eastern Pacific as far south as southern California; but hardly any are caught south of the Columbia River. They range north to Norton Sound in the Bering Sea, then to the Anadyr and Kuril Islands, Kamchatka, northeast Sakhalin, the southern part of the Sea of Okhotsk and are found in small numbers off northern Hokkaido, Japan.

COMMON NAMES
D: Pazifischer Heilbutt
DK: Stillehavshelleflynder
E: Fletán del Pacífico
F: Flétan du Pacifique
GR: Hálibat tou Irinikoú
I: Halibut del Pacifico
IS: Kyrrahafs lúda
J: Ohyô
N: Stillehavskveite
NL: Pacifische heilbot
P: Alabote do Pacífico
RU: Belokory paltus
S: Stillahavs-helgeflundra
SF: Tyynenmerenpallas
US: Pacific halibut

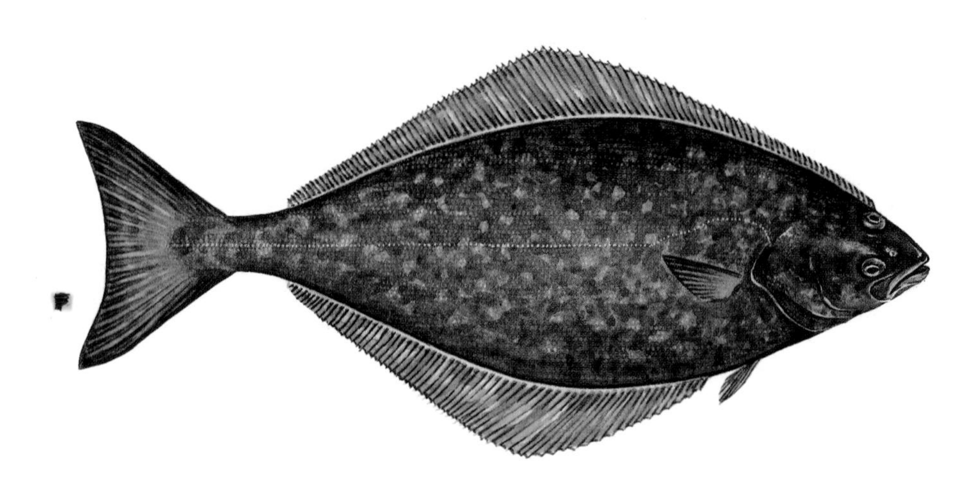

HIGHLIGHTS
Halibut fisheries in the eastern Pacific are tightly regulated, which may have helped to protect the resource in recent times. In the western Pacific, uncontrolled Russian fishing has greatly increased supplies of small halibut in recent years. It is not clear whether the western Pacific resources will support current levels of exploitation.

Major markets for Pacific halibut include Japan, North America and Western Europe. It is traded frozen, as headless and dressed fish, as fillets (sometimes called fletches) and as steaks.

Nutrition data:
(100 g edible weight)

Water	75.4 g
Calories	104 kcal
Protein	20.8 g
Total lipid (fat)	2.3 g
Omega-3	0.4 mg

PACIFIC HALIBUT *Hippoglossus stenolepis*

FISHING METHODS

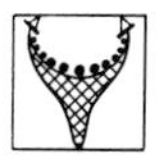 

MOST IMPORTANT FISHING NATIONS
United States, Canada, Russia

PREPARATION

USED FOR

EATING QUALITIES
One of the best of all white-meated fish, Pacific halibut is firm, sweet and well flavoured. It is popular in Canada as a superior fish-and-chips, which highlights the meaty texture and excellent flavour. Halibut stands up to all cooking methods, including broiling. Halibut soup thickens naturally with a gel contained in the bones. Fillets can be cut without bones, but if present on the plate the bones are easily removed.

IMPORTANCE
World catches of about 35,000 tons are recorded, most of it taken by the United States. Canada's fishery is also significant, while Japan now catches very little. However, Russia has recently increased its exploitation of halibut, selling large quantities of mainly smaller fish in world markets.

Halibut fishing has been restricted in the eastern Pacific since 1932, under an agreement between the USA, Canada and Japan. (Before the days of general 200-mile limits, Japan used to fish halibut in the region.) Canada and the US are both changing from the traditional management regime which imposed short seasons of only a few days. This resulted in uneven supplies to markets and considerable danger to fishermen who could not afford to miss limited time to fish, so would operate in any weather. The new system uses quotas which can be filled throughout the year; this has already increased the flow of chilled (rather than frozen) fish to market and improved returns to fishermen.

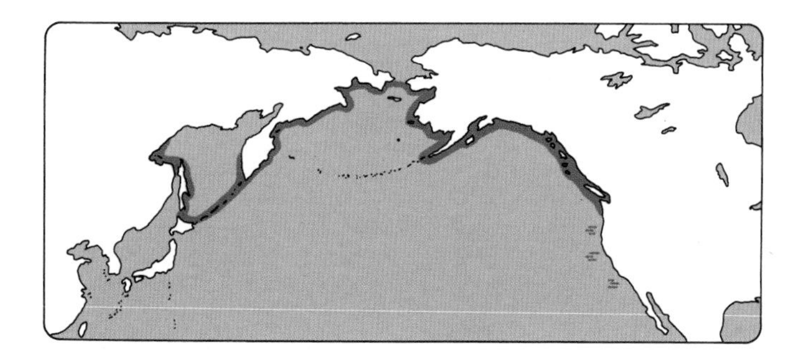

DK: Fiskeriet i det østlige Stillehav er nøje reguleret. I den vestlige del derimod, har ukontrolleret russisk fiskeri de seneste år reduceret bestanden væsentligt

D: Der Fischfang im östlichen Stillen Ozean ist genau geregelt. Im westlichen Teil dagegen hat unkontrollierte russische Fischerei in den letzten Jahren den Bestand erheblich dezimiert

E: La pesca del este del Océano Pacífico está regulada estrictamente. En el oeste, en cambio, la pesca incontrolada de los rusos durante recientes años ha reducido las existencias notablemente

F: La pêche dans l'océan Pacifique oriental est strictement réglementée. Par contre, dans la partie occidentale, la pêche russe non contrôlée a largement réduit la population

I: La pesca è rigorosamente regolata nel Pacifico orientale, mentre nella parte occidentale il patrimonio ha subito una forte riduzione in seguito alla pesca russa incontrollata degli ultimi anni

PACIFIC HERRING

Scientific name:
Clupea pallasii

Synonym: North Pacific herring
Family: Clupeidae — herrings, sardines
Typical size: 25 cm, up to 33 cm
See also Herring

Experts debate whether the Pacific herring is a separate species or a subspecies of the Atlantic herring (*Clupea harengus*). The question is largely academic as the two herring overlap only in a small area off the northern coast of Russia in the Arctic Sea. There are few differences between the Atlantic and Pacific species of herring. The most significant difference commercially is that the Pacific herring has larger eggs, making it significantly more valuable for roe products for Asian markets.

A small but important roe fishery utilizes the roe after it has been attached by the fish to kelp in protected bays and estuaries. The seaweed, together with the eggs, is harvested and processed as a very high-priced delicacy in Japan. Strict control over the fishing season for this product, called kazunoko-kombu, sometimes restricts harvesting to a matter of minutes each year.

COMMON NAMES
D: Pazifischer Hering
DK: Stillehavssild
E: Arenque del Pacífico
F: Hareng du Pacifique
GR: Rénga tou Irinikoú
I: Aringa del Pacifico
IS: Kyrrahafs-síld
J: Nishin
N: Sild
NL: Pacifische haring
P: Arenque do Pacífico
RU: Mongopozvonkovaya seld
S: Stillahavssill
SF: Vienansilli
TR: Pasifik ringasi
US: Pacific herring

DESCRIPTION

A small coastal pelagic species found in large schools. Pacific herring migrate inshore to breed, and they are then easily caught. The species is virtually identical to Atlantic herring in appearance.

HIGHLIGHTS

Pacific herring are traditionally fished when they migrate inshore to spawn, when they have the least oil in the flesh and are therefore least useful for curing, smoking or canning, as well as dry and tough to eat fresh or frozen.

If the fish is pursued further offshore by vessels capable of processing at sea, it is in much better condition for eating; oil content has been recorded at levels considerably higher than those usually found in Atlantic herring.

Nutrition data:
(100 g edible weight)

Water	71.5 g
Calories	191 kcal
Protein	16.4 g
Total lipid (fat)	13.9 g
Omega-3	1.7 mg

PACIFIC HERRING *Clupea pallasii*

FISHING METHODS

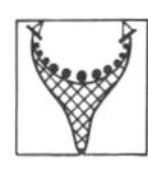  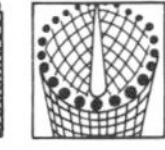

MOST IMPORTANT FISHING NATIONS
Russia, USA

PREPARATION
Roe

USED FOR

EATING QUALITIES
Pacific herring, like Atlantic herring, is a delicate fish with a small flake and many bones. In North America it is seldom eaten, being used for bait or roe. In Russia, China and Japan it is canned, salted or used fresh and frozen. It is as versatile as the Atlantic herring and in some locations has an even higher oil content, a feature greatly prized by smokers and curers because fish with more oil produce smoked and cured products that are moister and more tender.

IMPORTANCE
Pacific herring catches average about 200,000 tons yearly, though there are wide fluctuations from year to year in abundance and landings on both sides of the ocean.

In the eastern Pacific, herring is mostly caught for roe markets in Asia. Large quantities of spawning herring are caught in Pacific waters of the USA and Canada during a short season when the spawning fish is close to shore. The fish are quickly frozen, later thawed for processing. Because the fish are in spawning condition, oil content is very low. Most of the female carcasses and all the male fish are sold for bait, for fishmeal or are discarded. Some of the largest carcasses are salted and dried for migaki nisshin, a product used, like the roe, in Japan. Kazunoko, salted herring roe, is a highly regarded gift in Japan. Kazunoko-kombu is the herring eggs laid on kelp by the fish. It is salted and sold with the kelp and is an extremely expensive delicacy.

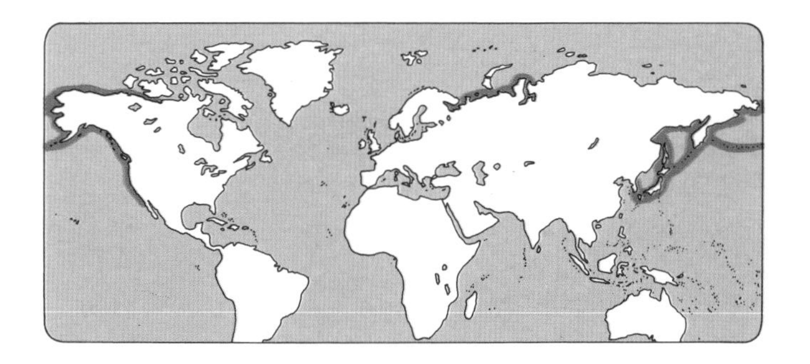

DK: Store mængder af gydende sild fanges i Stillehavet omkring USA og Canada i den korte gydeperiode, hvor fisken søger ind til kysten. Før forarbejdning, bør fisken hurtigt nedfryses

D: Grosse Mengen Hering werden im Stillen Ozean vor der amerikanischen und kanadischen Küste während der kurzen Laichzeit gefangen, wenn der Fisch sich in Küstennähe aufhält. Zur Verarbeitung sollte der Fisch schnell eingefroren werden

E: Se captura en gran número en la costa del Pacífico de EE.UU. y Canadá durante el desove. Hay que congelar el pescado rápidamente antes de prepararlo

F: Pêché en grand nombre dans le Pacific à hauteur des Etats-Unis et du Canada pendant la courte période de fraie où le poisson s'approche de la côte. Doit être congelé rapidement, avant la transformation

I: Si pescano grossi quantitativi di aringhe nel Pacifico lungo le coste degli USA e del Canada nel breve periodo di deposito delle uova. Il pesce va congelato rapidamente prima della lavorazione

PACIFIC OCEAN PERCH

Scientific name:
Sebastes alutus

Synonym: Pacific rockfish
Family: Scorpaenidae — scorpionfishes
Typical size: up to 50 cm, 1.4 kg

The Pacific ocean perch is a popular fish in North America and is beginning to be exported to Europe as well as to traditional markets in Japan. Fillets, with the skin on, are shipped fresh or frozen. If frozen, they are usually individually wrapped and graded in 2 ounce (55 g) steps.

DESCRIPTION
Pacific ocean perch is a red skinned species found mostly in deep water of more than 125 m from southern California to the Bering Sea. A closely related species, *S. paucispinosus*, is now considered to be the same fish, extending the range through the western Pacific to northern Honshu Island in Japan.

These rockfish are slow to mature and bear live young, in rather small numbers: only about 300,000 when the fish is 20 years old. Pacific ocean perch is slow growing and lives for perhaps as much as 30 years. The low level of fecundity and slow growth together mean that the species is susceptible to over-fishing. It is reported that it is an important constituent of the food supply of such valuable fishes as halibut and albacore tuna.

COMMON NAMES
D: Pazifischer Rotbarsch
DK: Stillehavsrødfisk
E: Gallineta del Pacífico
F: Sébaste du Pacifique
GR: Kokkinópsaro
I: Sebaste
IS: Blettakarfi
J: Arasukamenuke
NL: Pacifische roodbaars
P: Cantarilho do Pacífico
RU: Tikhookeanskij klyuvach
US: Ocean perch

HIGHLIGHTS
The Pacific ocean perch is commercially the most important rockfish species in the northeast Pacific. At one time, substantial quantities were exported to Japan, but developing markets in North America and increasing controls over the stocks of the species have effectively ended that trade.

Fillets are widely marketed throughout Canada and the United States and are popular in their own right as well as to substitute, less expensively, for lake perch.

Nutrition data:
(100 g edible weight)

Water	79.3 g
Calories	90 kcal
Protein	08.8 g
Total lipid (fat)	1.6 g
Omega-3	0.3 mg

PACIFIC OCEAN PERCH *Sebastes alutus*

FISHING METHODS

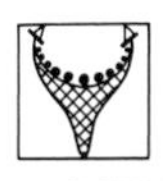 

MOST IMPORTANT FISHING NATIONS
Canada, United States

PREPARATION

USED FOR

EATING QUALITIES
Pacific ocean perch has pinkish-grey meat, which turns almost white when cooked. The flavour is mild and the texture moist. The species is excellent stuffed and baked. In North America, it is always sold with the skin on, partly to distinguish it from brown-skinned rockfishes, which are worth a little less in most markets.

An often more important reason for leaving the skin on is so that the fish can be sold, legally in California, Oregon and Washington but not elsewhere, as Pacific red snapper, confusing consumers into believing that they are buying true red snapper, which is a much more expensive — and quite different — fish. The red skin also serves to confuse consumers between the ocean perch and lake perch.

Pacific ocean perch, like other rockfish, should have firm flesh and taut, shiny skin. Flabbiness in either is a sign of deterioration.

IMPORTANCE
Pacific ocean perch landings average 40,000 to 50,000 tons a year, with about half the catches made by Canada. It is mostly trawled in deep water, but small quantities are caught on bottom-fishing long-lines. The hooked fish can be better, especially if brought alive to the boat so it can be bled. However, these perch have gas bladders and live at considerable depths; the bladders often burst if the fish are brought quickly to the surface, killing the fish before they reach the deck.

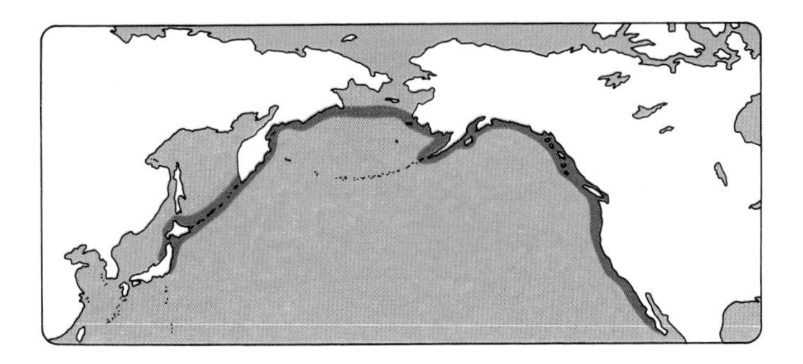

DK: Kommercielt en af de vigtigste rødfiskarter i det nordøstlige Stillehav. Filetterne, som også markedsføres som erstatning for den dyrere ferskvandsart, er ved at vinde indpas i Europa og Japan

D: Eine der wichtigsten kommerziellen Rotbarscharten im nordöstlichen Pazifik. Die Filets, die auch als Ersatz für die teurere Süsswasserart angeboten werden, gewinnen in Europa und Japan an Boden

E: Una de las más importantes gallinetas nórdicas del nordeste del Océano Pacífico. Sus filetes, también introducidos en el mercado en sustitución de la especie de agua dulce más cara, está ganando aceptación en Europa y Japón

F: Sur le plan commercial, la sébaste est l'une des plus importantes espèces du sud-est Pacifique. Les filets, commercialisés également en substitution à l'espèce d'eau douce, plus chère, gagnent du terrain, tant en Europe qu'au Japon

I: Fra gli scorfani, la specie commercialmente più importante del nord-est Pacifico. I filetti, commercializzati in sostituzione della specie d'acqua dolce più costosa, sta guadagnando terreno in Europa e nel Giappone

PACIFIC SALMON SHARK

Scientific name:
Lamna ditropis

Family: Lamnidae — mackerel sharks
Typical size: 180 cm, 175 kg; up to 305 cm

The population of salmon shark, a large North Pacific predator, is believed to be increasing, possibly due in recent years to the large increases in salmon stocks on which the species feeds.

Some researchers have indicated that the Pacific salmon shark prefers sockeye to other species of salmon. Resources of sockeye salmon have been particularly buoyant, with record runs being recorded in many rivers, especially in Alaska.

DESCRIPTION
The salmon shark is found close to shore as well as far out on the ocean, where it follows schools of salmon, squid and other prey. The species prefers cooler water and sometimes gather into small schools. They are fast swimmers, able to maintain body temperature well above that of the surrounding water.

Salmon shark are oviviparous, which means the eggs hatch in the uterus and the young are born alive. There are up to four pups in a litter. The young are known to cannibalise weaker members of the litter while still in the uterus.

COMMON NAMES
D: Pazifischer Heringshai
DK: Stillehavssildehaj
E: Marrajo salmón
F: Requin-taupe saumon
GR: Lámia tou Irinikoú
I: Smeriglio del Pacifico
IS: Kyrrahafshámeri
J: Nezumizame
NL: Pacifische haringhaai
P: Tubarao sardo do Japao
RU: Lososevaya akula
US: Salmon shark

HIGHLIGHTS
This big shark is found throughout the North Pacific. It is caught commercially by Japanese longliners and is targeted by sport fishermen in Canada and the USA. It does considerable damage to salmon nets and is believed to eat large quantities of salmon, explaining the name.

A common, large and very powerful shark, the species has not been associated with attacks on divers or bathers, although it is often seen in inshore waters.

Nutrition data:
(100 g edible weight)

Water	76.4 g
Calories	84 kcal
Protein	20.6 g
Total lipid (fat)	0.2 g
Omega-3	n.a. mg

PACIFIC SALMON SHARK *Lamna ditropis*

FISHING METHODS

MOST IMPORTANT FISHING NATIONS

Japan

PREPARATION

USED FOR

EATING QUALITIES

Salmon shark meat is fairly dark, but firm. Like all sharks, it is best if the fish is killed by stunning and bleeding, to remove as much blood as possible from the flesh. This helps to lighten the colour of the meat as well as extend the shelf life by reducing the amount of ammonia-forming urea in the system.

The meat is regularly used in northern Honshu, but elsewhere is not highly regarded. Fins, hides and livers are also utilised, with fins having particular value.

IMPORTANCE

Salmon shark fins are used when available. It is thought that the resource could support considerably greater fishing effort and that fins could provide useful revenue to fishermen. The fins of a salmon shark represent as much as 5 percent of the total weight of the live fish.

Shark fins are prepared by lightly salting the fins, then drying them for as long as a month in the air. At this stage of preparation, they can be sold to dealers who prepare them for consumption. The final preparation stages are complex and the quality variations between different fins, species and product quite detailed. Briefly, the fins are soaked in hot water, then the skins and cartilage separated from the rays or needles, which are the desired part. The needles are boiled, cooled slowly and arranged for packaging and sale.

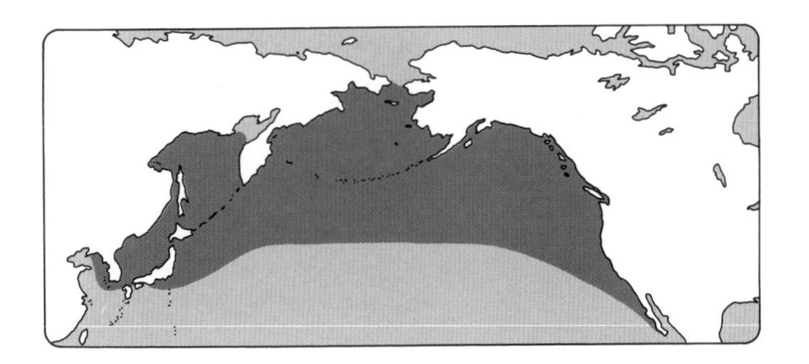

DK: Almindelig, stor og meget stærk haj, som ikke forbindes med angreb på mennesker. Konsumeres i Japan. Også finner, hud og lever bruges

D: Gewöhnlicher, grosser und sehr starker Hai, nicht für Angriffe auf Menschen bekannt. Wird in Japan gegessen. Flossen, Haut und Leber werden auch verarbeitet

E: Tiburón común, grande y muy fuerte, que no ataca a seres humanos. Se consume en Japón, y se usan también sus aletas, piel e hígado

F: Grand requin commun, très fort, qui ne s'attaque cependant pas à l'homme. Commercialisé au Japon. Les ailerons, la peau et le foie sont également utilisés

I: Squalo comune, grande e molto forte, che non è connesso con attacchi agli uomini. Si consuma in Giappone. Se ne usano anche le pinne, la pelle e il fegato

PACIFIC SAND LANCE

Scientific name:
Ammodytes hexapterus

Family: Ammodytidae — sand lances
Typical size: 20 cm

Closely related to the Atlantic sand lance (*A. americanus*), the Pacific species is sometimes targeted as a raw material for fishmeal by Japanese and Russian fishermen. Most of the enormous resource is eaten by other fish and marine mammals. It is one of the most important forage species in northern seas.

FISHING METHODS

MOST IMPORTANT FISHING NATIONS
Japan, Russia

PREPARATION
Fishmeal  

USED FOR

EATING QUALITIES
Dried for consumption in Japan, the sand lance can also be prepared like smelts, simply cleaned and fried, when it is said to be excellent and delicately flavoured.

COMMON NAMES
D: Pazifischer Sandaal
DK: Stillehavstobis
E: Lanzón del Pacífico
F: Lançon du Pacifique
GR: Ammóchelo tou Irinikoú
I: Cicerello del Pacifico
IS: Kyrrahafs sandsíli
J: Kita-ikanago
NL: Pacifische zandspiering
P: Galeota do Pacífico
RU: Tikhookeanskaya peschanka
US: Pacific sand lance

Nutrition data:
(100 g edible weight)

Water	78.5 g	Total lipid	
Calories	92 kcal	(fat)	2.5 g
Protein	17.5 g	Omega-3	0.6 mg

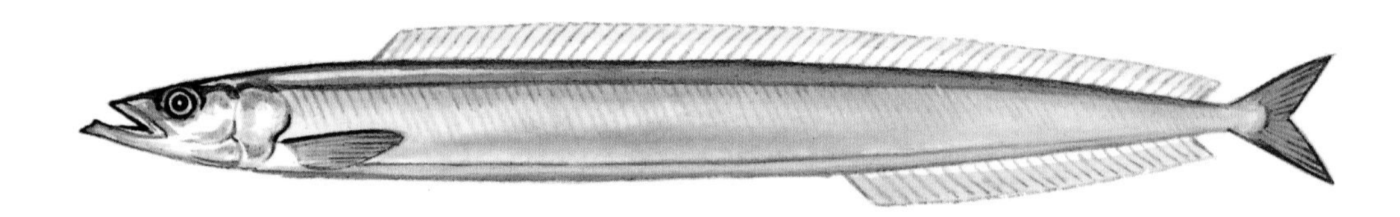

HIGHLIGHTS
Abundant throughout the north Pacific and adjacent Arctic waters, the sand lance is a small, smelt-like fish on which important commercial species such as salmon, cod and pollock depend for food. It is popular in Japan dried and salted.

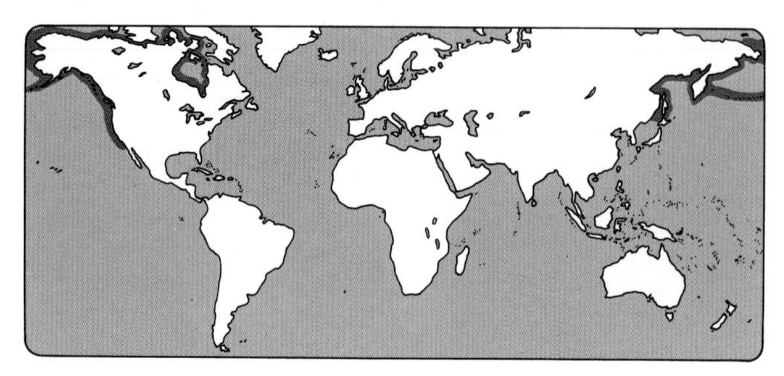

DK: En talrig, lille smelt-lignende fisk, der er føde for vigtige kommercielle arter som laks og torskefisk

D: Eine kleine, stintähnliche Art, die als Nahrung für wirtschaftlich bedeutende Arten wie Lachs und Dorsch dient

E: Pequeño y abundante pescado, parecido al eperlano, que sirve como alimento para especies comerciales importantes como el salmón y el bacalao

F: Très abondant, ce petit poisson genre éperlan sert principalement d'aliment pour les espèces commerciales importantes telles que le saumon et les gadidés

I: Piccolo pesce comune, simile allo sperlano. Alimento per altre specie commerciali importanti come i salmoni ed i merluzzi

PACIFIC SAURY

Scientific name:
Cololabis saira

Synonyms: Mackerel-pike, skipper
Family: Scomberesocidae — sauries
Typical size: 30 cm

Japan accounts for some 75 percent of the world's recorded prod-uction of 400,000 tons of Pacific saury. The species is seldom found close to shore, but ranges in large schools from Alaska southward off both sides of the Pacific.

FISHING METHODS

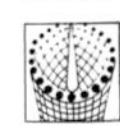

MOST IMPORTANT FISHING NATIONS
Japan, Russia, Korea, China

PREPARATION

USED FOR

EATING QUALITIES
The saury is an oily fish with dark meat, similar to jacks and mackerel. It has pronounced flavour. It is canned for human food; in this form it competes with canned mackerel and herring. It is also used for fishmeal and pet food.

COMMON NAMES
D: Kurzschnabel-Makrelenhecht
DK: Stillehavsmakrelgedde
E: Paparda del Pacífico
F: Balaou du Pacifique
GR: Loutsozargána tou Irinikoú
I: Costardella saira
J: Sanma
NL: Japanse makreelgeep
P: Agulhao do Japao
RU: Tikhookeanskaya saira
SF: Saira
TR: Zurna
US: Saury

LOCAL NAMES
KO: Ggong-chi

Nutrition data:
(100 g edible weight)

Water	72.3 g	Total lipid	
Calories	134 kcal	(fat)	5.5 g
Protein	21.0 g	Omega-3	1.1 mg

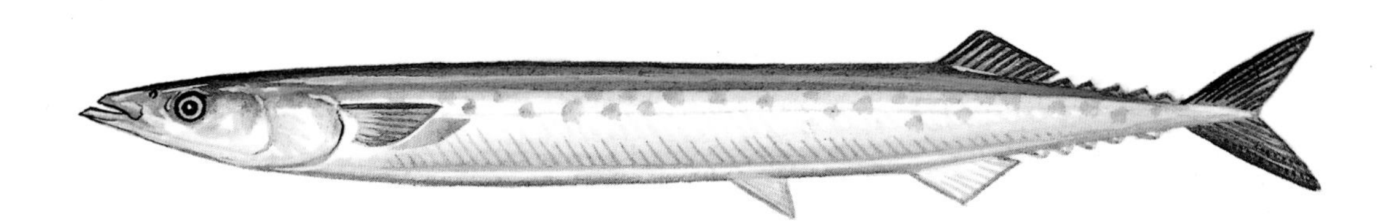

HIGHLIGHTS
Reportedly one of the most abundant species of the North Pacific, the Pacific saury is exploited off the Asian coast and is an important food species. Resources off North America, which are thought to be substantial, remain untouched.

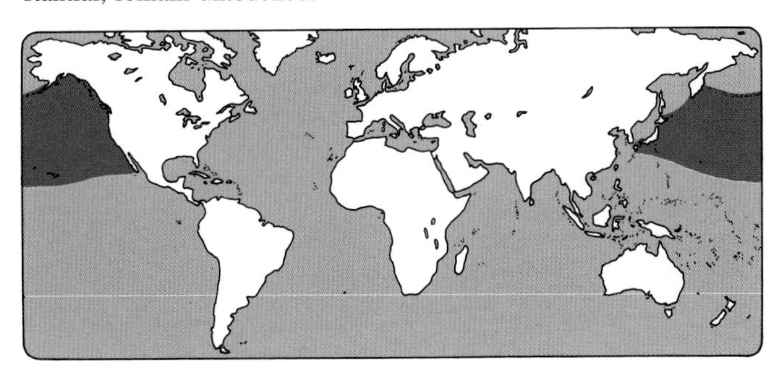

DK: Angiveligt en af det nordlige Stillehavs mest talrige arter. Fiskes langs Asiens kyster og er en vigtig konsumfisk

D: Angeblich eine der zahlreichsten Arten des nördlichen Stillen Ozeans. Wird entlang der Küste Asiens gefangen und ist ein wichtiger Speisefisch

E: Es una de las especies más abundantes del norte del Océano Pacífico. Se captura a lo largo de las costas de Asia y es un pescado de consumo importante

F: Sans doute l'une des espèces les plus abondantes du Pacifique Nord. Un poisson comestible important qui est pêché au large des côtes de l'Asie

I: Secondo le stime, una delle specie più abbondanti del Pacifico settentrionale. Viene pescato lungo le coste dell'Asia ed è un importante pesce di consumo

PERCH

Scientific name:
Perca fluviatilis

Family: Percidae — perches
Typical size: 35 to 50 cm, 1.2 kg

European perch is closely related to the yellow perch (*P. flavescens*) of North America. See under Yellow perch for more details. Together, the two species have an almost complete circumpolar range. The European perch was originally found from Ireland and Great Britain throughout northern Europe. It has been introduced to Asia and is now found across Russia to eastern Siberia.

Perch has also been introduced to more southerly countries in Europe and as far away as Australia, New Zealand and South Africa. Its popularity in England is one reason for these introductions: English settlers took the fish with them. An adaptable species, the perch has prospered and bred in many varying habitats far from its original range.

DESCRIPTION
Mostly a small fish of around 35 to 50 cm, the perch can grow as large as 50 cm and 4.75 kg. Its habitat is lakes and ponds, but it is also found in slow flowing rivers and is quite common in some of the brackish waters of the Baltic Sea.

COMMON NAMES
D: Flussbarsch, Barsch
DK: Aborre
E: Perca común, perca
F: Perche, perche de rivière
GR: Potamóperka
I: Pesce persico, perca
IS: Aborri
N: Abbor, åbor
NL: Baars
P: Perca
RU: Okun
S: Abborre
SF: Ahven
TR: Tatlisu levregi
US: Perch

LOCAL NAMES
PO: Okon

HIGHLIGHTS
The European perch has brightly coloured bands of yellow, green and brown on its sides. Even after the fish is dead, these colours remain quite bright, distinguishing the species from most others offered at the retail counter.

Although perch is caught in the Baltic Sea, it is mainly a lake fish. In landlocked Switzerland, it is in most years the most important fish species by volume produced, vying in other years with rainbow trout for that title.

Nutrition data:
(100 g edible weight)

Water	79.3 g
Calories	81 kcal
Protein	18.4 g
Total lipid (fat)	0.8 g
Omega-3	0.3 mg

PERCH — *Perca fluviatilis*

FISHING METHODS

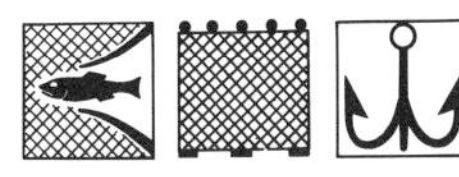

MOST IMPORTANT FISHING NATIONS
Switzerland, Germany, Poland, Sweden, Denmark

PREPARATION

USED FOR

EATING QUALITIES
European perch, like its American cousin the yellow perch (see that entry) has white meat with a small flake, delicate texture and mild flavour. An important food fish in central Europe and countries surrounding the Baltic, it is prepared in numerous ways and can also be used in recipes intended for soles, flounders and other fish with delicate meat.

Fillets are prepared from larger fish. These are usually fried. Dressed fish, whether small or larger, are pan-fried, grilled or baked.

IMPORTANCE
Although recorded caches of European perch are only about 4,000 tons a year, in the recent past they were three times that level. It is possible that some of the fall can be attributed to disruption in record keeping in eastern Europe after the collapse of Communism. In addition, recreational catches are not always recorded fully.

Perch was processed for export to North American markets in years when it was plentiful. Holland was the main producer. Markets in the mid-West of the United States frequently complained of an occasional bone remaining in the fillets.

Perch has significant recreational value throughout its wide range, taking baited hooks readily and providing good quality meat.

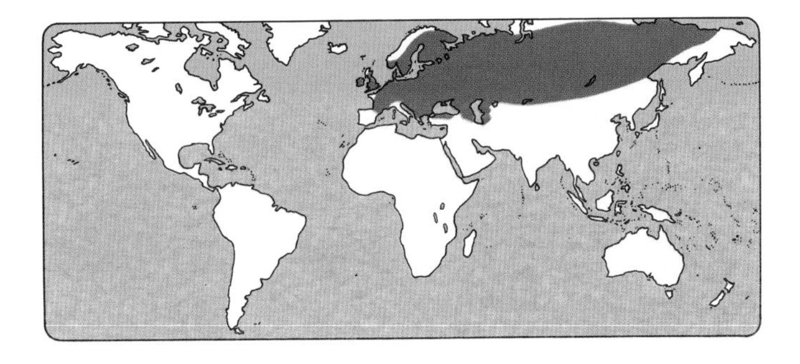

DK: Selvom aborren forekommer i Østersøen, er den generelt en ferskvandsfisk. I Schweiz er arten fangstmæssigt den største, kun i enkelte år overgået af regnbueørred

D: Auch wenn der Barsch in der Ostsee vorkommt, ist er eigentlich ein Süsswasserfisch. In der Schweiz ist die Art von der Fangmenge her am grössten, die nur in einzelnen Jahren von der Regenbogenforelle übertroffen wird

E: Aunque la perca vive en el Báltico, es ante todo un pescado de agua dulce. En Suiza esta especie es la más capturada, sólo superada en años aislados por la trucha arco iris

F: Même si la perche se rencontre dans la Baltique, elle vit généralement dans l'eau douce. En Suisse, la perche fait l'objet de la pêche la plus importante, surpassée seulement, certaines années, par la truite arc-en-ciel

I: Anche se il persico è catturato nel Baltico, è prevalentemente un pesce d'acqua dolce. In Svizzera costituisce la specie numero uno per volume, superata solo raramente dalla trota iridea

PERUVIAN ANCHOVY/ANCHOVETA

Scientific name:
Engraulis ringens

Family: Engraulididae — anchovies
Typical size: up to 20 cm

Anchoveta harvests fluctuate wildly, between almost nothing and 5 million tons in a year. It is one of the most intensively studied and managed fish populations, but the anchoveta's natural cycles have defeated the best efforts to control the fishery.

COMMON NAMES
D: Peru-Sardelle, Anchoveta
DK: Peruansk ansjos
E: Anchoveta peruana
F: Anchois du Pérou
GR: Gávros tou Peroú
I: Acciuga del Cile
IS: Perú-ansjósa
J: Iwashi
NL: Peruaanse ansjovis
P: Biqueirao do Peru
RU: Peruanskij anchous
SF: Perunsardelli
US: Anchovy

FISHING METHODS

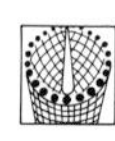

MOST IMPORTANT FISHING NATIONS
Peru, Chile

PREPARATION
Fishmeal, fishoil

USED FOR

EATING QUALITIES
The anchoveta is one of the world's most important resources for fishmeal and oil, but attempts to use the species for human consumption have made little progress because of the softness of the flesh and the extreme oiliness.

Nutrition data:
(100 g edible weight)

Water	65.5 g	Total lipid	
Calories	207 kcal	(fat)	15.0 g
Protein	18.0 g	Omega-3	3.0 mg

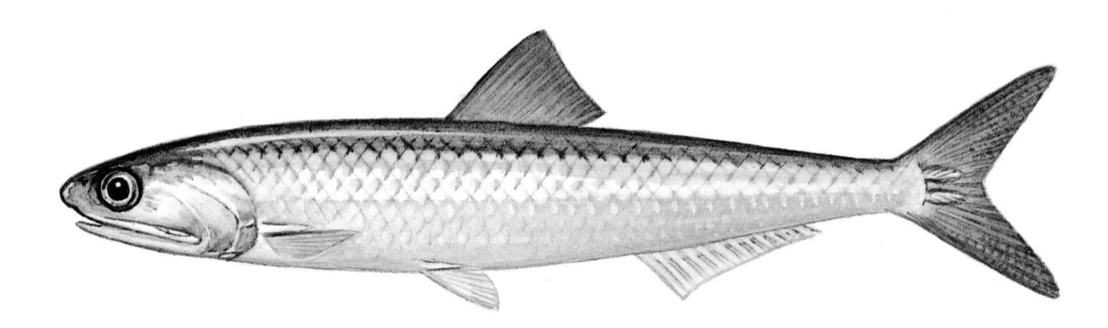

HIGHLIGHTS
In years when natural conditions are right, fishmeal made from the anchoveta is Peru's largest export. In other years, especially when El Niño drives the fish away from the coast, production is small. Huge populations of guano birds and pelicans also depend on the species.

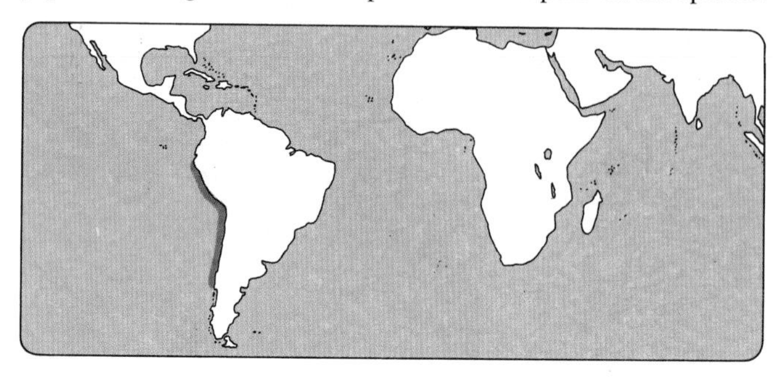

DK: I årene uden 'El Niño', og dermed under de rette naturlige betingelser, er fiskemel fremstillet af denne art Peru's største eksportartikel

D: In Jahren mit günstigen Strömungen und somit unter guten natürlichen Bedingungen ist diese zu Fischmehl verarbeitete Art Perus grösster Exportartikel

E: En los años en que las condiciones naturales son favorables - sin 'El Niño' - la harina de este pescado es el artículo de exportación más importante de Perú

F: Pendant les années de conditions naturelles favorables, la farine de poisson fabriquée de cette espèce constitue le plus grand article d'exportation du Pérou

I: Negli anni in cui le condizioni naturali sono favorevoli, la farina di pesce prodotta da questa specie è l'articolo d'esportazione più importante del Peru

PETRALE SOLE/BRILL

Scientific name:
Eopsetta jordani

Synonym: Petrale flounder
Family: Pleuronectidae — right-eye flounders
Typical size: up to 70 cm, 3 kg

Petrale sole, called brill in Canada, is found from Islas Los Coronados, Mexico north to the Bering Sea and the Aleutian Islands. It inhabits deep water most of the year, which puts it out of the reach of recreational fishermen.

FISHING METHODS

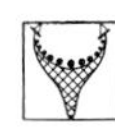

MOST IMPORTANT FISHING NATIONS
United States, Canada

PREPARATION

USED FOR

EATING QUALITIES
Light, white, very tender meat, with a moist texture and bland flavour make this excellent sole a favourite market fish on the west coast of North America. It is generally regarded as the premium Pacific coast sole. It is mostly available filleted, with the skin on.

COMMON NAMES
D: Kalifornische Scholle
DK: Kalifornisk flynder
E: Rodaballo de California
F: Plie de California
GR: Kalkáni tis Kalifórnias
I: Passera della California
IS: Kaliforníu koli
NL: Californische schol
P: Solha da Califórnia
RU: Kambala-romb
US: Petrale sole

Nutrition data:
(100 g edible weight)

Water	78.7 g	Total lipid	
Calories	87 kcal	(fat)	1.6 g
Protein	18.2 g	Omega-3	0.3 mg

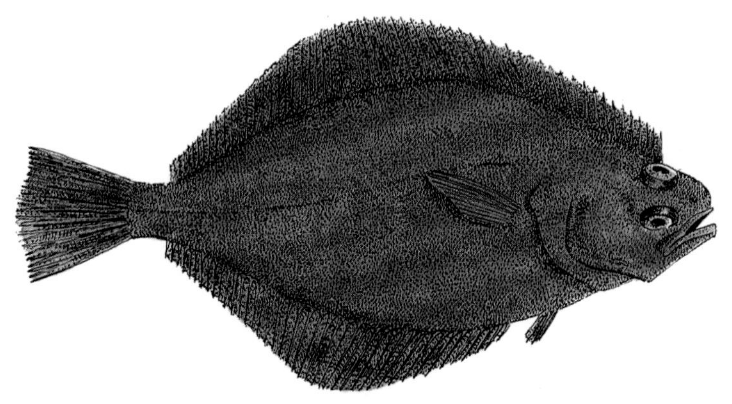

HIGHLIGHTS
The thick body of the petrale sole yields excellent fillets, which are regarded as second only to halibut among eastern Pacific flatfish for eating quality. The catch is mostly taken in deep water by trawlers, though there are some incidental catches on halibut longlines.

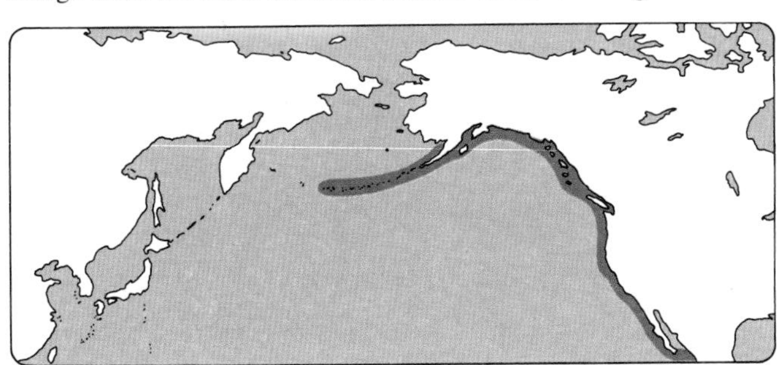

DK: Af fisken fås flotte filetter, der dog kvalitetsmæssigt, blandt det østlige Stillehavs fladfisk, overgås af helleflynder

D: Der Fisch ergibt schöne Filets, deren Qualität unter den Plattfischen des östlichen Stillen Ozeans jedoch vom Heilbutt übertroffen wird

E: De este pescado se obtienen excelentes filetes, y de entre los pleuronectos del este del Océano Pacífico sólo es superado en calidad por el hipogloso

F: Les filets de la plie de Californie sont excellents, et parmi les poissons plats de l'océan Pacifique oriental, seul le flétan la surpasse en qualité

I: Il pesce fornisce dei bei filetti la cui qualità, tra i pleuronettiformi del Pacifico orientale, viene superata soltanto da quella dell'halibut

PIKE

Scientific name: *Esox lucius*

Synonym: Northern pike
Family: Esocidae — pikes
Typical size: 1 m, 4.5 kg

The northern pike is valuable as a food fish and a sport fish. It is raised in ponds for both purposes, and wild stocks are generally robust. It is a tough fish, capable of adapting to new environments. The pike has been widely introduced. In some waters it is regarded as an important additional resource; in others it is blamed for the destruction of smaller native species.

DESCRIPTION
Pike are long and thin. They are aggressive predators, inhabiting shallow water in lakes and slow rivers throughout northern latitudes in America, Asia and Europe. The brightly spotted body appears to be flecked with gold, due to tiny gold spots on the tips of each scale.

The pike can be heavily infested with parasites, including the broad tapeworm which, if not killed by thorough cooking, can infect man. A cestode parasite which uses the pike as an intermediate host is responsible for large losses in usable catches of lake whitefish (*Coregonus clupeaformis*) in some areas. The pike also suffers from a trematode which causes unsightly cysts on the skin, though these can be removed by skinning the fish and are not harmful.

COMMON NAMES
D: Flusshecht, Hecht
DK: Gedde
E: Lucio
F: Brochet du Nord
GR: Toúrna
I: Luccio
IS: Gedda
J: Kawakamasu
N: Gjedde
NL: Snoek
P: Lúcio
RU: Schuka
S: Gädda
SF: Hauki
TR: Turna baligi
US: Pike

LOCAL NAMES
PO: Szczupak

HIGHLIGHTS
Although the angling world record pike weighed only 25 kg, there are many stories from medieval times of huge fish that pulled mules and milk maids into ponds. The Mannheim Hoax was about a pike supposedly released by the Emperor Friedrich II in 1230 and which, when caught in 1497 was 5.8 m long and weighed 250 kg. Unfortunately, the skeleton of this monster, long preserved in the Cathedral in Mannheim, is now proved to have been faked.

Pike grow larger in Europe than they do in North America.

Nutrition data:
(100 g edible weight)

Water	78.9 g
Calories	84 kcal
Protein	19.3 g
Total lipid (fat)	0.7 g
Omega-3	0.1 mg

PIKE — *Esox lucius*

FISHING METHODS

MOST IMPORTANT FISHING NATIONS
Canada, France, Poland, Russia, Czech Republic

PREPARATION
Minced  

USED FOR

EATING QUALITIES
Northern pike is excellent quality fish. The meat is white and sweet, with a firm flake. Small fish are often pan-fried, but it is even better baked, grilled or poached. Pike has a reputation for having a "muddy" flavour, especially in warmer weather, but this seems to be associated with inadequate de-sliming and scaling. The pike is covered in a thick mucous, which should be removed with boiling water before the fish is prepared for cooking. Alternatively, the fish can be skinned rather than simply scaled.

Pike, like shad, have numerous Y-shaped intramuscular bones, which detract from the enjoyment of the fish. These bones can be removed with tweezers from larger fish.

IMPORTANCE
Recorded world production of about 20,000 tons a year certainly understates the importance of this fish, which is farmed, raised for stocking for sport and is found naturally over a huge area. Canada is the major producer, but production in Russia is thought to be seriously under-reported. Many anglers' catches are also missed by the statistical collection process.

Pike is a very valuable game fish, sought for its fighting qualities as well as for its fine meat. It is part of the winter ice-fishery in Canada and has been introduced into many rivers far from its original range.

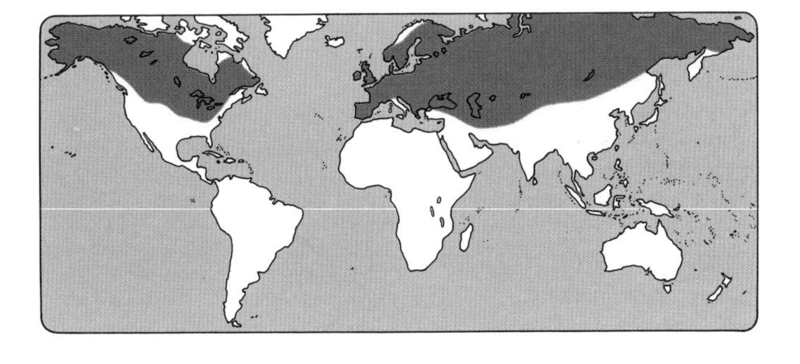

DK: En gedde formentlig udsat af Tysklands kejser Friedrich II i 1230 og fanget i 1497 var på hele 5,8 m og 250 kg. Kæmpens skelet, som opbevaredes i Mannheims katedral, er nu desværre afsløret som svindel

D: Ein Hecht - angeblich 1230 vom deutschen Kaiser Friedrich II ausgesetzt und 1497 gefangen - war 5,8 m lang und wog 250 kg. Das im Dom zu Mannheim aufbewahrte Gerippe des Riesen hat sich leider als Fälschung entpuppt

E: Un lucio, echado al agua en 1230 por el Emperador de Alemania, Friedrich II, y capturado en 1497, medía 5,8 m de longitud, y pesaba 250 kg. El esqueleto del gigante, preservado en la Catedral de Mannheim, ha resultado ser un engaño

F: Un brochet, supposé avoir été relâché en 1230 par l'empéreur Friedrich II et repêché en 1497, mesurait 5,8 m et pesait 250 kg. La squelette de ce monstre, conservé dans la cathédrale de Mannheim, s'est malheureusement avéré un faux

I: Un luccio rilasciato probabilmente dall'imperatore tedesco, Friedrich II, nel 1230 e catturato nel 1497 era di ben 5,8 m e 250 kg. Lo scheletro di questo gigante, conservato nella cattedrale di Mannheim, si è però rivelato una truffa

PINK SALMON

Scientific name: *Oncorhynchus gorbuscha*

Synonyms: Humpback salmon, gorbuscha
Family: Salmonidae — salmonids
Typical size: 2 kg

COMMON NAMES
D: Buckellachs, rosa Lachs
DK: Pukkellaks
E: Salmón rosado
F: Saumon rose
GR: Roz solomós
I: Salmone rosa
IS: Bleiklax, hnúdlax
J: Karafutomasu
N: Pukkellaks
NL: Pink zalm, roze zalm
P: Salmao rosa
RU: Gorbusha
S: Puckellax
SF: Kyttyrälohi
TR: Pembe alabalik
US: Pink salmon

Pink and sockeye salmon are the two bases of the Alaskan canned salmon industry. In most years, pinks are more abundant, though the value of sockeye production is much greater.

Line-caught (troll) pink salmon is a premium product, although only a small part of the catch is taken on lines and most of this comes from Canadian waters. The better quality that can be achieved with trolled salmon could be an important factor in expanding markets for the species in general. Worldwide markets for canned salmon are sliding slowly downward, increasing the industry's need to develop new products and new markets for pink salmon.

DESCRIPTION
Pink salmon is the smallest of the six Pacific salmon species. The largest fish recorded was 6.4 kg, It is a silver fish when in the ocean, with large spots on the back and tail which help to identify it. In spawning dress, the males develop a deep hump and dark red skin, while the females may be red or black. The species is found throughout the North Pacific, though little is harvested in the more southerly parts of its range.

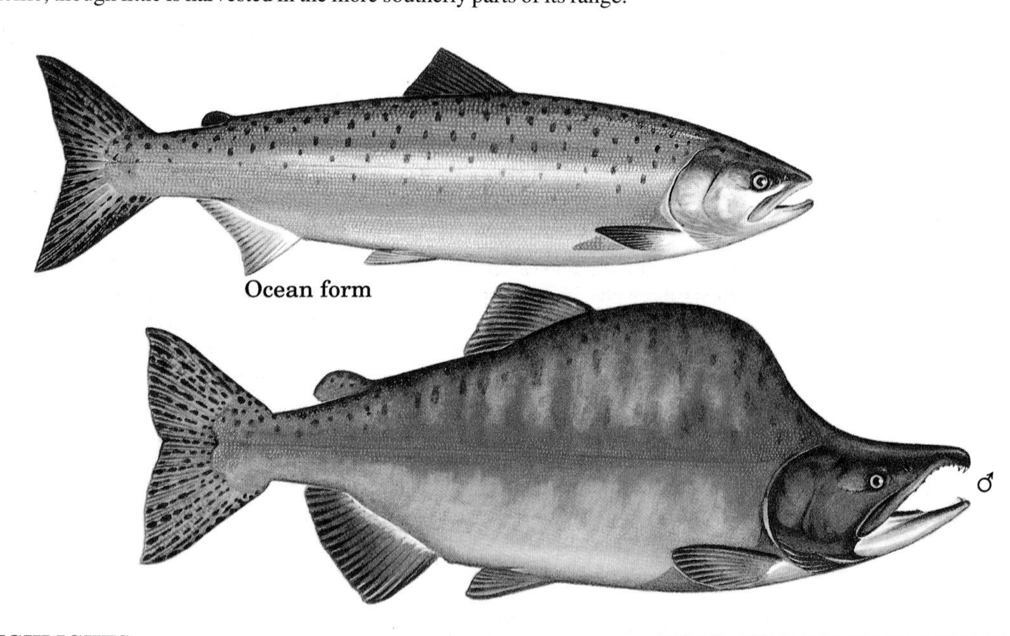

HIGHLIGHTS
As ranched production of pink salmon increases in Alaska and Siberia, processors are being forced to find new ways of presenting and marketing the species. Consumer acceptance of this inexpensive salmon seems to be good, but harvesters and processors have quality and handling problems to solve before markets will give full value to the species.

Pink salmon roe is highly valued, especially in Japan but increasingly in western Europe, where salmon caviar is gaining new markets.

Nutrition data:
(100 g edible weight)

Water	76.4 g
Calories	111 kcal
Protein	19.9 g
Total lipid (fat)	3.5 g
Omega-3	1.0 mg

FISHING METHODS

 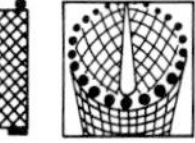 

MOST IMPORTANT FISHING NATIONS
United States, Russia, Canada, Japan

PREPARATION
Caviar

USED FOR

EATING QUALITIES
When fresh, pink salmon has good flavour and texture, with a small flake. Fish frozen at its freshest is also excellent. Unfortunately, most pink salmon is destined for the can; much of this product is poorly handled, partly because the fish is caught in enormous volumes which make it difficult to give proper attention to quality. The edibility characteristics of such fish in fresh or frozen form is definitely inferior. The meat becomes slightly rancid rather quickly and the colour, already pale, fades to a yellowish tinge.

Canned pink salmon is an inexpensive food, widely used for sandwiches and recipes such as fish cakes.

IMPORTANCE
Salmon catches fluctuate widely. Landings of pinks average around 300,000 tons, with substantial increases in Alaska in recent years due to the success of hatcheries in ranching the species by stocking natal streams to which the fish return after growing in the ocean.

The United States is the largest producer in most years, but Russian catches sometimes exceed American. Canada and Japan, the only other producers, have much smaller stocks. Hatcheries are now being built on some Russian rivers and it is expected that world production of pink salmon will continue to expand, provided markets can be found for the additional catch.

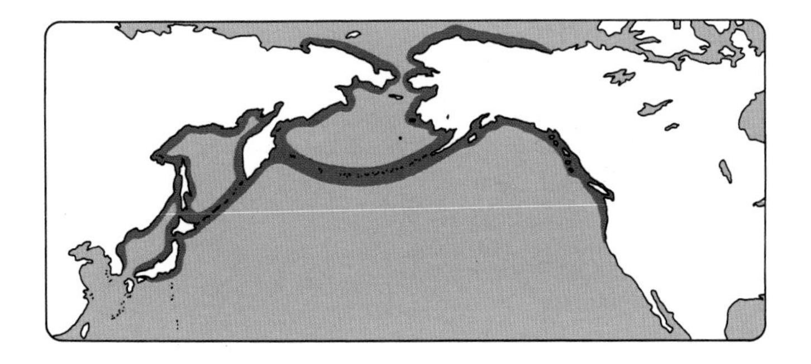

DK: I takt med det stigende opdræt af pukkellaks i Alaska og Sibirien, tvinges producenterne til at udvikle nye markedsføringsstrategier. Rognen har et stort marked i specielt Japan

D: Mit der zunehmenden Aufzucht von Buckellachs in Alaska und Sibirien sind die Produzenten gezwungen, neue Marketingstrategien zu entwickeln. Der Rogen findet besonders in Japan grossen Absatz

E: El aumento de la cría del salmón jorobado en Alaska y Siberia ha forzado a los productores a desarrollar nuevas estrategias de comercialización. Las huevas tienen un gran mercado, especialmente en Japón

F: Au rythme que l'élevage du saumon rose augmente en Alaska et en Sibérie, les producteurs sont forcés à développer de nouvelles stratégies de marketing. Les oeufs sont très demandés, en particulier sur le marché japonais

I: Il crescente allevamento del salmone rosa nell'Alasca costringe i produttori a sviluppare nuove strategie di marketing. Le uova hanno un grande mercato, specialmente nel Giappone

PLAICE

Scientific name:
Pleuronectes platessa

Family: Pleuronectidae — right-eye flounders
Typical size: 50 cm, 2 kg

Plaice is the most popular flatfish in Europe. It is found from the western Mediterranean and southern Spain along the Continental Shelf to the Barents Sea. It is also found around Iceland and occasionally off eastern Greenland. It prefers shallow water; small plaice are often seen on bathing beaches. Even the oldest and largest fish are seldom found in depths greater than 100 m.

The commercial fishery is tightly regulated. The Netherlands is the largest producer, followed by the United Kingdom. Most catches of plaice are taken in the North Sea and Irish Sea. There is a significant seasonal fishery in Iceland.

DESCRIPTION
Plaice are deep brown with orange or red spots on the upper side. The under side is creamy white.

Plaice has reasonable shelf life if well iced, but freshly caught fish has considerably more and sweeter flavour than the normally available product which is several days old. Frozen plaice, like other frozen flatfish, are often superior to aging iced product.

COMMON NAMES
D: Scholle, Goldbutt
DK: Rødspætte
E: Solla, platija, platura
F: Plie, carrelet
GR: Glóssa, Europai. chomatída
I: Passera di mare
IS: Skarkoli
N: Gullflyndre, rødspette
NL: Schol, plaat, pladijs
P: Solha, solha avessa
RU: Morskaya kambala
S: Rödspätta, rödspotta
SF: Punakampela
TR: Pisi baligi
US: European plaice

LOCAL NAMES
PO: Gladzica

HIGHLIGHTS
Plaice is the most important flatfish in Europe, providing substantial quantities of product every year. It is especially popular in northern Europe. Consumers look for its distinctive orange spots, which are regarded as the certain way to distinguish the well regarded plaice from less esteemed flatfish like European flounder.

Despite their perceived market differences, plaice and flounder are sufficiently similar that they breed together. Hybrids of the two species are quite common in some areas.

Nutrition data:
(100 g edible weight)

Water	79.5 g
Calories	79 kcal
Protein	16.7 g
Total lipid (fat)	1.4 g
Omega-3	0.1 mg

FISHING METHODS

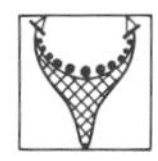

MOST IMPORTANT FISHING NATIONS

Netherlands, United Kingdom, Denmark, Belgium, Iceland

PREPARATION

USED FOR

EATING QUALITIES

Plaice has tender but firm, white meat with an excellent flavour. The flake, like that on most flounders and soles, is small and the meat holds together quite well.

Small plaice are generally sold dressed, ready for the pan, while larger fish are filleted. The fillets from both the top and under sides are sometimes halved, so that each fish supplies four small fillets. These are particularly tasty battered and fried, or rolled around stuffing and baked with a light sauce.

Whole plaice are good grilled or poached. Cooking these fish on the bone seems to result in better flavour. The meat is easily removed from the backbone. The tiny pinbones usual in flounders are often small enough to ignore.

In Germany and Denmark, plaice are hot smoked, either whole or in slices.

IMPORTANCE

The most important commercial flatfish species in European fisheries, the plaice is also an esteemed fish for sporting anglers. Fish as large as 7 kg have been taken by recreational fishermen, although the normal commercial size of the fish is about 2 kg. Catches have been stable in recent years, between 150,000 and 200,000 tons.

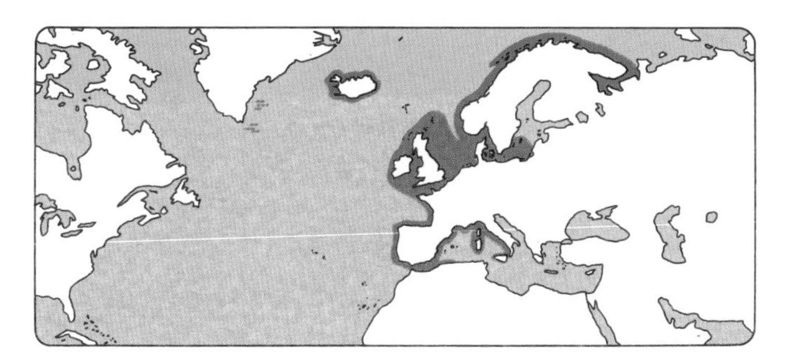

DK: Europas vigtigste fladfisk. Trods ydre forskelle sammenlignet med flyndere, er de dog så nært beslægtede, at 'bastarder' er ret almindeligt forekommende

D: Europas wichtigster Plattfisch. Trotz äusserer Unterschiede so eng mit der Flunder verwandt, dass Kreuzungen nicht ungewöhnlich sind

E: El pleuronecto más importante de Europa. Aunque su aspecto difiere de las peludas, son especies tan afines que los 'bastardos' son bastante corrientes

F: Le poisson plat le plus important de l'Europe. Malgré les différences externes qui le distinguent des flets, les deux familles sont tellement proches que des hybrides des deux espèces sont fréquemment rencontrés

I: Il pesce piatto più importante dell'Europa. Nonostante le differenze esteriori rispetto alle specie affini, sono abbastanza comuni gli ibridi

POLLACK/SAITHE

Scientific name:
Pollachius virens

Synonyms: Coalfish, black pollack
Family: Gadidae — cods
Typical size: 30 to 70 cm, 5 to 10 kg

This species is variously known as saithe, pollack, pollock and coley. It is a coldwater fish, but migrates more freely than some other close relatives such as cod, often visiting coastal waters in spring and returning to deeper waters in winter.

Saithe is sometimes used as a substitute for more expensive species.

DESCRIPTION
Very similar to cod and haddock, pollack or saithe is found on both sides of the Atlantic, although populations and catches are far greater on the eastern side. The species commonly grows to about 70 cm, but the oldest and largest fish may reach 130 cm, with a weight of 30 kg. The average size of pollack harvested appears to have been falling, possibly as a result of either fishing pressure or temperature changes. In the 1930s, fish of 110 cm and 18 kg were not unusual.

Pollack are known as voracious and aggressive feeders, sometimes chasing smaller fish up to the beach, where the pollack can be easily caught in traps. A Scandinavian observer in 1892 wrote of one 3.5 cm saithe consuming 77 herrings of 1 cm each at a single meal.

COMMON NAMES
D: Seelachs, Köhler
DK: Sej, gråsej
E: Carbonero, fogonero
F: Lieu noir, colin noir
GR: Mávros bakaliáros
I: Merluzzo carbonaro
IS: Ufsi
J: Porakku
N: Sei
NL: Koolvis
P: Escamudo, paloco
RU: Sajda
S: Sej, gråsej
SF: Seiti
US: Pollock

LOCAL NAMES
AU: Coley
PO: Czarniak

HIGHLIGHTS
Saithe, known as Boston bluefish in Canada and pollack or pollock in the United States, remains a valued species for fish-and-chips and similar recipes where its darker meat colour is concealed by the coating.

The preference of consumers in Europe and North America for white meated fish has not saved the pollack from intensive harvesting. Resources of this important demersal species are under pressure on both sides of the Atlantic.

Nutrition data:
(100 g edible weight)

Water	80.2 g
Calories	82 kcal
Protein	18.3 g
Total lipid (fat)	1.0 g
Omega-3	0.2 mg

POLLACK/SAITHE *Pollachius virens*

FISHING METHODS

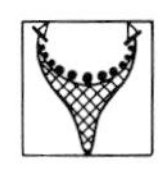 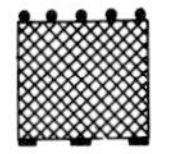 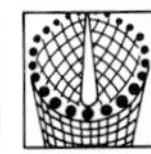

MOST IMPORTANT FISHING NATIONS

Norway, Iceland, Faroe Islands, France, Canada, United Kingdom

PREPARATION

USED FOR

EATING QUALITIES

Saithe has darker and slightly softer meat than cod, with more fat. Otherwise, it is very similar. When cooked, it lightens in colour, though it is not as pearly white as cod or haddock. The flavour is more pronounced than its whiter-meated cousins. The species has good quality meat which can be used for most preparations suitable for cod or haddock.

Pollock is often used for stockfish or salt fish. It can be smoked. Seelachs is a German preparation made from smoked pollock which resembles inferior smoked salmon. It is sometimes sliced and canned in oil.

IMPORTANCE

North Atlantic landings of pollack average close to 500,000 tons, of which at least 80 percent is from the eastern Atlantic. Catches by Canada and the United States have fallen considerably. Norway and Iceland are the largest producers.

The very similar *Pollachius pollachius* provides harvests of a further 15,000 to 20,000 tons, mainly by Norway, France, Spain and the UK.

Saithe is a less expensive alternative to cod and, sometimes, haddock. It is used for portions, sticks, fingers and similar products. It is especially suitable for breaded items; with these, the consumer sees the meat only after it is cooked, when it is a lighter colour.

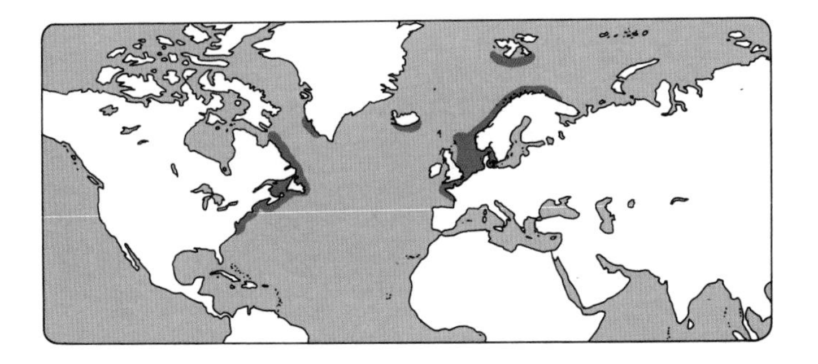

DK: Selvom det europæiske og det nordamerikanske marked foretrækker fisk med lyst kød, fiskes der intensivt efter den mørkere sej. Bestanden af denne vigtige bundfisk er derfor ved at være truet

D: Auch wenn auf dem europäischen und nordamerikanischen Markt helles Fleisch bevorzugt wird, so wird der dunklere Köhler doch intensiv gefangen. Der Bestand dieses wichtigen Bodenfisches steht deshalb kurz vor der Gefährdung

E: Aunque los consumidores europeos y norteamericanos prefieren pescados de carne blanca, es intensiva la pesca del carbonero de carne más obscura. Por eso, se ve amenazada la existencia de este importante pescado de fondo

F: Bien que les marchés en Europe et aux Etats-Unis préfèrent les poissons à chair blanche, le lieu noir fait l'objet d'une pêche intensive, et la population de cet important poisson des fonds est en voie de devenir une espèce menacée

I: Anche se i mercati europeo e nordamericano preferiscono il pesce dalle carni bianche, questa importante specie dalle carni scure è soggetta ad una pesca intensiva che minaccia la sua popolazione

PORBEAGLE

Scientific name: *Lamna nasus*

Synonym: Mackerel shark
Family: Lamnidae — mackerel sharks
Typical size: 150 to 180 cm, 135 kg

The decline of porbeagle stocks has made this once quite commonly seen shark a comparative rarity now. Because of its size, the porbeagle is generally regarded as dangerous, though there are few if any fully authenticated accounts of attacks on man.

FISHING METHODS

 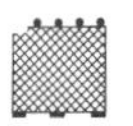 

MOST IMPORTANT FISHING NATIONS
Faroe Islands, France, Norway

PREPARATION

 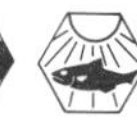 

USED FOR

EATING QUALITIES
Porbeagle sharks provide firm, darkish coloured meat, similar to mako shark in texture and flavour. It works well for broiling or grilling, less well for frying or baking unless thoroughly marinated in brine, milk or other acidic solution before cooking.

COMMON NAMES
D: Heringshai
DK: Sildehaj
E: Marrajo sardinero
F: Requin-taupe commun
GR: Karcharías, skylópsaro
I: Smeriglio, talpa
IS: Hámeri
J: Môkazame, nishikudazame
N: Håbrann
NL: Haringhaai, neushaai
P: Tubarao sardo
RU: Seldevaya akula, lamna
S: Håbrand, sillhaj
SF: Sillihai
TR: Dikburun karkarias
US: Porbeagle

LOCAL NAMES
MO: Lkars

Nutrition data:
(100 g edible weight)

Water	74.8 g	Total lipid	
Calories	102 kcal	(fat)	1.1 g
Protein	23.0 g	Omega-3	0.2 mg

HIGHLIGHTS
Once an important Atlantic and Indian Ocean resource, porbeagle stocks and catches have declined. The meat is well regarded, firm and tasty, similar to mako shark. The species is also sought for its valuable fins and liver.

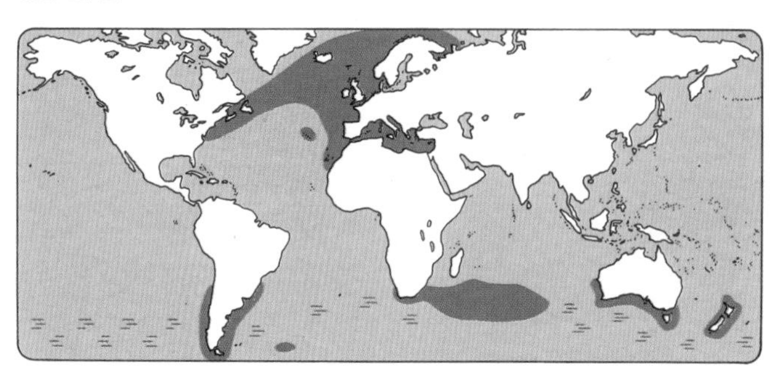

DK: Bestandene i Atlanterhavet og Det indiske Ocean er stærkt dalende. Fint fast kød, i smag meget lig Mako-hajens. Finner og lever benyttes

D: Bestände im Atlantik und Indischen Ozean gehen stark zurück. Festes Fleisch, das geschmacklich dem des Makos ähnelt. Verarbeitung von Flossen und Leber

E: Las existencias del Atlántico Norte y del Océano Indico han bajado mucho. Su carne es exquisita y firme, similar al tiburón Mako. Se usan sus aletas e hígado

F: Les populations de l'Océan Indien et de l'Atlantique sont en déclin rapide. La chair ferme ressemble en goût à celle du mako. On utilise les ailerons et le foie

I: Le risorse dell'Atlantico e dell'Oceano Indiano sono in forte declino. Le carni sode sono simili a quelle dello squalo mako. Si usano le pinne e il fegato

PUFFER

Scientific name:
Fugu vermiculare porphyreum

Family: Tetraodontidae — puffers
Typical size: 35 cm

The skin, gonads, liver, intestines and even the blood of puffers contains a deadly toxin. If carefully prepared, excluding all traces of these parts, the meat is edible. In Japan, chefs are required to train for several years before being tested as fugu chefs: the main test is that they cook and eat several of the fish themselves.

COMMON NAMES
J: Shosaifugu
NL: Egelvis
US: Sea squab, fugu

FISHING METHODS

MOST IMPORTANT FISHING NATIONS
Japan, Korea

PREPARATION

USED FOR

EATING QUALITIES
Puffer (or fugu) is said to be delicious, but the authors have neither the courage nor the funds to check this out personally. Fugu is used, in specialist restaurants in Japan, for a wide variety of preparations.

Nutrition data:
(100 g edible weight)

Water	n.a.	Total lipid	
Calories	n.a.	(fat)	n.a.
Protein	n.a.	Omega-3	n.a.

HIGHLIGHTS
Puffers protect themselves from predators with spines on their skin and by swelling their bodies, with air or water, to twice the normal size. These defenses are inadequate: man, especially in Japan, still eats them, even though the meat may be deadly.

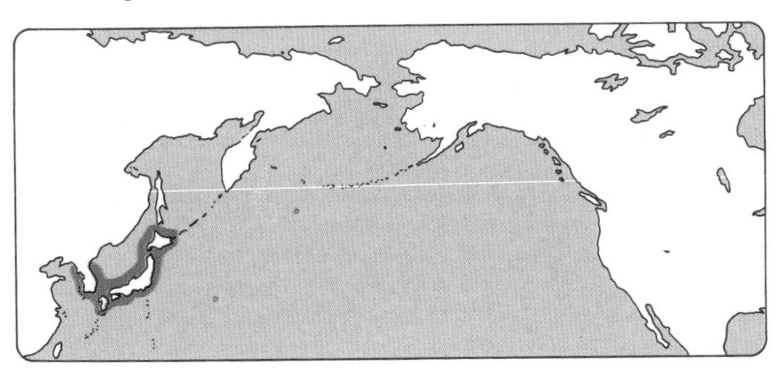

DK: Trods forsvarsmekanismer som pigge og opsvulmning af kroppen samt det faktum at kødet kan være dødelig giftigt, fiskes og spises den alligevel

D: Der Fisch verteidigt sich mit Dornen und durch Anschwellen seines Körpers. Er wird gefangen und gegessen, obwohl sein Fleisch tödlich giftig sein kann

E: A pesar de sus defensas como espinas e hinchar del cuerpo y el hecho de que la carne puede ser mortalmente venenosa, se captura y se come todavía

F: En dépit de ses défenses telles qu'épines et gonflement du corps, et le fait que sa consommation risque d'être mortelle, il est pêché et même mangé

I: Nonostante i meccanismi di difesa, aculei e il gonfiamento del corpo, e il fatto che le carni possono essere mortalmente velenose, è pescato e mangiato

RAINBOW TROUT/STEELHEAD

Scientific name: *Oncorhynchus mykiss*

Family: Salmonidae — salmonids
Typical size: trout 25 cm, 500 g; steelhead 50 cm, 5 kg

This species is called rainbow trout when it grows in fresh water. The anadromous form is called steelhead. Steelhead is found in small quantities along the Pacific coast of North America and is mostly reserved for recreational fisheries. Rainbow trout, originally from lakes and streams around the North Pacific basin from the Amur to Mexico, has been domesticated and transplanted to almost every part of the world. Hybrids of the rainbow and steelhead races have been developed which grow fast and have red meat. These are favourites with farmers and consumers, especially in Japan.

DESCRIPTION
Rainbow trout are easily distinguished by the rainbow-like colours on their flanks which give them their name. Trout are usually harvested at about 500 g and used in dressed form; larger trout may be filleted. Modern machinery can remove every bone from a dressed or filleted trout, an important factor in the growth of the trout industry.

Steelhead are not easy to distinguish from Atlantic salmon, a feature which has several times caused problems when a steelhead was wrongly thought to be an Atlantic salmon escaped from a farm.

COMMON NAMES
D: Regenbogenforelle
DK: Regnbueørred
E: Trucha arco iris
F: Truite arc-en-ciel
GR: Iridízousa péstrofa
I: Trota iridea
IS: Regnboga-silungur
J: Nijimasu
N: Regnbueaure, regnbueørret
NL: Regenboogforel
P: Truta arco íris
RU: Raduznuju forel
S: Regnbåge, regnbågslax
SF: Kirjolohi
TR: Alabalik türü
US: Rainbow trout, steelhead

LOCAL NAMES
PO: Pstrag teczowy

Ocean form

HIGHLIGHTS
Rainbow trout and the red-meated hybrids of rainbow and steelhead are excellent aquaculture fish, fast growing and well established in their markets. Production can easily be increased to meet demand. Rainbows grown in sea water are popular in Japan and Norway, where they are regarded as similar to Atlantic salmon but are sold for a slightly lower price.

The rainbow is an important game fish; wild populations are well established far beyond the original natural range of the species.

Nutrition data:
(100 g edible weight)

Water	76.7 g
Calories	125 kcal
Protein	19.6 g
Total lipid (fat)	5.2 g
Omega-3	1.0 mg

FISHING METHODS

MOST IMPORTANT FISHING NATIONS

Denmark, France, Germany, USA, Japan, United Kingdom

PREPARATION

USED FOR

EATING QUALITIES

Rainbow trout has delicate meat, pale brown to almost white. The flake is small and the bones easily removed. Fresh and frozen rainbow trout are popular with consumers around the world. Smoked trout is also a common product, especially in Europe.

Steelhead meat is pinkish red, with a distinctively salmon-like flavour and texture. The fish is very similar in most respects to Atlantic salmon. Farmed hybrid rainbows often have very red meat and deep bodies, both characteristics developed for the Japanese market which favours similar features in sockeye salmon (*O. nerka*).

IMPORTANCE

Rainbow trout is farmed in many countries, from Norway to Lesotho and Argentina. World production is reported to be over 200,000 tons a year and growing, but the figure is probably understated since many rainbows are taken by recreational anglers who do not have to report their catches. Rainbows are often hatched and stocked into rivers and lakes specifically to attract recreational fishermen.

Rainbow trout are one of the most securely domesticated fish. Production is carried out in large units with low costs and regular crop cycles, able to offer fresh fish year round to their customers.

Steelhead catches are almost entirely recreational.

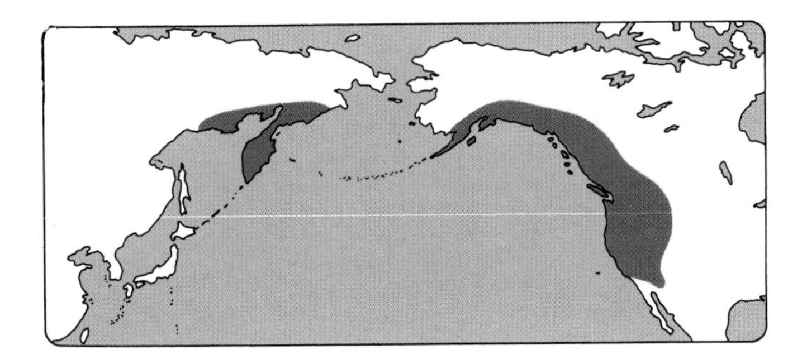

DK: En vigtig art inden for sportsfiskeri og akvakultur. Regnbueørred og krydsningen af denne og steelhead er stærkt efterspurgt, og med deres hurtige vækst vil en produktionsforøgelse nemt kunne foretages

D: Wichtige Art für Sportfischerei und Aquakultur. Die Regenbogenforelle und ihre Kreuzungen sind sehr gefragt und auf Grund ihrer hohen Wachstumsgeschwindigkeit ist eine Produktionserweiterung leicht machbar

E: Especie importante para la pesca deportiva y para la acuicultura. La trucha arco iris y los cruces con otras truchas son muy demandadas. Gracias a su rápido crecimiento será fácil aumentar la producción

F: Importante espèce pour la pêche sportive et la pisciculture. La truite arc-en-ciel et le croisement avec d'autres truites sont très demandés, et grâce à leur croissance rapide, il est facile de réaliser une augmentation de la production

I: Una specie importante per la pesca sportiva e l'acquicoltura. La trota iridea e gli incroci con altre trote sono fortemente richiesti. Con la sua rapida crescita sarà facile incrementarne la produzione

RAY'S BREAM

Synonyms: Smallscale pomfret, angel fish
Family: Bramidae — pomfrets
Typical size: 40 to 60 cm

A pomfret-like fish, Ray's bream is highly prized in certain countries for its fine eating qualities.

DESCRIPTION

Ray's bream is a deeply compressed fish, with coppery colours on the sides which fade soon after it is caught.

In the eastern Atlantic, it is found from the Bay of Biscay along the coasts of Spain and Portugal to North Africa. It migrates into the North Sea in some summers and is also found in the Mediterranean. Most of the fish is in Spanish waters and Spanish fishermen are the main harvesters of the species.

There are isolated populations off South Africa, New Zealand, Australia and Chile. None of these other populations is large enough to support a fishery.

The species is also reported from the western Atlantic, but there have been only a few occurrences and these sightings are probably of strays, rather than evidence of a self-sustaining resource.

Scientific name:
Brama brama

COMMON NAMES

D: Brachsenmakrele
DK: Havbrasen
E: Japuta del Atlántico
F: Brème de mer, hirondelle
GR: Lestí, lestíka
I: Pesce castagna
IS: Stóri bramafiskur
J: Echiopia, shimagatsuo
N: Havbrasme
NL: Braam
P: Xaputa, chaputa
RU: Morskoj lesch
S: Rays havsbraxen
SF: Merilahna
US: Atlantic pomfret

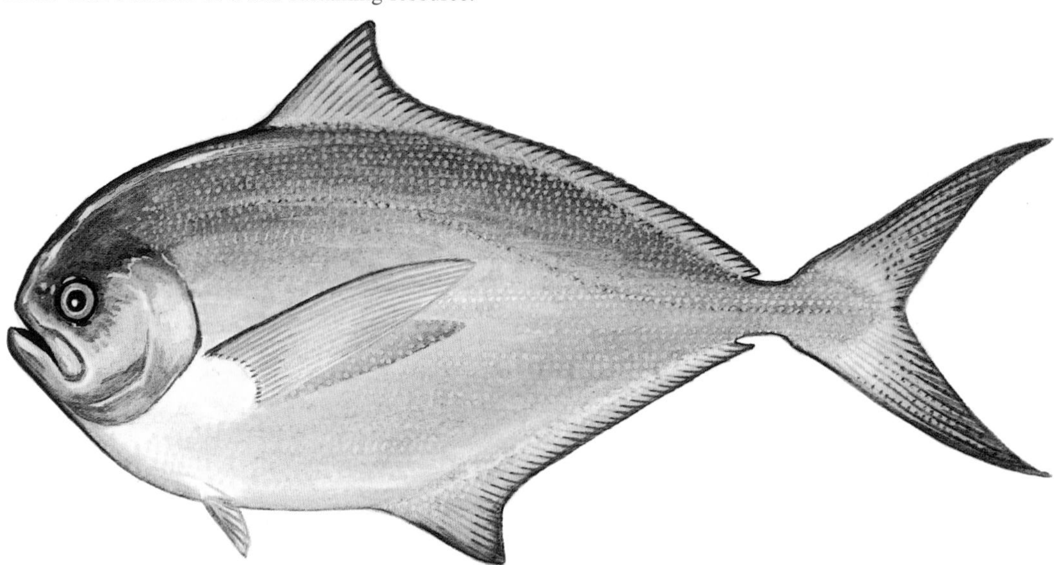

HIGHLIGHTS

An excellent table fish, Ray's bream or Atlantic pomfret (the name favoured by FAO, and used also in North America) is particularly prized in Spain and France.

The name of Ray's bream honours the great naturalist John Ray who first described the species in 1681. Ray also reported an irregular phenomenon, which still occurs in some years, when large numbers of these fish strand themselves along the North Sea beaches of Britain and Europe in late summer and early autumn.

Nutrition data:
(100 g edible weight)

Water	75.6 g
Calories	98 kcal
Protein	21.0 g
Total lipid (fat)	1.5 g
Omega-3	0.3 mg

RAY'S BREAM *Brama brama*

FISHING METHODS

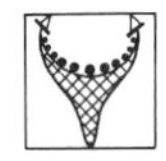 

MOST IMPORTANT FISHING NATIONS

Spain

PREPARATION

USED FOR

EATING QUALITIES

Ray's bream is a fine eating fish. The meat is white with a slightly pink colouration and the texture resembles that of skate, with long, tender strings rather than flakes. It has excellent flavour and stands up to most cooking methods.

Commercially, it is usually filleted, although it is excellent baked whole (dressed). Small quantities were canned in past years; the flesh withstands this treatment, but the return on the fish is greater if it is sold in fresh or frozen form. The fish often contain parasites, but although unaesthetic these are not harmful to humans.

IMPORTANCE

Landings of Ray's bream fluctuate considerably from year to year, mostly between 5,000 and 11,000 tons. Spain is the major producer, but vessels from Russia and South Africa occasionally target the species, increasing catches for those years.

A pelagic species, it is caught with submerged long-lines set at depths of around 100 m. It is also trawled, especially during its seasonal migrations northwards: in some years, it is found as far north as Murmansk, Russia. In other years it may not even enter the North Sea.

The species may grow as large as 70 to 75 cm, but the majority of the fish are caught between 40 and 60 cm.

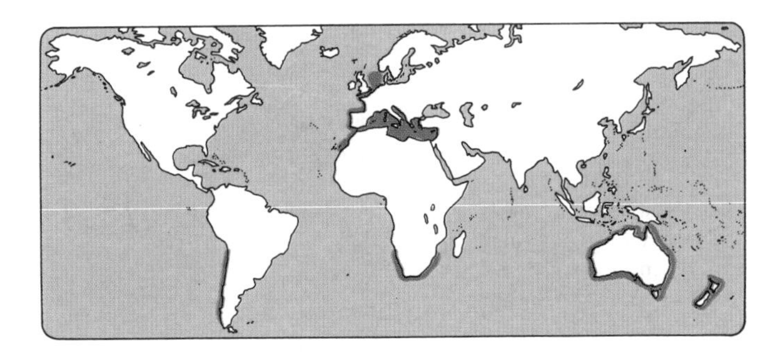

DK: En fortrinlig spisefisk, opkaldt efter den store naturalist John Ray, der som den første observerede det fænomen, at fisken fra juni til september lader sig strande på Nordsøens kyster

D: Ein vorzüglicher Speisefisch, benannt nach dem grossen Naturalisten John Ray, der als erster das Phänomen beobachtete, dass der Fisch sich von Juni bis September an den Küsten des Nordsees anschwemmen lässt

E: Exquisito pescado comestible, que debe su nombre al gran naturalista, John Ray, que fue el primero en observar el fenómeno que de junio a septiembre el propio pescado se deja encallar en las costas del Atlántico Norte

F: Un excellent poisson comestible, dénommé d'après le grand naturaliste John Ray, qui fut le premier à observer le phénomène qu'entre juin et septembre, le poisson s'échoue aux côtes de la mer du Nord

I: Un ottimo pesce commestibile, a cui è stato messo il nome del famoso naturalista John Ray, che fu il primo ad osservare lo strano fenomeno che il pesce resta in secca da giugno a settembre sulle spiagge del Mare del Nord

RED COD/RED CODLING

Scientific name:
Pseudophycis bachus

Family: Moridae — codlings or morid cods
Typical size: 40 to 70 cm, 1.5 to 2.5 kg; up to 90 cm

Similar to the southern bastard codling (*P. barbata*) with which it is often caught, the red cod provides good quality meat to a limited coastal fishery in New Zealand, where it is frozen for domestic consumption as well as for export.

FISHING METHODS

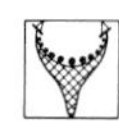

MOST IMPORTANT FISHING NATIONS
New Zealand

PREPARATION

USED FOR

  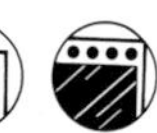

EATING QUALITIES
The red codling has white, moist flesh which flakes easily. The texture is delicate and the fat content low. Fillets are packed in layer packs or are used to make fish blocks for the manufacture of portions and coated items.

COMMON NAMES
D: Neuseeland-Eisfisch
E: Brotolilla
F: Morue rouge de N. Zélande
I: Busbana neozelandese
J: Akadara
P: Abrótea da Nova Zelândia
RU: Krasnaya treska
US: Morid cod

LOCAL NAMES
NZ: Hoka

Nutrition data:
(100 g edible weight)

Water	81.0 g	Total lipid	
Calories	73 kcal	(fat)	0.6 g
Protein	16.9 g	Omega-3	0.3 mg

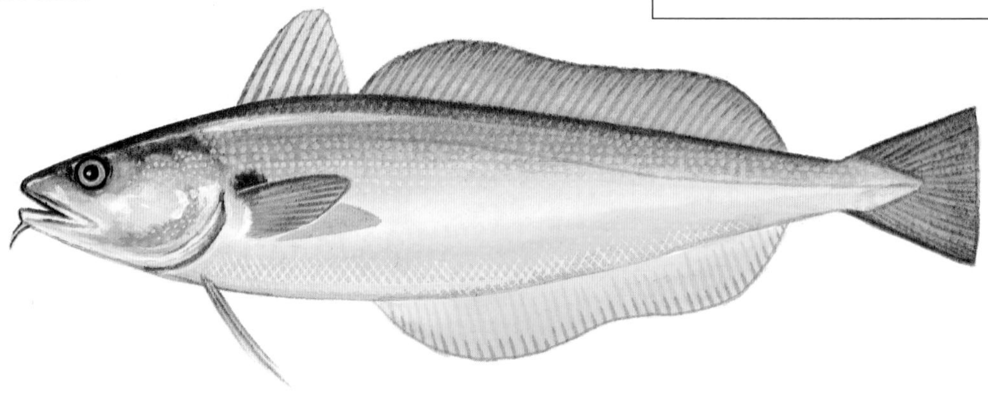

HIGHLIGHTS
A coastal species trawled in New Zealand, the red codling is also found in southern waters of Australia, although it is seldom caught there. Frozen fillets are produced, as well as fish blocks. The resource is moderate, supporting fisheries of around 7,000 tons annually.

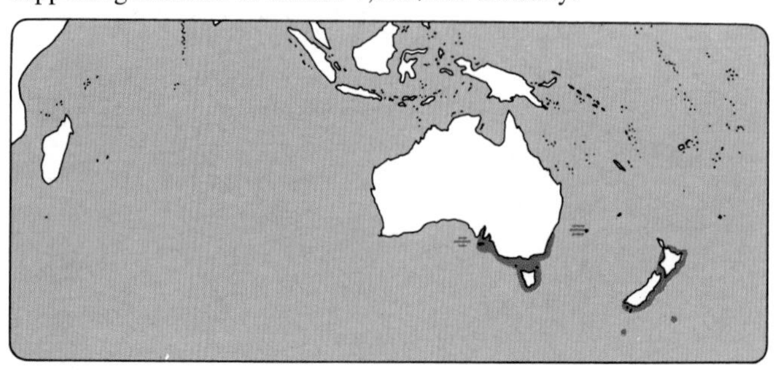

DK: Begrænset kystfiskeri ved New Zealand. Såvel hjemme- som eksportmarkedet aftager denne fremragende fisk i frossen form

D: Unbedeutende Küstenfischerei von Neuseeland. Dieser ausgezeichnete Fisch in gefrorenem Zustand findet seine Abnehmer im In- und Ausland

E: Pesca de bajura limitada en Nueva Zelanda. Tanto el mercado doméstico como el de exportación compran este pescado exquisito en estado congelado

F: Pêche côtière limitée près de la Nouvelle-Zélande. Poisson délicieux vendu à l'état surgelé tant sur le marché intérieur que pour l'exportation

I: Pesca limitata presso la costa della Nuova Zelandia. Pesce squisito, venduto congelato sia sul mercato nazionale che su quello internazionale

RED DRUM

Scientific name:
Sciaenops ocellatus

Synonym: Channel bass
Family: Sciaenidae — drums
Typical size: 50 cm; up to 1.52 m, 45 kg

Commercial fishing for red drum virtually ceased as pressure on the resource intensified. Regulators in the USA have largely reserved the species for recreational fishermen, who dominate the management process in the southeast and Gulf states.

FISHING METHODS

MOST IMPORTANT FISHING NATIONS
United States, Mexico

PREPARATION

USED FOR

EATING QUALITIES
Blackened redfish, fillets seared with hot spices in a cast iron pan at high temperature, became a fashionable and highly popular dish in the USA in the early 1980s. The meat is white and firm when cooked, with a mild flavour which blends well with the spices.

COMMON NAMES
D: Augenfleck-Umberfisch
DK: Rød trommefisk
E: Corvinón ocelado
F: Tambour rouge
GR: Stiktomylokópi
I: Ombrina ocellata
NL: Rode ombervis
P: Corvinao de pintas
SF: Punarumpukala
US: Redfish

Nutrition data:
(100 g edible weight)

Water	78.5 g	Total lipid	
Calories	85 kcal	(fat)	1.0 g
Protein	19.0 g	Omega-3	0.2 mg

HIGHLIGHTS
Red drum stocks are reported to be making a comeback in the wild and the species is now being farmed in Texas, with small quantities offered for commercial sale. It remains a popular fish, with considerable unfilled demand, especially from restaurants.

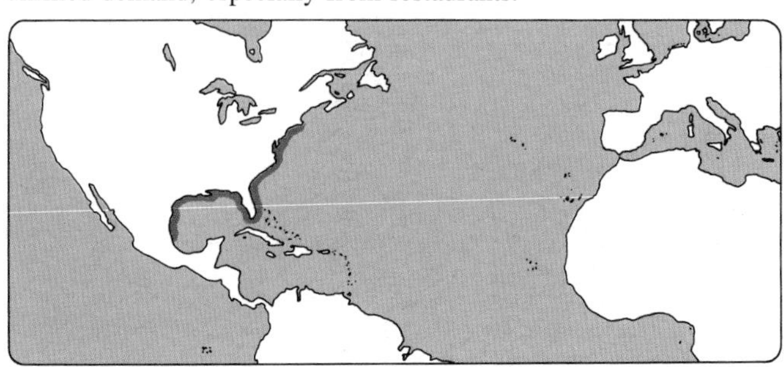

DK: Den naturlige bestand er igen i stigning. Fisken opdrættes i Texas, men kun et lille kvantum fisk sælges kommercielt

D: Der natürliche Bestand nimmt wieder zu. Der Fisch wird in Texas gezüchtet, aber nur eine geringe Menge lässt sich vermarkten

E: El número de las especies en estado salvaje está aumentando. El pescado se cría en Texas, pero sólo se venden pequeñas cantidades para uso comercial

F: La population naturelle est de nouveau en progression. Dans le Texas, le tambour fait l'objet d'élevage dont seulement une moindre quantité est commercialisée

I: Il patrimonio naturale sta nuovamente crescendo. Il pesce viene allevato in Texas, ma solo piccole quantità vengono commercializzate

RED GURNARD

Scientific name:
Chelidonichthys kumu

Synonym: Bluefin gurnard
Family: Triglidae — searobins
Typical size: 30 to 50 cm, 500 to 1,500 g

The red gurnard, which may grow as long as 60 cm, is larger than most other searobins, providing fillets that are usable commercially. A coastal species, it is quite common in New Zealand waters and is also found in Australia, South Africa and parts of Japan and China. New Zealand produces about 60 percent of world landings, which exceed 4,000 tons yearly.

FISHING METHODS

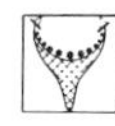  

MOST IMPORTANT FISHING NATIONS
New Zealand, Japan, Korea

PREPARATION

USED FOR

EATING QUALITIES
Red gurnard has pink-coloured, firm meat, which is sometimes a little dry. Low in fat, it keeps quite well and has a mild flavour.

COMMON NAMES
D: Blauflossen-Knurrhahn
DK: Blåfinnet knurhane
E: Rubi kumu
F: Grondin aile bleue
GR: Kapóni
I: Capone, gallinella
J: Minamihôbô
NL: Kumupoon
P: Carbra kumu
RU: Morskoj petukh
US: Searobin

LOCAL NAMES
KO: Seong-dae

Nutrition data:
(100 g edible weight)

Water	78.1 g	Total lipid	
Calories	87 kcal	(fat)	0.9 g
Protein	19.8 g	Omega-3	0.2 mg

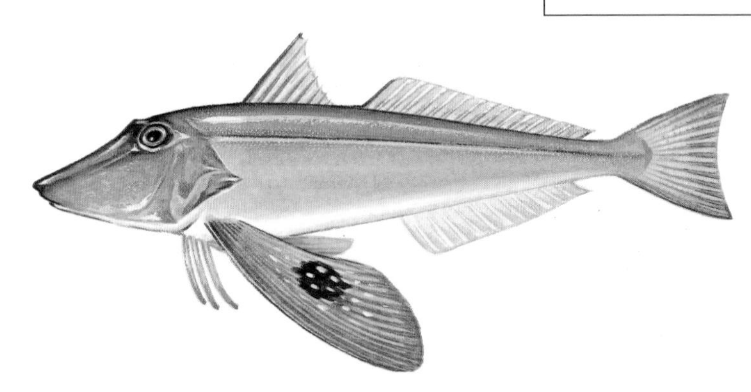

HIGHLIGHTS
Mainly a by-catch species, resources of red gurnard, especially in Australia, are believed to be able to support significantly increased fishing. Frozen fillets are generally inexpensive and quite well accepted on world markets.

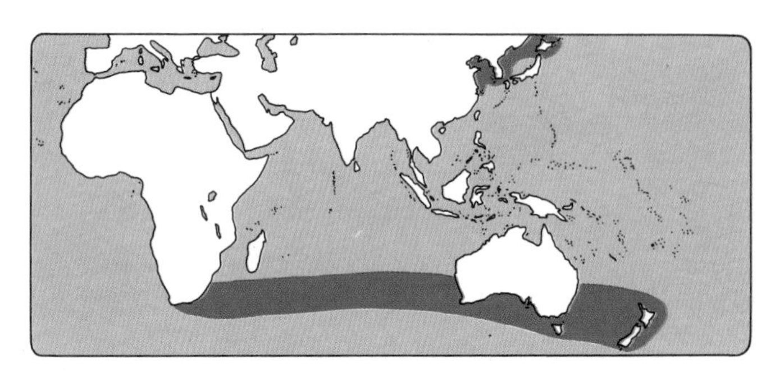

DK: Tages primært som bifangst, men forekomsterne ved specielt Australien skønnes at kunne bære et mere intensiveret fiskeri

D: Hauptsächlich Beifang, aber schätzungsweise können die Bestände besonders in australischen Gewässern einen intensivierten Fang verkraften

E: Se captura principalmente como pescado secundario, pero especialmente en Australia se estima que las existencias pueden soportar una pesca más intensiva

F: Pêché principalement comme prise secondaire, mais les ressources près de l'Australie semblent cependant pouvoir supporter une pêche intensivée

I: Preso normalmente come cattura secondaria, ma le risorse presenti soprattutto in Australia possono probabilmente sopportare una pesca più intensa

RED HAKE

Scientific name:
Urophycis chuss

Synonym: Squirrel hake
Family: Gadidae — cods
Typical size: 50 cm, 1.5 kg

Red hake and white hake (*Urophycis tenuis*) are often not distinguished in fisheries or in markets. Canadian landings of hake, for example, are listed simply as "hake." These two northern hake species can be regarded as a single commodity in commercial terms.

FISHING METHODS

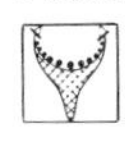

MOST IMPORTANT FISHING NATIONS

United States, Canada

PREPARATION

Fishmeal

USED FOR

EATING QUALITIES

White meat, rather soft but very sweet, characterises the red hake. It is essential to cook it with the skin on, or the flakes of meat disintegrate. The species has poor keeping qualities. Small fish are used for fishmeal, because it is uneconomic to process them for food.

COMMON NAMES

D: Roter Gabeldorsch
DK: Rød skægbrosme
E: Locha roja
F: Merluche écureuil
GR: Kókkinos bakaliáros
I: Musdea atlantica
IS: Leirbrosma
J: Reddoheiku
NL: Atlantis. gaffelkabeljauw
P: Abrótea vermelha, linguiça
RU: Krasnyj nalim
SF: Suomuturska
US: Red hake

LOCAL NAMES

PO: Meitus czerwony

Nutrition data:
(100 g edible weight)

Water	81.0 g	Total lipid	
Calories	72 kcal	(fat)	0.4 g
Protein	17.2 g	Omega-3	0.1 mg

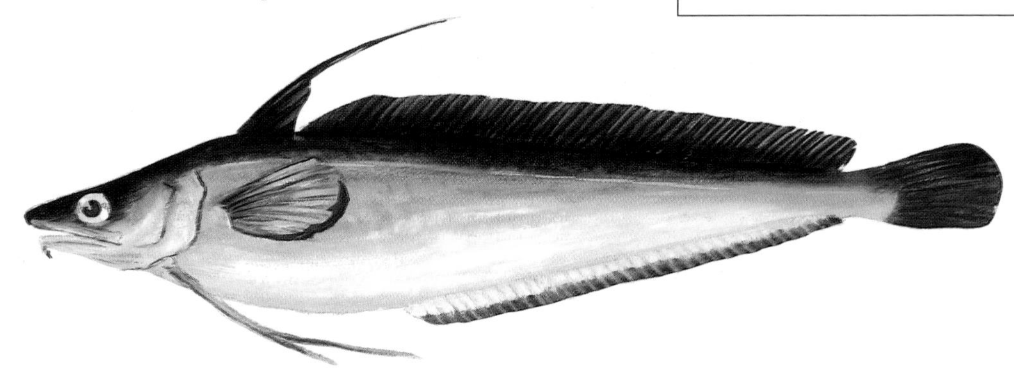

HIGHLIGHTS

Total catches of about 2,000 tons yearly are greatly reduced from the 18,000 tons recorded fifty years ago. Red hake live inside scallops, and remain close to the scallop beds until they mature. They are taken on lines by anglers, although they offer little fight.

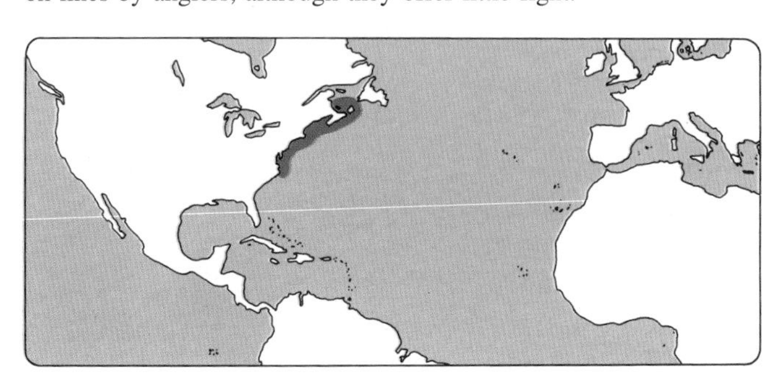

DK: Det lyse, meget søde kød er karakteristisk for denne fisk. Kødet er blødt og bør koges med skindet på for at undgå, at det falder fra hinanden

D: Das weisse, sehr süsse Fleisch ist für diesen Fisch charakteristisch. Das Fleisch ist weich und sollte mit Haut gekocht werden, damit es nicht zerfällt

E: Este pescado se caracteriza por su carne blanca y muy dulce. La carne es blanda, y se recomienda cocerla con piel para evitar que se desmenuce

F: La chair blanche, très douce est caractéristique de ce poisson. La chair assez molle doit être préparée avec la peau pour éviter qu'elle ne tombe en morceaux

I: Caratteristico per le sue carni bianche e molte dolci. Essendo di carne morbida va bollito con la pelle per evitare che cada in pezzi

RED SEA BREAM

Scientific name:
Chrysophrys major

Family: Sparidae — porgies
Typical size: 100 cm

The classification of the red sea bream is confused. This species may be found commonly under three different scientific names: in addition to *Chrysophrys major*, it is also known as *Pagrus major* and *Sparus major*. Of the three names, *Pagrus major* appears to be the one favoured currently. Whatever it is called, it is the same fish and one that is highly esteemed in the marketplace, where it fetches premium prices.

DESCRIPTION
Red sea bream, as its name implies, has red skin. The fish looks similar to a red snapper and in some areas it, and the closely related *Chrysophrys* (or *Pagrus*) *auratus* are caught and marketed as snapper.
Found over rocky bottoms and reefs, it is slow growing. While slow growing species are not usually good candidates for aquaculture, this one is sufficiently valuable that farmers consider it profitable even though they have to wait for their harvests.

A porgy, red sea bream is very similar to the red porgy (*Pagrus pagrus*) of European waters.

COMMON NAMES
D: Roter Tai, Akadei
DK: Japansk guldbrasen
E: Dorada gigante
F: Daurade japonaise
GR: Fangrí tis laponías
I: Orata del Giappone
J: Madai
NL: Japanse goudbrasem
P: Dourada do Japao
RU: Krasnyj morskoj karas
SF: Punahammasahven
TR: Kirmizi fangri
US: Porgy

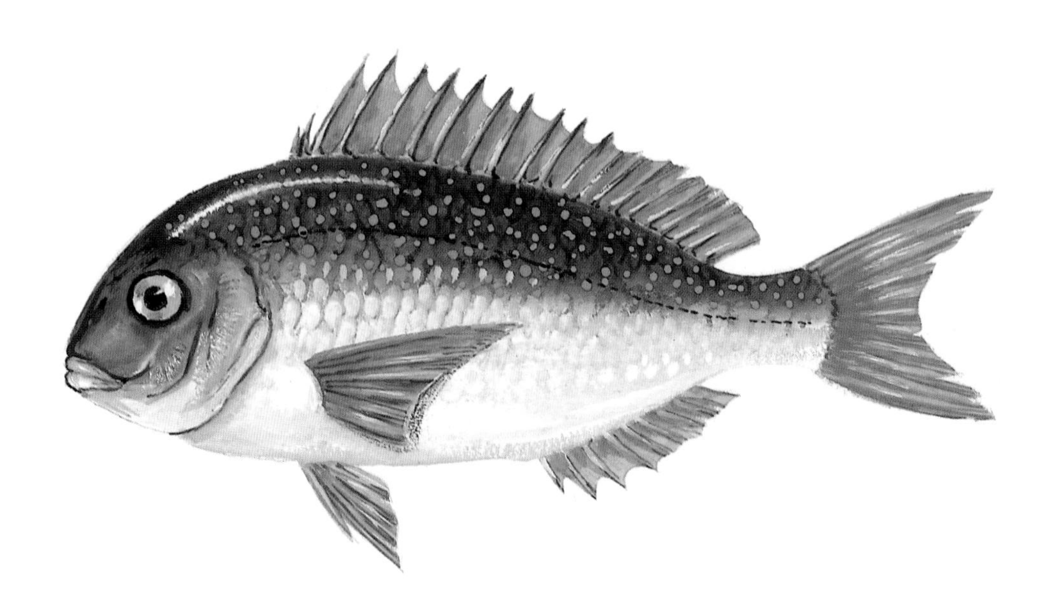

HIGHLIGHTS
This highly regarded marine food fish is farmed in Japan, where it commands a high price, especially if sold live. Development of techniques to culture this and related species are being carried out in many other countries, from Hong Kong to Australia.

An attractive, brightly coloured fish when taken from the water, wild stocks are too small to support much recreational fishing, but the species is targeted by anglers who appreciate its strength as well as its eating quality.

Nutrition data:
(100 g edible weight)

Water	77.1 g
Calories	94 kcal
Protein	20.0 g
Total lipid (fat)	1.6 g
Omega-3	0.4 mg

RED SEA BREAM *Chrysophrys major*

FISHING METHODS

MOST IMPORTANT FISHING NATIONS

Japan

PREPARATION

Live   

USED FOR

EATING QUALITIES

Red sea bream is a superb table fish. The pinkish meat whitens when cooked. It has firm but moist texture and an excellent, full flavour. It is most versatile and can be used in many different fish recipes, including broiling, frying and steaming. Thick, meaty fillets can be cut without bones.

Red sea bream is similar to red snapper (*Lutjanus campechanus*) in eating characteristics and meat quality.

In Japan, where the fish is particularly prized live or fresh, red sea bream is stunned and incapacitated by inserting a small spike into a precisely located part of the fish's brain. This lowers the metabolism of the fish, which remains alive but inert, making it possible to transport it over considerable distances. This technique is increasingly used with a wider variety of fish species, ensuring maximum freshness for the consumer.

IMPORTANCE

Most of the world's production of around 70,000 tons is farmed in Japan. Wild harvest are made in Japan as well as Korea and Hong Kong. China is believed to be farming the species, although no production has so far been recorded from that country.

A high value fish, the red sea bream is farmed in cages in the sea from where it can be readily shipped live to market.

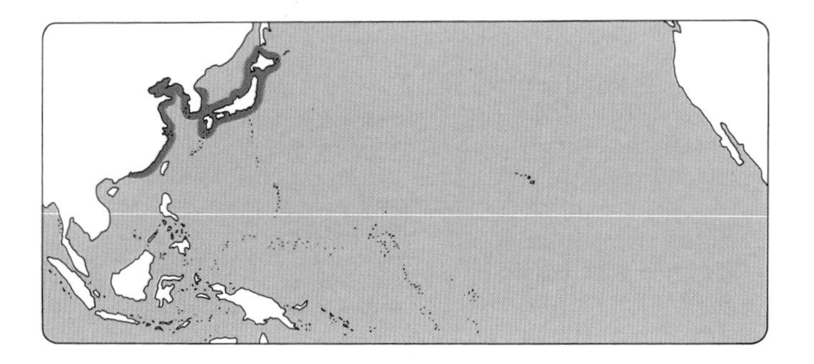

DK: En højt estimeret spisefisk opdrættet i Japan, hvor især den levende fisk giver høje priser. Andre lande forsøger opdræt af denne og beslægtede arter

D: Ein hoch geschätzter Speisefisch. Wird in Japan gezüchtet, wo besonders der lebende Fisch hohe Preise einträgt. In anderen Ländern wird versucht, diese und verwandte Arten zu züchten

E: Muy estimado pescado comestible, criado en Japón, donde se vende a buen precio, especialmente si el pescado se vende vivo. Otros países tratan de criar esta y otras especies afines

F: Un poisson de table très apprécié, cultivé au Japon où son prix est élevé, en particulier pour le poisson vivant. Dans plusieurs pays des essais d'élevage, tant de la daurade japonaise que d'autres espèces apparentées, sont en cours

I: Un pesce commestibile altamente apprezzato. Viene allevato nel Giappone, dove soprattutto il pesce vivo è venduto a prezzi alti. In altri paesi vengono attuati sperimenti di allevamento di questa e di altre specie affini

RED SNAPPER

Scientific name:
Lutjanus campechanus

Synonym: Northern red snapper
Family: Lutjanidae — snappers
Typical size: 50 cm, 2 kg; grows to 100 cm, 15 kg

Red snapper, called northern red snapper by FAO, is an important food and game fish from the Gulf of Mexico and the southeast Atlantic coast of the United States. Although it ranges as far north as Massachusetts, it is in practice hardly ever found north of the Carolinas. Florida is the main catching and consuming area, but imports supply most of the demand.

Highly prized for its eating qualities, its fame has been taken by tourists visiting Florida back to their homes throughout the USA. There is far too little of the species to satisfy the demand, which has to be supplied with other snappers, some from the Gulf of Mexico and adjacent areas, but many from much more distant waters, including Brazil and Thailand.

COMMON NAMES
D: Nördlicher Schnapper
DK: Nordlig snapper
E: Pargo del Golfo
F: Vivaneau campèche
GR: Kókkinos loutiános
I: Lutiano rosso
IS: Raudglefsari
NL: Red snapper
P: Luciano do Golfo
RU: Krasny lutsian
US: Red snapper

LOCAL NAMES
ME: Guachinango del Golfo

DESCRIPTION
Red snapper has rose-red skin and carmine fins. Precise identification is difficult. There are no obvious features to distinguish this fish from several other species. Buyers must rely on the expertise of their suppliers, or have elaborate tests done by a laboratory if they suspect their fish is not truly red snapper.

HIGHLIGHTS
Red snapper is one of the most desired food fish in the USA, but it is scarce and the resources, long fished very hard, appear to be declining. At one time, fishermen kept red snappers alive in wells on the boats. Now, freshness is assured by clean handling and the liberal use of ice.

Florida law requires that red snapper be sold with the skin on, to aid identification. Most fish are sold drawn, with the gills left in. Red snapper is often filleted at the restaurant or retail store.

Nutrition data:
(100 g edible weight)

Water	78.3 g
Calories	90 kcal
Protein	19.6 g
Total lipid (fat)	1.3 g
Omega-3	0.3 mg

RED SNAPPER

Lutjanus campechanus

FISHING METHODS

MOST IMPORTANT FISHING NATIONS

Mexico, United States

PREPARATION

USED FOR

EATING QUALITIES

Red snapper has moist, pinkish meat with a mild, delicate flavour. The delicate meat is best cooked with the skin on; the skin also helps to identify the fish as genuine red snapper.

Smaller fish are more valuable than larger ones, because they provide fillets which can be served whole. Larger fillets must be cut, giving a less attractive appearance. Large fillets are also more difficult to cook, as they are thicker and more likely to dry out on top.

IMPORTANCE

Red snapper is a very important fish in Florida and the Gulf states of the USA, but about 75 percent of the total catch of 5,000 tons is taken by Mexico. Red snapper has been heavily fished in American waters and the species is now closely protected. Much of the remaining resource is reserved for sport fishermen. Some observers expect that commercial fishing for red snapper will be banned. Shrimp fishing is also affected by the wide popularity of this species: shrimpers have been accused of destroying young snappers and shrimp fishing is restricted as a result.

There are many snappers that closely resemble red snapper. Substitution of these lower priced fish is a frequent problem for buyers seeking genuine red snapper. It is not clear why two or more snapper species that look, taste and perform the same should be differently valued by the market, but the fact is that they are.

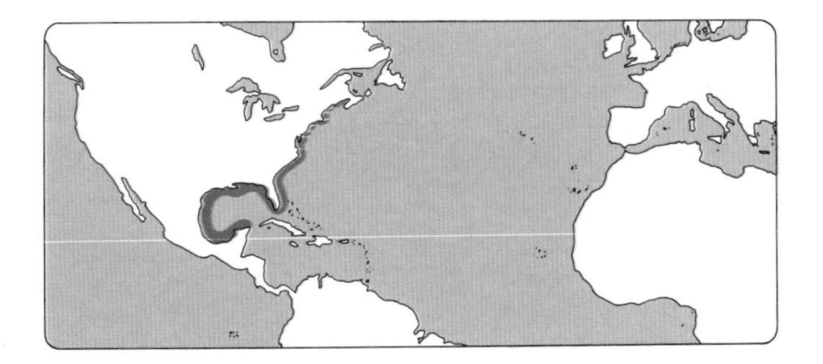

DK: Faldende forekomster af denne fisk, som i USA er en af de populæreste konsumfisk. Den rensede fisk sælges mest i fersk, hel form. Filetteres ofte først lige inden serveringen

D: Der Bestand dieses Fisches der in den USA einer der begehrtesten Speisefische ist, schrumpft. Der gesäuberte Fisch wird hauptsächlich frisch und im ganzen Stück verkauft. Das Filetieren erfolgt meistens erst kurz vor dem Auftragen

E: Van bajando las existencias de este pescado que es uno de los pescados de consumo más populares en EE.UU. Normalmente, el pescado limpiado se vende fresco y entero. Es normal filetear el pescado inmediatemente antes de servirlo

F: Très apprécié aux Etats-Unis, ce vivaneau est actuellement en déclin. En général, le poisson nettoyé est vendu à l'état frais et en entier, pour être découpé en filets seulement au moment de le servir

I: Popolazione decresente di questo pesce che è uno dei pesci commestibili più popolari negli Stati Uniti. Il pesce pulito è venduto fresco e intero. Normalmente è tagliato in filetti solo al momento di essere servito

REDFISH

Scientific name:
Sebastes marinus

Synonyms: Rosefish, golden redfish, Norway haddock
Family: Scorpaenidae — scorpionfishes
Typical size: 25 to 50 cm, up to 100 cm

There are four species of almost identical redfish in the North Atlantic. Trawls may bring up two or three together and they are difficult to tell apart. This entry covers all these species, which, in addition to *S. marinus* are: *S. fasciatus*, Acadian redfish, which lives in shallower water and may range further south; *S. mentella*, deepwater redfish, which ranges further north and into deeper water; and *S. viviparus*, small redfish, which reaches a maximum size of only about 30 cm and is found only in the eastern Atlantic.

It should be noted that there is still some disagreement among fish scientists on these classifications, which are not universally accepted. So far as markets are concerned, fillets from any of them are either redfish or ocean perch according to local usage and are entirely interchangeable, if not identical.

DESCRIPTION

Redfish inhabit cold deep water in the North Atlantic. They have red skin which, although it slowly fades after capture, remains an attractive feature and one which enables the fish to be sold in North America as a substitute for the more expensive lake perch.

COMMON NAMES

D: Grosser Rotbarsch
DK: Stor rødfisk
E: Gallineta nórdica
F: Sébaste, rascasse
GR: Kokkinópsaro tis Norvigías
I: Scorfano di Norvegia
IS: Gullkarfi
J: Menuke
N: Uer, rødfisk
NL: Roodbaars, noorse schelvis
P: Peixe vermelho
RU: Morskoj okun
S: Rödfisk, kungsfisk
SF: Punasimppu, puna-ahven
TR: Kirmizi balik
US: Ocean perch

LOCAL NAMES

PO: Karmazyn

HIGHLIGHTS

These redfish are live-bearers, and produce comparatively few young. They also grow slowly, living for up to 48 years. The oldest fish may be quite large: there is a skeleton in the Royal Ontario Museum of a redfish that weighed 11.2 kg.

Consumers in northern Europe and North America like this fish, which is adaptable in cooking and attractive on the plate with its red skin. Supplies have been limited in recent years, but the taste for the fish appears to be continuing.

Nutrition data:
(100 g edible weight)

Water	78.1 g
Calories	98 kcal
Protein	18.4 g
Total lipid (fat)	2.7 g
Omega-3	0.3 mg

REDFISH *Sebastes marinus*

FISHING METHODS

MOST IMPORTANT FISHING NATIONS

Faroe Islands, France, Norway, Iceland, Canada

PREPARATION

USED FOR

EATING QUALITIES

Redfish has a firm textured meat which whitens when cooked. The skin fades after a time, making fresh fillets tricky to handle and distribute as they must be moved through the distribution system quickly. Most redfish, or ocean perch as it is called in the United States, is frozen, almost always as fillets with the skin on, though there are many different packs and size grades.

Redfish is best baked or poached. Gentle cooking brings out the delicate flavour. The skin has a tendency to shrink: it is worth scoring the skin before cooking the fish.

IMPORTANCE

Redfish supports a considerable fishery, although the slow growing species is easily stressed by fishing, causing numerous fluctuations in supply from different areas. It is normally trawled in deep water.

The individual species in the commercial catch are not identified, but it is thought that deepwater redfish is the most important off Newfoundland, while Acadian redfish provides the bulk of the catch from Georges Bank and the Gulf of Maine.

In northern European waters, redfish is probably the dominant species. Note that the map shows aggregate distribution for all four redfish.

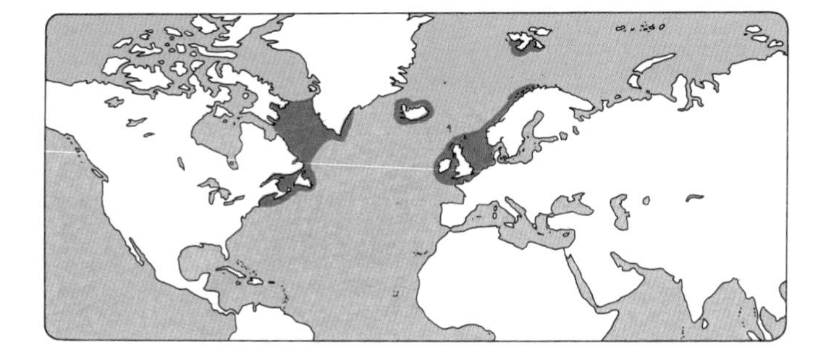

DK: Rødfiskene får som levendefødende langt færre unger end de æglæggende fisk. De vokser langsomt i løbet af deres 48 leveår. De ældste fisk bliver ret store, hvilket bekræftes af et eksemplar på 11,2 kg fanget i Canada

D: Die lebendgebärenden Grossen Rotbarsche bekommen weit weniger Junge als eierlegende Fische. Sie wachsen langsam in ihren 48 Lebensjahren. Die ältesten Fische werden ziemlich gross, wie ein Exemplar von 11,2 kg in Kanada belegt

E: Como vivíparos las gallinetas nórdicas no tienen tantos alevines como los ovíparos. Crecen lentamente durante sus 48 años de vida. Los mayores son bastante grandes. Lo confirma un ejemplar de 11,2 kg capturado en Canadá

F: En tant que vivipare, la procréation chez le sébaste est bien moindre que chez les ovipares. Les sébastes grandissent leur vie durant et peuvent atteindre 48 ans, ce dont témoigne un exemplaire de 11,2 kg, pêché au Canada

I: Essendo vivipari, gli scorfani di Norvegia producono meno giovani dei pesci ovipari. Crescono lentamente e vivono fino a 48 anni. I più vecchi sono abbastanza grandi; il maggiore catturato in Canada era di 11,2 kg

REX SOLE

Scientific name:
Glyptocephalus zachirus

Synonym: Long-finned sole
Family: Pleuronectidae — right-eye flounders
Typical size: Up to 60 cm, 1.5 kg

Rex sole is an abundant species which could be more fully utilized. Larger fish are found in Alaska. It is possible to fillet these mechanically, making an economical product possible. Smaller rex soles from other areas are less desirable because of the difficulty of processing them.

COMMON NAMES
D: Amerikanische Scholle
DK: Amerikansk skærising
E: Lenguado americano
F: Plie cynoglosse royale
GR: Kalkáni tis Amerikís
I: Passera del Pacific
NL: Amerikaanse schol
P: Solhao americano
RU: Dlinnoperaya kambala
US: Rex sole

FISHING METHODS

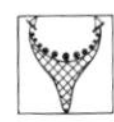

MOST IMPORTANT FISHING NATIONS
United States, Canada

PREPARATION

USED FOR

EATING QUALITIES
Rex sole has good flavour and delicately textured, white meat, but the fillets are very thin, which detracts from the mouth feel of the meat and makes the fish difficult to process. It is best cooked on the bone.

Nutrition data:
(100 g edible weight)

Water	81.3 g	Total lipid	
Calories	72 kcal	(fat)	0.7 g
Protein	16.5 g	Omega-3	0.1 mg

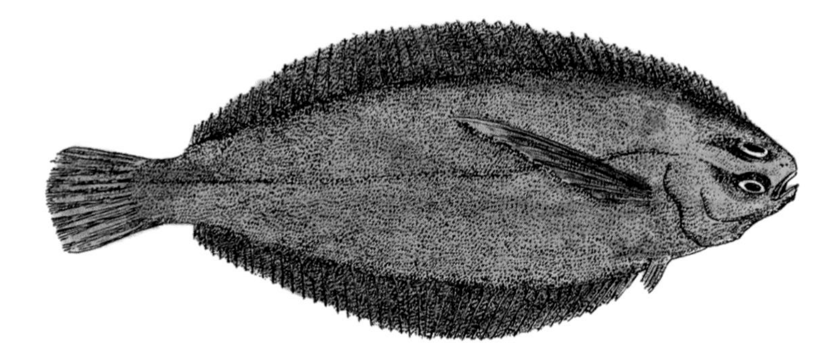

HIGHLIGHTS
Rex sole is found in the eastern Pacific from the Bering Sea to northern Mexico. It is closely related to the Atlantic witch, but is much less valuable because of the thinness of the muscle. It is delicate, requiring careful handling; shelf life may be limited.

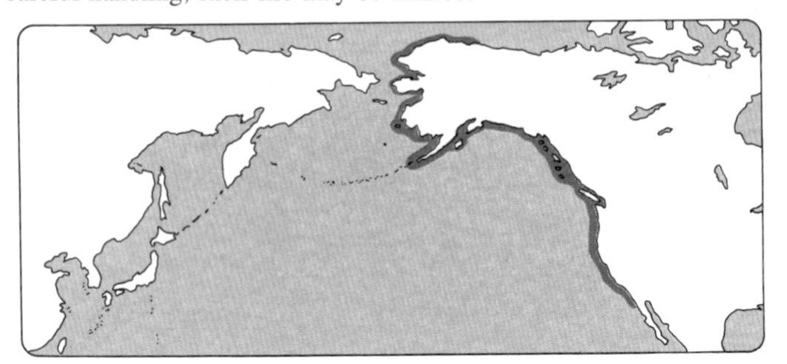

DK: En talrig men underudnyttet art. De største eksemplarer, som findes i Alaska, kan maskinfiletteres til et prismæssigt interessant produkt

D: Eine zahlreiche, aber nicht voll ausgenutzte Art. Die grössten, in Alaska vorkommenden Fische können preisgünstig maschinell filetiert werden

E: Especie abundante, no aprovechada plenamente. En Alaska, los ejemplares más grandes son fileteados a máquina, y es un producto económicamente interesante

F: Espèce très nombreuse mais mal exploitée. Les plus grands exempl. vivent en Alaska. Découpés en filets à la machine, on obtient un produit de prix intéressant

I: Una specie numerosa, ma poco utilizzata. Gli esemplari più grandi, che si trovano in Alasca, sono filettati a macchina, dando un prodotto economico

RIVER HERRING

Scientific name:
Alosa pseudoharengus

Synonym: Branch herring
Family: Clupeidae — herrings
Typical size: 25 to 30 cm, 300 to 400 g
Also known as *Pomolobus pseudoharengus*
Alewives are anadromous, swimming into Atlantic rivers of North America from Newfoundland to North Carolina. It was introduced to the Great Lakes, where at one time it became overabundant, dying in large and nuisance numbers along the shores.

COMMON NAMES
D: Atlantischer Maifisch
DK: Flodsild
E: Pinchagua
F: Gaspareau
GR: Potamórenga
I: Falsa aringa atlantica
NL: Bastaardelft, rivierharing
P: Alosa cinzenta
RU: Seldeobrazny pomolobus
SF: Harmaasilli
US: Alewife

FISHING METHODS

 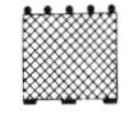

MOST IMPORTANT FISHING NATIONS
Canada, United States

PREPARATION

USED FOR

EATING QUALITIES
Although alewife meat is light and white, it tends to be dry and there are numerous small bones which make the fish difficult to eat and unacceptable to many tastes. Alewives are also used for crab and lobster bait and sometimes for pet food.

Nutrition data:
(100 g edible weight)

Water	68.6 g	Total lipid	
Calories	185 kcal	(fat)	13.0 g
Protein	16.9 g	Omega-3	2.6 mg

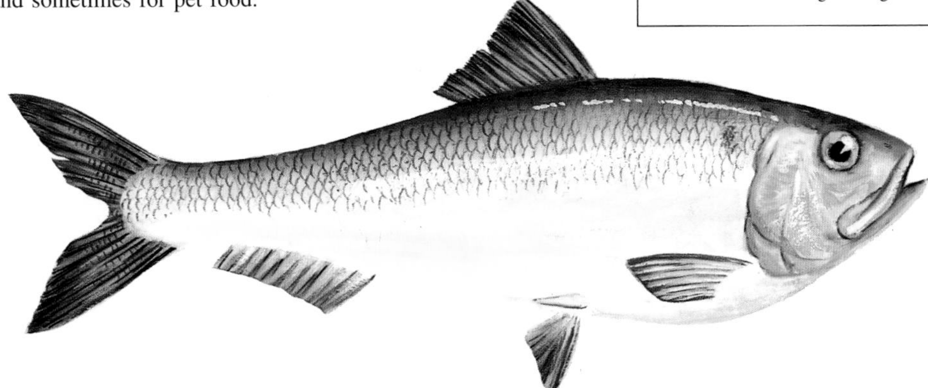

HIGHLIGHTS
Early North American colonists found alewives in enormous quant-ities, and salted them for preservation. Pollution of spawning rivers has reduced numbers, while markets have also declined. The species contributes some 15,000 tons to annual harvests.

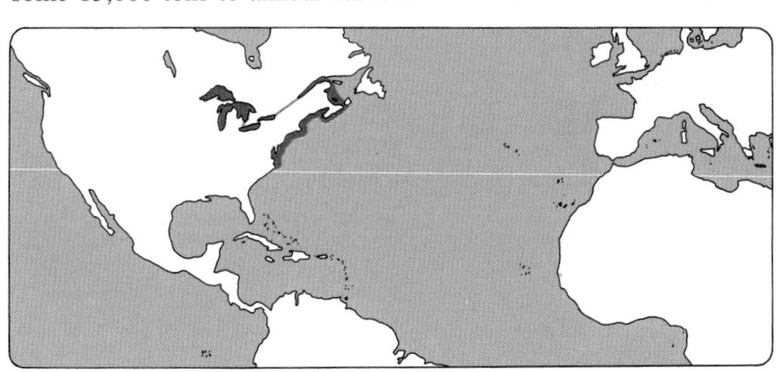

DK: De første kolonister fiskede enorme mængder flodsild, som de saltede. Forurening af gydefloderne har reduceret bestanden og dermed markedet

D: Die ersten Siedler fingen Riesenmengen des Atlantischen Maifisches, die sie einsalzten. Die Verschmutzung der Laichflüsse hat den Bestand dezimiert

E: Los colonizadores capturaron enormes cantidades de pinchaguas, que salaron. La polución de los ríos de desove ha reducido las existencias y el mercado

F: Les premiers colons pêchaient d'énormes quantités de gaspareaux pour les saler. La pollution des rivières de fraie a réduit leur nombre et le marché sensiblement

I: I primi colonizzatori pescarono enormi quantità di false aringhe atlantiche, conservandole sotto sale. L'inquinamento dei fiumi ha ridotto la quantità e il mercato

ROACH

Scientific name:
Rutilus rutilus

Family: Cyprinidae — carps and minnows
Typical size: up to 46 cm

Roach can live in poor quality, even polluted water. They are often abundant in rivers, lakes, canals and reservoirs. There is little commercial fishing for roach, but the economic value of the species for recreation is immense.

FISHING METHODS

  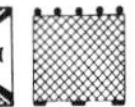 

MOST IMPORTANT FISHING NATIONS
France

PREPARATION

USED FOR

EATING QUALITIES
Roach is eaten mainly in eastern Europe, where it is salted or smoked for flavour as well as preservation. The meat is bland and there are numerous small bones, which are softened by curing so that they become edible.

COMMON NAMES
D: Plötze, Rotauge
DK: Skalle
E: Bermejuela, calandino
F: Gardon, guidon blanc
GR: Platítsa, asprítsa
I: Triotto rosso
IS: Rodagægir
N: Mort
NL: Blankvoorn
P: Ruivaca
RU: Plotva
S: Mört
SF: Särki
TR: Kizilgöz, kizil sazan
US: Roach

LOCAL NAMES
PO: Ploc

Nutrition data:
(100 g edible weight)

Water	76.3 g	Total lipid	
Calories	112 kcal	(fat)	5.6 g
Protein	17.8 g	Omega-3	0.4 mg

HIGHLIGHTS
The roach is one of the most important recreational species in Europe. The fish bite freely and boldly and may be found in canals and streams in the middle of industrial areas, where families can easily target them for an evening's sport.

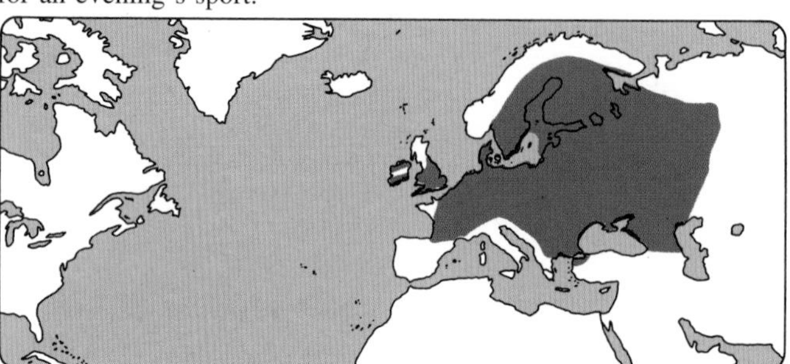

DK: Da fisken forekommer i industriområder og tilmed nemt tages med krog, er den en af de populæreste arter hos europæiske lystfiskere

D: Da dieser Fisch leicht und häufig anbeisst, gehört er zu den beliebtesten Arten bei europäischen Sportfischern. Kommt in Industriegebieten vor

E: Como vive en aguas próximas a zonas industriales y muerde fácilmente el anzuelo, es una de las especies más populares entre los pescadores deportivos europeos

F: Le gardon vit dans les zones industrielles. Comme il mord facilement à l'hameçon, il est une des espèces les plus populaires chez le pêcheur sportif européen

I: Dato che vive nelle zone industrializzate e per lo più è facile prendere con l'amo, è una delle specie più popolari fra i pescatori sportivi

SABLEFISH

Scientific name:
Anoplopoma fimbria

Synonym: Black cod
Family: Anoplomatidae — sablefishes
Typical size: 80 cm, 6 kg; up to 120 cm, 14 kg

Almost all the 40,000 tons of sablefish landed are caught in N. America, but sold in Japan. Only small fish normally remain for the US and Canadian markets, despite the popularity of this highly palatable fish.

FISHING METHODS

MOST IMPORTANT FISHING NATIONS
United States, Canada

PREPARATION

USED FOR

EATING QUALITIES
Sablefish has pearly white meat with a large flake. It has a high oil content and particularly sweet and distinctive flavour. Kasu-cod (marinated in sake) is a Japanese recipe which has become very popular in N. America.

COMMON NAMES
D: Kohlenfisch
DK: Fakkelfisk
E: Bacalao negro
F: Morue charbonnière
I: Merluzzo dell'Alasca
IS: Drungi, svartthorskur
J: Gindara
NL: Zandvis
P: Peixe Carvao do Pacífico
RU: Anoplopoma ugolnaya ryba
SF: Silosimppu
US: Sablefish

Nutrition data:
(100 g edible weight)

Water	71.0 g	Total lipid (fat)	15.3 g
Calories	191 kcal		
Protein	13.4 g	Omega-3	1.4 mg

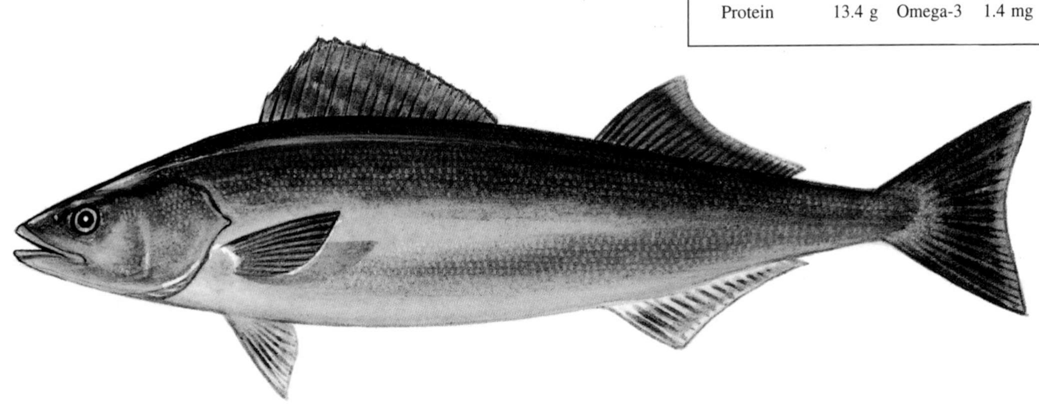

HIGHLIGHTS
Sablefish, or black cod as it is more often called, is frequently hot smoked in chunks or fillets. It is excellent baked, broiled or cold smoked. The sablefish name derives from the soft, almost furry, feel of the dark skin.

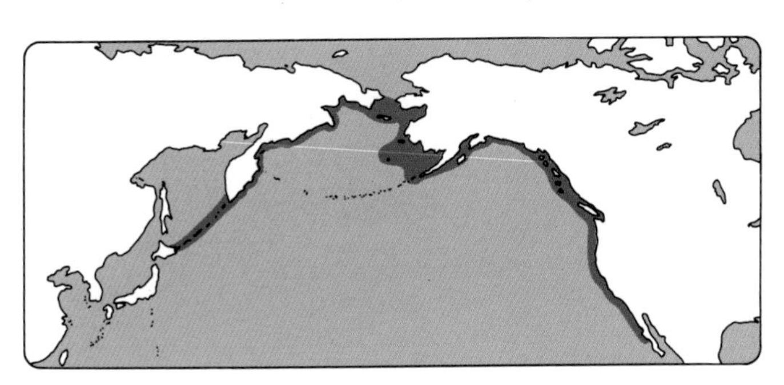

DK: Fakkelfisk, eller 'black cod' som den normalt kaldes, fås ofte varmrøget i skiver eller filetter. Er også fortrinlig indbagt, stegt og koldrøget

D: Der Kohlenfisch wird meistens warmgeräuchert in Scheiben oder Filets angeboten. Ebenfalls schmackhaft im Schlafrock, gebraten und kaltgeräuchert

E: El bacalao negro se consume a menudo ahumado en caliente en rodajas o filetes. También es excelente empanado, frito y ahumado en frío

F: La morue charbonnière est souvent servie fumée à chaud en tranches ou en filets. Excellente également préparée en croûte, frite ou saurée

I: Il merluzzo dell'Alasca si consuma spesso affumicato a caldo in fette o filetti. E' ottimo anche in pasta, arrostito e affumicato a freddo

SARDINE

Scientific name:
Sardina pilchardus

Synonym: Pilchard
Family: Clupeidae — herrings, sardines
Typical size: 20 to 25 cm

The European sardine or pilchard (the name sardine is reserved for this species in Britain, but is used for immature herring in North America) is found along the coasts of the Eastern North Atlantic, from Iceland to West Africa, as well as in the Mediterranean.

At one time, Russian factory vessels turned large quantities of sardines into fishmeal, or froze them in blocks for processing later into inexpensive food products. The resource fluctuates, but is consistently important to the coastal states which target sardines.

DESCRIPTION
Sardines are small, silvery pelagic fish very similar in appearance to small herring. Alive, the back is greenish and the flanks have a yellow tinge, but the colours fade quickly after death, leaving the whole fish silvery.

Large shoals are found at shallow depths, often less than 40 m. They swim even closer to the surface at night, when they can be attracted with lights and caught in seines, such as the traditional lampara nets of the Mediterranean.

COMMON NAMES
D: Sardine, Pilchard
DK: Sardin
E: Sardina europea, parrocha
F: Sardine européenne
GR: Sardélla
I: Sardina, sardella
IS: Sardína
J: Iwashi, maiwashi
N: Sardin
NL: Sardien, pelser
P: Sardinha
RU: Sredizemnomorskaya sardina
S: Sardin
SF: Sardiini
TR: Sardalya
US: Sardine

LOCAL NAMES
KO: Chang o ri
IN: Tembang
PH: Tunsoy, tamban
TH: Pla kureb, pla lang khieo
MA: Tamban
IL: Sardin zefoni

HIGHLIGHTS
Canned sardines are an important seafood commodity, but the fish is very versatile and used in many other ways throughout Europe.

The fish are caught in vast numbers during seasonal migrations. High-speed machines are used to remove heads and viscera and process the fish for cans. Canned sardines come in many forms, including skinless and boneless, smoked and fried. The fish are packed in olive or soybean oils, with tomato, mustard or other sauces or even, as is increasingly popular for health reasons, in plain water.

Nutrition data:
(100 g edible weight)

Water	67.7 g
Calories	165 kcal
Protein	20.6 g
Total lipid (fat)	9.2 g
Omega-3	2.0 mg

SARDINE *Sardina pilchardus*

FISHING METHODS

 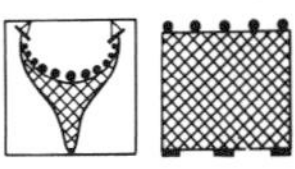 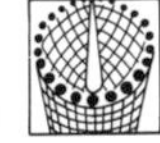

MOST IMPORTANT FISHING NATIONS

Russia, Spain, Morocco, Portugal

PREPARATION

USED FOR

EATING QUALITIES

Sardines are delectable canned, smoked, fried or simply grilled. The oily flesh cooks almost white, the flavour is delicate and the meat flakes easily from the many small bones. If the fish is canned or smoked, the bones are softened and can be eaten without a problem. In fact, nutritionists often recommend canned sardines as a source of calcium because the bones are eaten.

There are numerous traditional preparations for sardines throughout Europe. The species has been important for many centuries. It is smoked, salted or pickled, prepared in many of the ways used for herring and several that are unique to this small, tasty fish.

IMPORTANCE

The European pilchard or sardine supports major commercial fisheries in the Mediterranean Sea and the Atlantic. Worldwide, the species usually ranks in the top five clupeid (herring-like) species by volume of landings. Landings fluctuate between 1 and 1.5 million tons. Russian factory vessels target sardines off North Africa, accounting for almost half the catch in some seasons. Spain, Morocco and Portugal have important canning industries based on fish from the same general region, where there have been numerous disputes over fishing rights and borders.

Elsewhere, catches in the western Mediterranean of about 250,000 tons and in the northeast Atlantic of 175,000 contribute to the total.

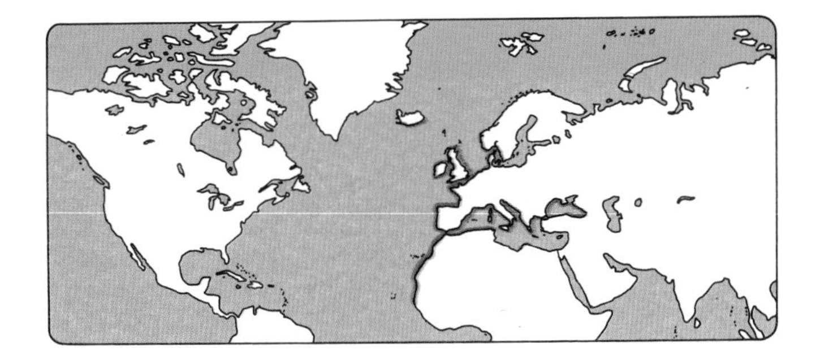

DK: Stor anvendelse inden for konservesindustrien. Den rensede fisk forekommer i adskillige variationer: med/uden skind, med/uden ben, røget, stegt, i olie, sauce eller vand

D: Weit verbreitet in der Konservenindustrie. Der gesäuberte Fisch kommt in verschiedenen Varianten vor: mit/ohne Haut, mit/ohne Gräten, geräuchert, gebraten, eingelegt in Öl, Sosse oder Wasser

E: Se usa mucho en la industria conservera. El pescado limpiado se ofrece en varias presentaciones: con o sin piel, con o sin espinas, ahumado, asado, en aceite, salsa o agua

F: Poisson de très grande importance pour l'industrie des conserves. La sardine est préparée dans de multiples variations: avec ou sans peau, avec ou sans arêtes, fumée, grillée, à l'huile, à l'eau ou en sauce

I: Pesce molto usato dall'industria conserviera. Dopo la lavorazione si presenta in tanti modi: con/senza pelle, con/senza spine, affumicato, arrostito, sott'olio, in salsa o acqua

SCUP

Scientific name:
Stenotomus chrysops

Synonym: Northern porgy
Family: Sparidae — porgies
Typical size: 25 cm, 750 gm

COMMON NAMES
D: Skap
DK: Nordlig skælfisk
E: Sargo de América del Norte
F: Spare doré
GR: Sargós tis Amerikís
I: Sarago americano
IS: Grænflekkur
P: Sargo da América do Norte
RU: Skap-khrizops
SF: Amerikanhammasahven
US: Scup

Scup are found from southern Maine to the Gulf of Mexico, though in the southerly part of this range, from South Carolina onwards, the species is largely replaced by the almost identical longspine porgy, *S. caprinus*. Scup school inshore in Chesapeake Bay in April, moving up the coast to Narragansett Bay where they support important traditional fisheries. They are seldom caught north of Cape Cod. They winter offshore, where they can be caught with trawls.

Anglers have mixed feelings about scup: some enjoy the easy capture and surprising strength; others regard them as a nuisance, taking bait intended for more desirable species.

DESCRIPTION
Scup are small, deep-bodied, fish with spiny fins and dull, silvery skin. They may grow to 35 cm and 2 kg, but such large fish are exceptional.

The first recorded failure of the scup fishery, which has a history going back to the early settlements in New England, was between 1896 and 1902. Its abundance since then has been erratic.

HIGHLIGHTS
Scup used to be caught in large numbers in traps in southern Massachusetts and Rhode Island. These fish were fine quality, as they were taken from the water alive, with the minimum of bruising or other damage. Trap fishing has now almost ceased.

This and related species range into the Gulf of Mexico, where it is thought that catches could be increased if markets were developed. The boniness of the fish is likely to deter most Americans from buying it, however, so export markets offer the best prospects.

Nutrition data:
(100 g edible weight)

Water	77.7 g
Calories	96 kcal
Protein	17.5 g
Total lipid (fat)	2.9 g
Omega-3	0.6 mg

FISHING METHODS

 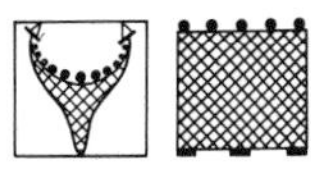

MOST IMPORTANT FISHING NATIONS

United States

PREPARATION

USED FOR

EATING QUALITIES

Scup has tender, white meat with a sweet, delicate flavour. Unfortunately, it also has many small bones, which are difficult to remove and detract from the enjoyment of the fish. It should be scaled when it is dressed; if the scales are not removed promptly after the fish is caught, the skin tends to adhere to the meat and becomes difficult to detach.

Scup are excellent hot smoked. The process softens the small bones, making them more or less edible.

IMPORTANCE

The United States lands about 6,000 tons of scup a year, but catches are highly variable. The species appears to suffer major boom-and-bust cycles of abundance. These have been charted for at least a century. They may be based on climate variations from year to year, but there is no generally agreed explanation.

Before the days of 200-mile fishing limits, vessels from a number of European countries targeted scup off the USA. Now, the fishery is mainly directed at Japanese buyers, who require good quality product carefully graded by size. Unfortunately, much of the scup available on the US domestic market has been rejected for export, or is in some other way second rate. Domestic markets are limited to ethnic groups familiar with similar porgies.

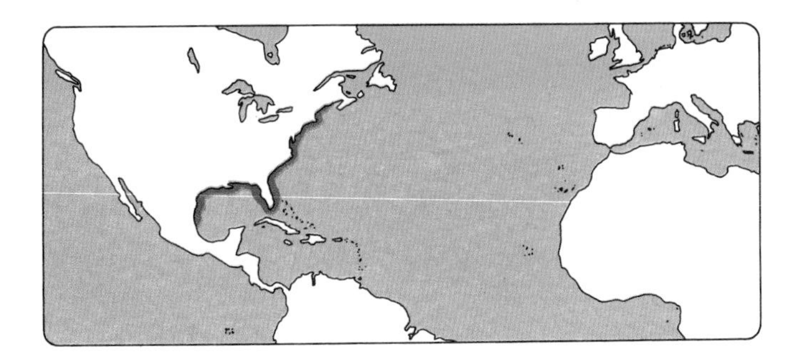

DK: De førhen så store mængder nordlig skælfisk fanget i fælder, var af god kvalitet, da fiskene blev taget levende i vandet med et minimum af skrammer. Fældefiskeriet er nu næsten ophørt

D: Der früher so zahlreiche, mit Fallen gefangene Skap war von guter Qualität, da er lebend und mit wenigen Schrammen aus dem Wasser geholt wurde. Die Fallenfischerei hat jetzt fast aufgehört

E: El sargo de América del Norte se capturaba antes con trampas en grandes cantidades, y era de buena calidad, porque salía del agua vivo, sin sufrir casi daños. La pesca con trampas casi ha desaparecido ya

F: Les grandes quantités de spares dorés qui étaient pêchés autrefois dans des pièges étaient de bonne qualité car les poissons étaient pris vivants avec un minimum de balafres. Actuellement, la pêche à pièges a presque cessé

I: Prima i saraghi americani, catturati in trappole e tirati su dall'acqua vivi con un minimo di graffi, erano di buona qualità. Ora è praticamente cessata la pesca con le trappole

SEAROBIN

Scientific name:
Prionotus carolinus

Synonym: Gurnard
Family: Triglidae — searobins
Typical size: 35 cm

The tapered, almost triangular body of the searobin is ornamented with large fins and spines. It makes loud, drumming noises by vibrating its swim bladder. These fish are spiny and awkward to handle, and yield only small amounts of meat.

COMMON NAMES
E: Rubio carolino
F: Grondin carolin
RU: Trigla
SF: Amerikankurnusimppu
US: Searobin

FISHING METHODS

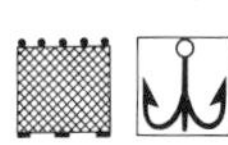

MOST IMPORTANT FISHING NATIONS
Not available

PREPARATION
Caviar 

USED FOR

EATING QUALITIES
Regarded as a trash fish in the United States, searobins have firm, rather dry meat but are good if stuffed and baked. The fish are rather small for filleting, reducing chances of popularizing the species. Fillets, if prepared, are good battered and fried.

Nutrition data:
(100 g edible weight)

Water	78.2 g	Total lipid	
Calories	90 kcal	(fat)	1.7 g
Protein	18.8 g	Omega-3	0.3 mg

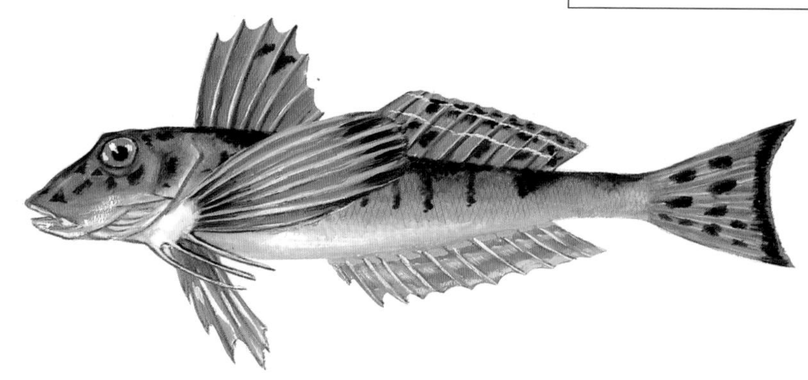

HIGHLIGHTS
Searobins inhabit the continental shelf from the Gulf of Maine to South Carolina. Closely related searobins are found southward to Venezuela. Searobins are taken as a by-catch in commercial trawl fisheries and regarded as a nuisance by anglers.

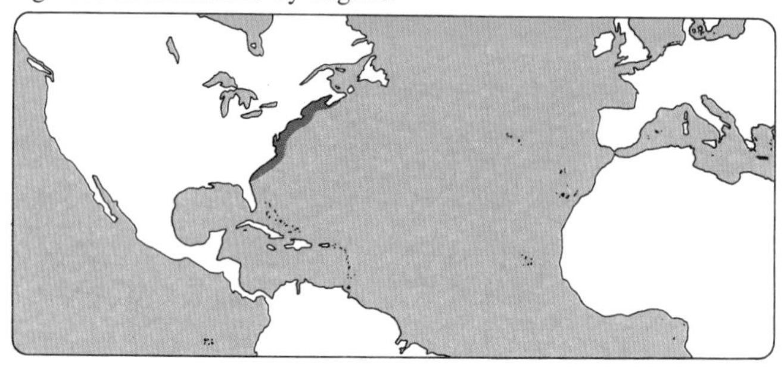

DK: Dens vibrerende svømmeblære laver høje trommelyde. Fisken er pigget, ikke særlig kødfuld og fra et industrielt synspunkt uegnet til forarbejdning

D: Seine vibrierende Schwimmblase erzeugt starke Trommellaute. Er ist stachelig, nicht sonderlich fleischig und zur industriellen Verarbeitung ungeeignet

E: Su vejiga natatoria vibrante produce altos sonidos como de tambor. Es erizado, tiene poca carne y no sirve para producción industrial

F: A l'aide de la vessie natatoire vibrante, il émet de hauts sons de tambour. Couvert d'aiguillons et pas très charnu - peu intéressant pour la pêche commerciale

I: Produce alti suoni tamburellanti con la vescica natatoria vibrante. Il pesce è spinoso con poca carne e quindi poco adatto per la lavorazione industriale

SEVRUGA STURGEON

Scientific name:
Acipenser stellatus

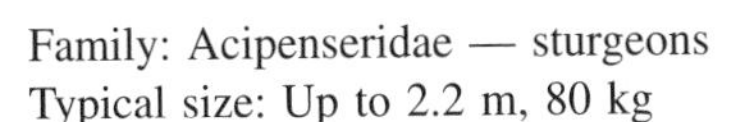

Family: Acipenseridae — sturgeons
Typical size: Up to 2.2 m, 80 kg

The stellate sturgeon, better known as sevruga, is one of the three most important species for caviar with beluga (*Huso huso*) and Russian or osetra sturgeon (*A. gueldenstaedti*). The resources are supported by artificial propagation (ranching).

FISHING METHODS

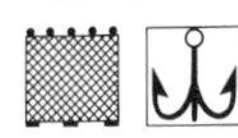

MOST IMPORTANT FISHING NATIONS
Russia, Kazakhstan,

PREPARATION
Caviar 

USED FOR

EATING QUALITIES
Sevruga caviar is packed under red lids. The eggs are small and dark and make, in the opinion of many experts, the tastiest of all sturgeon caviars. The meat is very firm and well flavoured, excellent smoked, broiled or baked.

COMMON NAMES
D: Sternhausen, Sterg
DK: Stjernehus
E: Esturión estrellado
F: Esturgeon étoilé, sevruga
GR: Stourióni
I: Storione stellato
IS: Stjörnustyrja
J: Chôzame
NL: Stersteur
P: Esturjao estrelado
RU: Sevruga
S: Stjärnstör, sevruga
SF: Tähtisampi
TR: Mersin
US: Star sturgeon

Nutrition data:
(100 g edible weight)

Water	78.7 g	Total lipid	
Calories	97 kcal	(fat)	3.6 g
Protein	16.2 g	Omega-3	0.7 mg

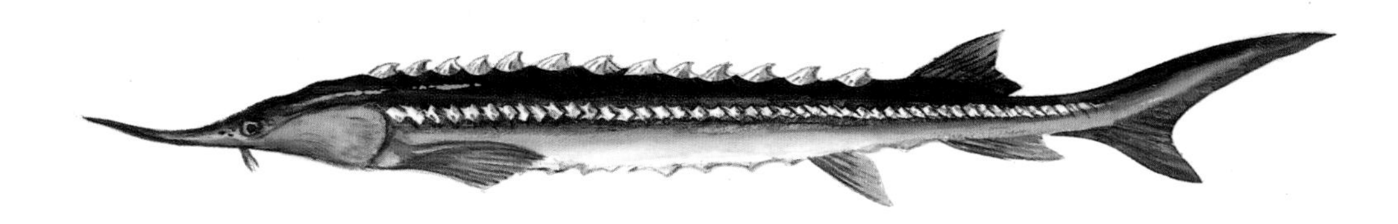

HIGHLIGHTS
This anadromous sturgeon from the Caspian, Azov and Black Seas is a major source of top quality caviar, producing up to 700,000 dark grey eggs up to 2.5 mm in diameter. Poaching, following the breakup of the former USSR, may be adding stress to the resource.

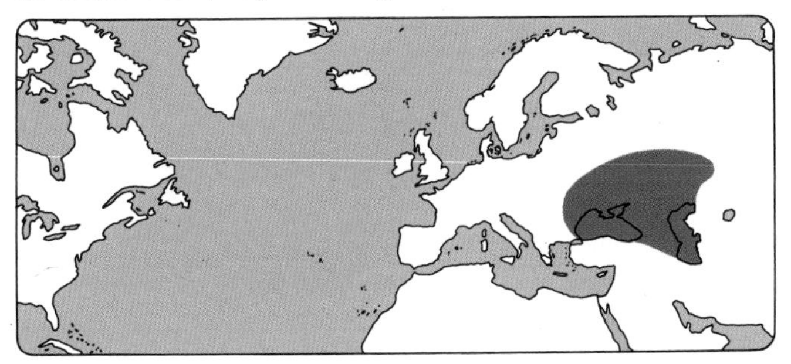

DK: Af æggene fås en førsteklasses kaviar. Ulovligt fiskeri, eskaleret efter USSR's opløsning, formodes at belaste bestanden fremover

D: Seine Eier ergeben erstklassigen Kaviar. Ungesetzliche Fischerei nach dem Zusammenbruch der Sowjetunion wird den Bestand zukünftig vermutlich belasten

E: De las huevas se obtiene un caviar exquisito. La pesca ilegal, que ha aumentado después de la disolución de la antigua URSS, reducirá las existencias

F: Les oeufs donnent un caviar de première qualité. Depuis la dissolution de l'URSS, la pêche illégale qui ne cesse d'augmenter représente une menace grave

I: Dalle uova si ottiene un caviale di ottima qualità. Se prevede che la pesca illegale, aumentata dopo lo scioglimento dell'URSS, farà diminuire le risorse

SHAD

Scientific name:
Alosa sapidissima

Family: Clupeidae — herrings and sardines
Typical size: 50 cm, 2 to 3 kg

The American or white shad was successfully transplanted from the Atlantic coast of the United States to the Pacific in both North America and Kamchatka. Although large spawning runs take place in a number of rivers along the west coast of the United States, the fish is not highly regarded locally. Most of the commercial harvest is from the Atlantic coast, where the roe, and sometimes the meat, is eaten. There is a small, mainly recreational, fishery in Pacific rivers.

DESCRIPTION
The species may grow as large as 5 to 6 kilos. A herring-like fish with very oily meat, it enters rivers on its spawning migrations based on water temperatures. The earliest migrations are in Florida in mid-November. As the water warms to the north, the fish follow; the latest migrations are in May or even June in the St. Lawrence.

Like salmon, shad return to the rivers where they hatched. Construction of hatcheries has assisted the recovery of some shad populations, but although the fish is sought by a small group of fishermen and by consumers of roe, it does not have the wide appeal of other species, so re-stocking has not been extensive.

COMMON NAMES
D: Amerikanischer Maifisch
DK: Amerikansk stamsild
E: Sábalo americano
F: Alose savoureuse
GR: Trichiós tis Amerikís
I: Alaccia americana
IS: Skjaddi, amerísk augnasíld
J: Shyado
NL: Amerikaanse elft
P: Sável americano
RU: Shed
S: Shad
SF: Amerikankantasilli
US: Shad

LOCAL NAMES
PO: Alosa

HIGHLIGHTS
This large anadromous herring is fished for its roe as it returns to rivers to spawn. It represents one of the few successful attempts to introduce a fish species to a remote location, without disrupting native fauna.

Anglers like shad, which can be caught by fly-fishing as well as by trolling, spinning and bait-casting, but there is little commercial use outside the Atlantic coastal states. Commercial fishing is quite heavily restricted in most states to protect stocks for recreational use.

Nutrition data:
(100 g edible weight)

Water	68.2 g
Calories	192 kcal
Protein	16.9 g
Total lipid (fat)	13.8 g
Omega-3	2.8 mg

SHAD *Alosa sapidissima*

FISHING METHODS

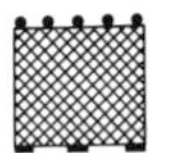 

MOST IMPORTANT FISHING NATIONS

United States

PREPARATION

Roe 

USED FOR

EATING QUALITIES

Prized in the United States for its roe, the shad has very bony meat and is difficult to fillet. Once cleaned, the meat is very good: delicate and light, with plenty of flavour. However, removing the bones requires considerable skill and experience. Most people do not bother, simply discarding the meat and the male fish while retaining the very tasty roe from the female fish.

As an alternative to removing the bones, the fish can be cooked very slowly in the oven, wrapped tightly in foil with lemon juice or other acid. This process softens the bones enough so that they can be eaten.

Roes are generally sold as pairs, in whole skeins containing the eggs. Most of the production is sold fresh. Small amounts are frozen, although the very high fat content of the roe means that it cannot be kept for more than two or three months before it starts to turn rancid.

The roes are best if dipped in hot water briefly to firm them before they are fried. They are an important seasonal item on the menus of many fine fish restaurants along the eastern seaboard of the United States.

IMPORTANCE

The once huge populations of shad started to decline in the 1850s and continued to fall for the next hundred years. Stocks have been recovering, thanks to harvesting controls and improvements to spawning habitats.

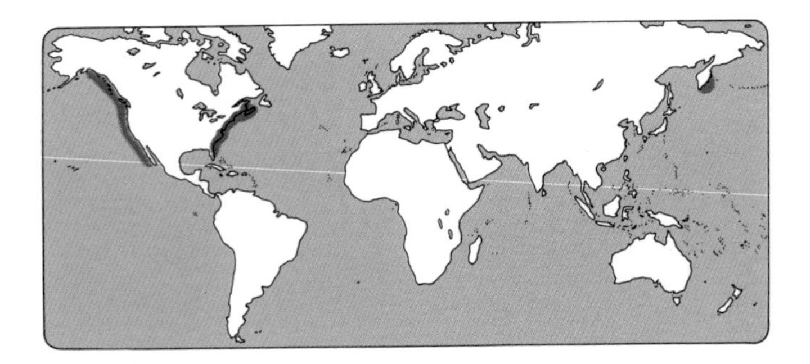

DK: Stor anadrom sild, der fiskes for at udnytte rognen, når den vender tilbage til floden for at gyde. Spises ikke på grund af benet kød

D: Grosser anadromer Hering, der des Rogens wegen gefangen wird, wenn er zu den Laichplätzen die Flüsse hinaufsteigt. Das gräten-reiche Fleisch wird nicht gegessen

E: Gran arenque anádromo, que es capturado por sus huevas, cuando vuelve al río para desovar. No es popular por sus muchas espinas

F: Grand hareng anadrome que l'on pêche pour les oeufs lorsqu'il revient vers les rivières pour jeter le frai. A cause de la chair pleine d'arêtes, on ne le mange pas

I: Grande aringa anadroma, catturata per le uova quando all'epoca della ri-produzione risale i fiumi. La carne non si mangia, perché piena di spine

SHEEPSHEAD

Scientific name:
Archosargus probatocephalus

Synonym: Sheepshead sea bream
Family: Sparidae — porgies
Typical size: up to 70 cm, 9 kg

Very similar to scup, sheepshead grows much larger but is much less abundant. It ranges the Atlantic coast of North America from Cape Cod south to Texas; it is also found off northeast South America. U.S. catches average about 1,500 tons annually.

FISHING METHODS

MOST IMPORTANT FISHING NATIONS
United States

PREPARATION

USED FOR

EATING QUALITIES
Sheepshead have white, sweet meat which can be dry if overcooked but is excellent if prepared well. The species has large bones, which are easily removed from a cooked fillet. It is a popular food fish with anglers, who catch it from jetties or rocky shores.

COMMON NAMES
D: Schafskopf-Brassen
DK: Havrude
E: Sargo chopa
F: Rondeau mouton
GR: Amerikanikós sargós
I: Sarago americano
IS: Kindarhaus
NL: Schaapskopbrasem
P: Sargo choupa
RU: Karas baranjya golova
US: Seabream

Nutrition data:
(100 g edible weight)

Water	77.8 g	Total lipid	
Calories	102 kcal	(fat)	2.4 g
Protein	20.2 g	Omega-3	0.3 mg

HIGHLIGHTS
Most of the sheepshead catch is taken by anglers in Florida, where it is a prized food fish. Sheepshead may exceed 10 kg. Commercial catches of this good quality fish are small, but easily sold throughout the eastern seaboard of the United States.

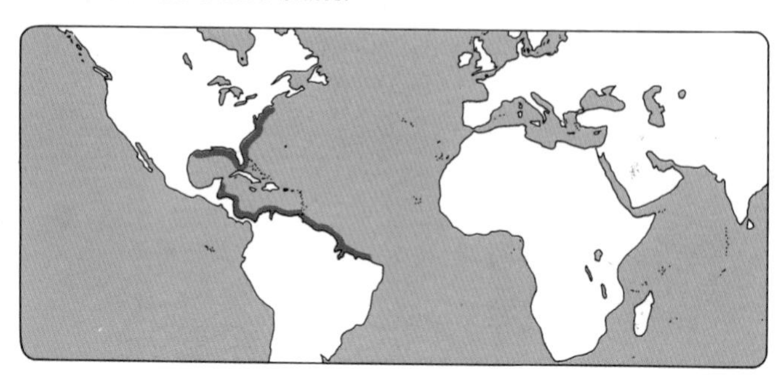

DK: Størsteparten af den årlige fangst på 1.500 tons fiskes af sportfiskere i Florida. Benene fjernes nemt fra det i øvrigt fortrinlige kød

D: Der grösste Teil der jährlichen Fänge von 1500 Tonnen wird von Sportfischern in Florida gefangen. Die Gräten sind leicht aus dem vorzüglichen Fleisch zu entfernen

E: La mayor parte de las 1.500 toneladas al año son capturadas por pescadores deportivos en Florida. Las espinas se separan fácilmente de la exquisita carne

F: La majeure partie des prises annuelles de 1.500 t est réalisée par des pêcheurs sportifs en Floride. Les arêtes s'enlèvent facilement de la chair excellente

I: Della pesca annuale di 1.500 tonnellate, la maggior parte è catturata dai pescatori sportivi in Florida. La carne pregiata è facile da togliere dalle spine

SHEEPSHEAD/FRESHWATER DRUM

Scientific name:
Aplodinotus grunniens

Family: Sciaenidae — drums
Typical size: 45 cm, 1 kg

The sheepshead or freshwater drum is the only freshwater member of this large family. It is fished commercially in Canada, but is mainly taken by anglers, who respect its fighting abilities and generally do not care to eat their catch.

FISHING METHODS

MOST IMPORTANT FISHING NATIONS
Canada, United States

PREPARATION

USED FOR

EATING QUALITIES
The quality of the meat varies in different lakes and streams, but generally the smaller fish are regarded as more palatable. Sheepshead has a large flake and a coarse texture. Commercially caught fish are used mainly for animal feed, especially for mink.

COMMON NAMES
D: Süsswasser-Trommelfisch
DK: Flodtrommefisk
E: Umbrina de agua dulce
F: Malachigan d'eau douce
GR: Mylokópi tou glykoú neroú
I: Ombrina d'acqua dolce
IS: Vatnabaulari
NL: Zoetwatertrommelvis
P: Corvina de água doce
RU: Presnovodny barabanschik
SF: Jokirumpukala
US: Freshwater drum

Nutrition data:
(100 g edible weight)

Water	77.3 g	Total lipid (fat)	4.9 g
Calories	114 kcal		
Protein	17.5 g	Omega-3	1.0 mg

HIGHLIGHTS
Widely distributed in the Mississippi drainage and in other lakes and rivers of eastern North America, freshwater drum in Lake Erie and other Great Lakes once supported a substantial commercial industry. Now, the fish is regarded mainly as a recreational species.

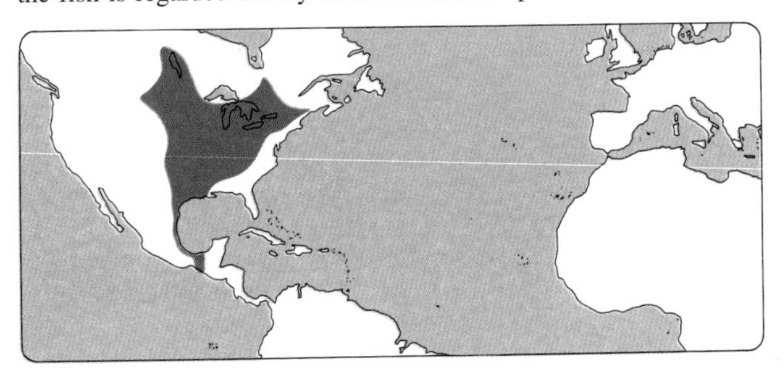

DK: Familiens eneste ferskvandseksemplar. Storflaget og groft kød. Kvaliteten varierer fra område til område, men små fisk er generelt de bedste

D: Das einzige Süsswasserfisch seiner Familie. Grobes Fleisch in dicken Lagen. Qualität schwankt von Gegend zu Gegend, kleine Fische sind im allgemeinen am besten

E: Unico ejemplar de agua dulce de la familia. Tiene lonchas grandes y carne gruesa. La calidad varía de zona a zona. Los pescados pequeños son los mejores

F: L'unique exemplaire d'eau douce de la famille. La qualité de la chair en grosses lamelles varie suivant les lieux; les petits poissons sont les meilleurs

I: L'unico esemplare d'acqua dolce della famiglia. La qualità della carne, a grosse falde, varia da zona a zona, essendo migliori gli esemplari piccoli

SHORTFIN MAKO

Scientific name:
Isurus oxyrinchus

Synonyms: Mackerel shark, mako, bonito shark
Family: Lamnidae — mackerel sharks
Typical size: 180 to 250 cm, 60 to 135 kg

Shortfin mako sharks are found widely in temperate and tropical waters throughout the world, both on the high seas and inshore. It prefers temperatures above about 16°C. It is believed to be the fastest swimming shark and one of the fastest and most active of all fishes. It can produce very fast bursts of speed when hunting or when hooked.

Anglers are excited by the fish's speed as well as by its habit of leaping out of the water when hooked. An exciting fish to catch, it is also good to eat.

DESCRIPTION

A slender, fast-moving shark, the mako is a deep metallic blue on the back and snow white below. After death, the blue colour fades to a brownish grey. The skin, like that of most sharks, is rough and abrasive, although it contains no scales.

Mako females may grow as large as 380 cm and weigh 570 kg. Little is known of reproduction, but young are born alive in litters of eight to ten pups.

COMMON NAMES

D: Blauhai, Mako, Makrelenhai
DK: Makrelhaj, mako-haj
E: Marrajo dientuso, atunero
F: Lamie à nez pointu
GR: Skombrokarcharías
I: Squalo mako
IS: Makrílháfur
J: Aozame
N: Makrellhai
NL: Makreelhaai, mako
P: Tubarao anequim
RU: Cherno-rylaya akula
S: Makrillhaj
SF: Makrillihai, makohai
TR: Dikburun
US: Mako shark

HIGHLIGHTS

One of the most important and sought-after game fishes, the shortfin mako is also a desirable commercial species, targeted for its fine quality meat as well as its fins and skin. Increasing quantities have been caught commercially in recent years as American and European consumers developed greater taste for shark meat.

The prime value of the species remains as a game fish. It is a very powerful, fast fish, making great leaps out of the water, sometimes landing in the boat, even damaging boats on occasion.

Nutrition data:
(100 g edible weight)

Water	77.1 g
Calories	88 kcal
Protein	21.0 g
Total lipid (fat)	0.4 g
Omega-3	0.1 mg

FISHING METHODS

MOST IMPORTANT FISHING NATIONS
United States, Italy, Spain, Japan, Cuba

PREPARATION

USED FOR

EATING QUALITIES
Mako is widely considered to be one of the best sharks to eat. Similar to swordfish in appearance, it has often been substituted for swordfish in retail markets and restaurants. Now that the meat is appreciated for its own qualities, it is approaching swordfish in cost, so substitution is less profitable and, presumably, less common.

Mako meat is generally a little darker than swordfish and is always moister, with less of a "meaty" texture. The large, boneless steaks are easy to prepare and are particularly good broiled or grilled, with or without prior marinading.

IMPORTANCE
Figures indicating the catches of shortfin mako are lacking, but Italy, the USA and many other countries have small landings. Total catches worldwide are probably substantial.

In addition to the high quality meat, makos provide fins for shark-fin soup and skins for leather. Jaws and teeth are sometimes sold as ornaments and trophies. Note that the skins of fish intended for human consumption often retain little value for leather, as the fish must be handled differently for this purpose.

Fins can be removed and utilized whatever alternative purposes are intended for the carcass. Shark fins have high value and ready markets.

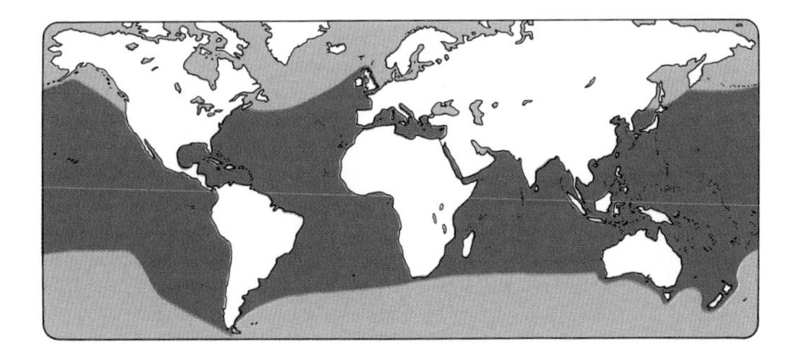

DK: En af de vigtigste og mest eftertragtede sportsfisk. Sildehajen har samtidig stor kommerciel betydning på grund af kvalitetskødet, finnerne og skindet

D: Der Heringshai ist für Sportfischer einer der wichtigsten und gefragtesten Fische. Er besitzt gleichzeitig grosse wirtschaftliche Bedeutung wegen seines Qualitätsfleisches, der Flossen und der Haut

E: Uno de los pescados deportivos más importantes y más populares. La lamia también tiene gran importancia comercial gracias a su carne de calidad, sus aletas y su piel

F: L'un des plus importants poissons sportifs très recherché. Représente en plus un grand intérêt commercial en raison de sa chair de qualité, les ailerons et la peau

I: Uno dei pesci sportivi più importanti e ricercati. Lo smeriglio ha inoltre grande importanza commerciale per la carne di qualità, le pinne e la pelle

SILVER HAKE

Scientific name:
Merluccius bilinearis

Synonyms: Hake, Atlantic hake, offshore hake
Family: Merlucciidae — merluccid hakes
Typical size: 35 cm; up to 76 cm, 2.3 kg

Silver hake is an abundant species on the Atlantic continental shelf of Canada and the United States. Fresh fish is now air freighted from the United States to European markets. Frozen headless and dressed fish competes with imported whiting in American markets.

FISHING METHODS

MOST IMPORTANT FISHING NATIONS
United States, Russia

PREPARATION

USED FOR

EATING QUALITIES
Silver hake has tasty, white meat with a firmer texture than most other hakes and whitings, despite its comparatively high oil content. The meat is versatile and stands up well to most cooking methods, including hot smoking.

COMMON NAMES
D: Nordamerikanisch. Seehecht
DK: Nordvestatlantisk kulmule
E: Merluza norteamericana
F: Merlu argenté
GR: Bakaliáros tou Atlantikoú
I: Nasello atlantico
IS: Silfurlysingur
N: Lysing
NL: Noordwestatlantische heek
P: Pescada prateada
RU: Serebristy khek
S: Brosme
SF: Hopeakummeliturska
US: Whiting, silver hake

LOCAL NAMES
PO: Morszczuk

Nutrition data:
(100 g edible weight)

Water	80.3 g	Total lipid	
Calories	85 kcal	(fat)	1.3 g
Protein	18.3 g	Omega-3	0.2 mg

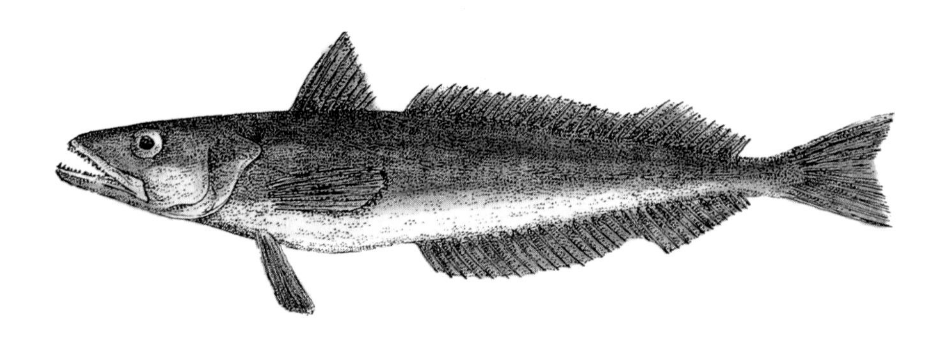

HIGHLIGHTS
Production of 50,000 to 100,000 tons of this species remains far below the estimated potential, which is 300,000 to 500,000 tons. Silver hake is increasing markets slowly as its good quality meat becomes more widely appreciated.

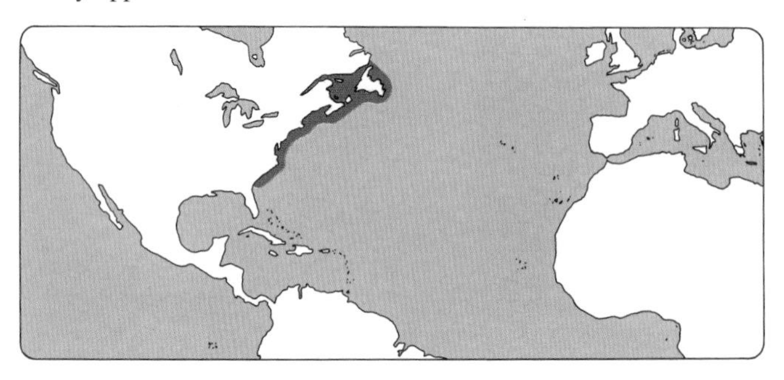

DK: Der fiskes årligt kun ca. 20% af den anslåede, mulige fangstmængde på 300.000 til 500.000 tons. Dog stigende efter-spørgsel

D: Es werden jährlich nur rund 20% der veranschlagten möglichen Fangmenge von 300.000 bis 500.000 Tonnen gefangen. Die Nachfrage ist jedoch im Anstieg begriffen

E: Al año se captura tan sólo un 20% de la cantidad potencial estimada en 300.000 a 500.000 toneladas, pero la demanda sigue aumentando

F: La pêche annuelle ne représente que 20% envir. des quantités potentielles évaluées à 3 à 500.000 tonnes. On note cependant une demande croissante

I: Se ne pesca soltanto circa il 20 per cento della quantità potenziale stimata di 300.000 a 500.000 tonnellate. La domanda è comunque crescente

SLENDER ALFONSINO

Scientific name:
Beryx splendens

Synonyms: Red bream, scarlet bream
Family: Berycidae — alfonsinos
Typical size: 60 cm

The slender alfonsino is found in many parts of the world, but the greatest resources appear to be around New Zealand, off Portugal and northwest Africa and in northern Japan. The fish has also been reported off South Africa. Recorded production is only 2,000 to 5,000 tons yearly, most of it from New Zealand.

COMMON NAMES

D: Südlicher Kaiserbarsch
E: Alfonsino besugo
F: Béryx long
I: Berice rosso
IS: Fagurserkur
J: Kinmedai
NL: Echte beryx, alfonsino
P: Imperador costa estreita
RU: Obyknovenny beriks
SF: Hohtolimapää
US: Alfonsino

LOCAL NAMES

AU: Alfonsino
NZ: Slender beryx

FISHING METHODS

MOST IMPORTANT FISHING NATIONS

New Zealand

PREPARATION

USED FOR

   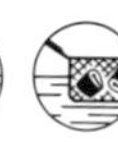  

EATING QUALITIES

The slender alfonsino has white, firm meat with a mild flavour, not unlike orange roughy. It can be used in a wide variety of fish recipes and takes sauces well.

Nutrition data:
(100 g edible weight)

Water	77.3 g	Total lipid	
Calories	98 kcal	(fat)	2.6 g
Protein	18.6 g	Omega-3	0.7 mg

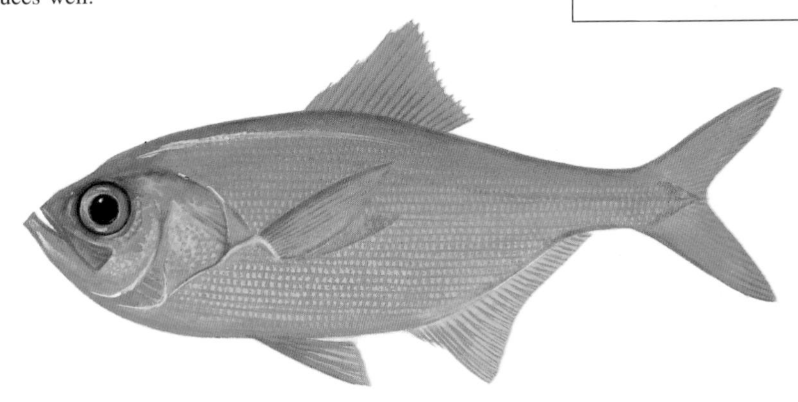

HIGHLIGHTS

Widely found but nowhere abundant, the slender alfonsino is well regarded as a food fish wherever it is available. It lives in moderately deep water down to 800 m and is invariably caught in trawls, often by factory vessels which fillet and freeze on board for best quality.

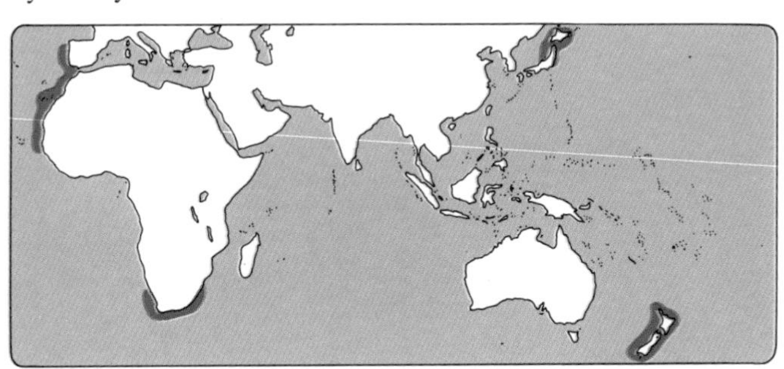

DK: Små forekomster af denne fisk, der dog er vidt udbredt og lokalt anses for en god spisefisk. Lever i vanddybder på indtil 800 m

D: Geringe Bestände dieses Fisches, der jedoch weit verbreitet ist und örtlich als guter Speisefisch gilt. Lebt in Meerestiefen bis zu 800 m

E: Se encuentra en muchos lugares, pero nunca en abundancia. Es apreciado como pescado comestible. Vive en aguas profundas de hasta 800 m

F: Poisson largement répandu mais pas très abondant et localement considéré comme un excellent poisson de table. Il vit à des profondeurs de jusqu'à 800 m

I: Le risorse di questo pesce sono modeste, ma largamente diffuse. Ritenuto un pesce commestibile buono. Vive in profondità fino a 800 metri

SMALLMOUTH BUFFALO

Scientific name:
Ictiobus bubalus

Family: Catastomidae — suckers
Typical size: up to 90 cm, 16 kg

COMMON NAMES
RU: Melkorotaya bufala
US: Smallmouth buffalo

Found in large rivers and lakes, the smallmouth buffalo stays close to the bottom where it vacuums up shellfish and algae, grinding the food with bony plates in its throat.

FISHING METHODS

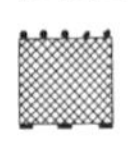

MOST IMPORTANT FISHING NATIONS
United States

PREPARATION

USED FOR

EATING QUALITIES
Considered the best of the suckers and similar to carp, the smallmouth buffalo has firm, white, sweet meat. It also has numerous small, intramuscular bones running the full length of the body. Chefs score the meat to ensure these bones are well heated and softened. Buffalo should be cooked with the skin on, or the flakes will fall apart.

Nutrition data:
(100 g edible weight)

Water	77.2 g	Total lipid	
Calories	117 kcal	(fat)	6.3 g
Protein	15.0 g	Omega-3	1.3 mg

HIGHLIGHTS
Both the roes and milts of the smallmouth buffalo are delicacies. They are rolled in cornmeal and fried, or lightly cooked then mixed with eggs and scrambled. Commercial catches have collapsed, but the species appears to be a good candidate for aquaculture.

DK: Fiskens rogn regnes for en delikatesse - serveres paneret eller letkogt blandet i røræg

D: Der Rogen des Fisches gilt als Delikatesse - paniert gebraten oder kurz gekocht und mit Eiern verrührt

E: Las huevas del pescado son consideradas un manjar exquisito. Se sirven empanadas o ligeramente cocidas, mezcladas con huevos revueltos

F: Les oeufs du poisson sont considérés comme une grande friandise. Ils sont servis cuits à la poële en panure ou mélangés avec des oeufs brouillés

I: Le uova di questo pesce sono considerate un piatto prelibato. Si preparano impanate oppure leggermente bollite mischiate con uova strapazzate

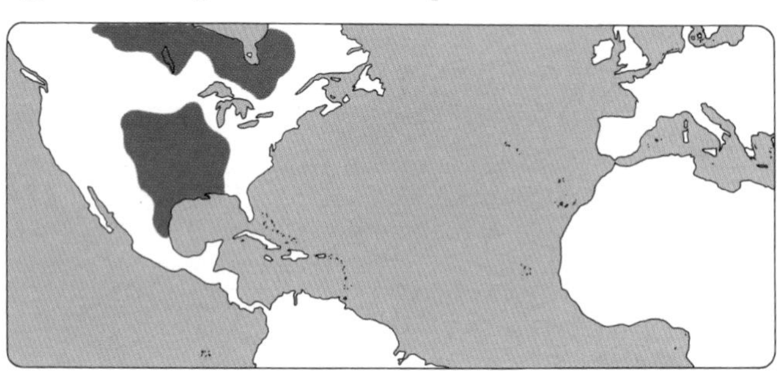

SNOEK

Scientific name:
Thyrsites atun

Family: Gempylidae — snake mackerels
Typical size: 50 to 100 cm, 1 to 3 kg; up to 150 cm, 6 kg

Known as barracouta in New Zealand, this is an important, inexpensive species. World catches fluctuate between 50,000 and 100,000 tons, depending on the intensity of fishing by Russian trawlers off southern Africa. Resources are believed to be capable of sustaining considerably increased fishing effort.

FISHING METHODS

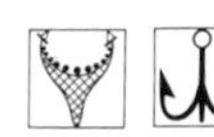

MOST IMPORTANT FISHING NATIONS

New Zealand, Russia, South Africa, Australia, Argentina, Chile

PREPARATION

USED FOR

EATING QUALITIES

Snoek has dark meat, which whitens when cooked. It is firm and tasty and is very good smoked. Canned snoek has some similarity to canned mackerel, though the meat tends to be drier.

COMMON NAMES

D: Snoek
DK: Slangemakrel
E: Sierra, sierra del sur
F: Escolier, escolar
I: Tirsite
J: Barakuta
NL: Snoekmakreel
P: Senuca, foguete
RU: Snek, barakuta
SF: Kuta
US: Barracouta

LOCAL NAMES

NZ: Barracouta

Nutrition data:
(100 g edible weight)

Water	73.9 g	Total lipid (fat)	4.5 g
Calories	114 kcal		
Protein	18.4 g	Omega-3	0.9 mg

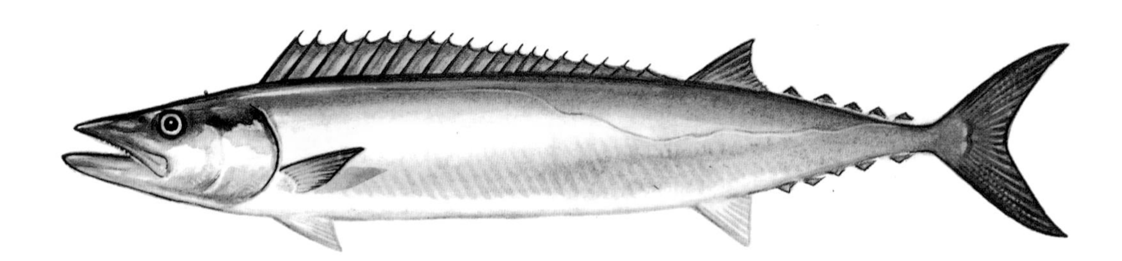

HIGHLIGHTS

Skinless frozen fillets, mainly from New Zealand and Chile, are a tasty and affordable fish product. They make excellent fish-and-chips. In Japan, snoek is sometimes used for surimi. Canned snoek has fallen from favour; it was important during World War II.

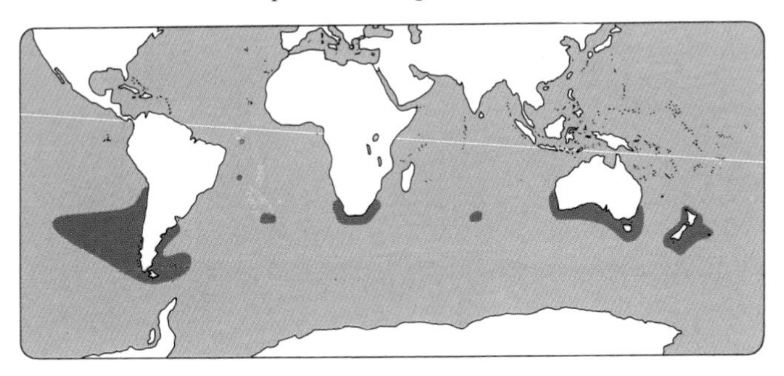

DK: De frosne filetter er velsmagende og prismæssigt et rimeligt produkt velegnet til panering. I Japan anvendes fisken ofte til surimi

D: Die eingefrorenen Filets sind schmackhaft und preiswert sowie gut zum Panieren geeignet. In Japan wird der Fisch häufig zu 'Surimi' verarbeitet

E: Los filetes congelados son sabrosos y el precio es razonable. Apropiado para empanar. En Japón sirve a menudo para 'surimi'

F: Les filets surgelés ont très bon goût, et quant au prix, c'est un produit raisonnable qui convient pour paner. Au Japon, il sert souvent pour préparer du surimi

I: I filetti congelati sono gustosi e il prezzo ragionevole. Adatti per essere impanati. Nel Giappone il pesce è spesso usato per surimi

SOCKEYE SALMON

Scientific name:
Oncorhynchus nerka

Synonyms: Red salmon, kokanee, blueback
Family: Salmonidae — salmonids
Typical size: 45 cm, 4 kg

Sockeye salmon is probably the most valuable wild salmon resource in the oceans of the world; it is an abundant fish which fetches a high price. Alaska's Bristol Bay hosts the largest fisheries. The only substantial sockeye runs south of Alaska are in the Skeena and Fraser Rivers of British Columbia.

DESCRIPTION
Sockeye salmon is a silvery fish with speckles on its back, very similar to chum salmon (*O. keta*). These two species cannot be easily distinguished, although sockeye's deep red flesh colour is often used as an indication. In breeding dress the fish turn deep red and the males develop kypes, strongly hooked lower jaws. Sockeye from a particular run tend to be remarkably similar in size. Most fish are between 3 and 5 kg. The largest sockeye recorded was 7 kg.

Sockeye spend their early lives in freshwater lakes, entering the ocean in their second or third spring. They generally return to their natal stream at the age of four. Like pink salmon and unlike the other Pacific salmon species, they conform quite strictly to the pattern evolved for their particular river.

COMMON NAMES
D: Nerkalachs, Rotlachs
DK: Kokanee-laks
E: Salmón sockeye
F: Saumon rouge
GR: Kókkinos solomós
I: Salmone rosso
IS: Raudlax
J: Beni-zake
N: Indian-laks, sockeye
NL: Rode zalm, sockeye zalm
P: Salmao vermelho do Pacíf.
RU: Krasnaya nerka
S: Indianlax, sockeye
SF: Punalohi, intiaanilohi
US: Sockeye, red salmon

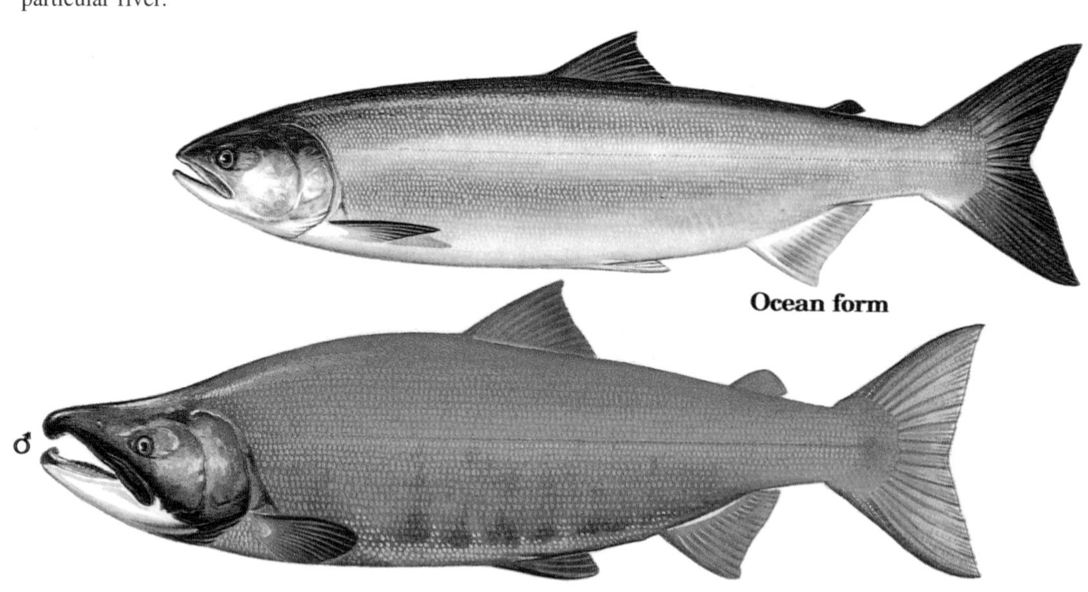

HIGHLIGHTS
Although sockeye salmon has not been successfully domesticated for farming, supplies have been increasing as a result of enhancement and habitat improvement programs. Markets from Japan to Europe value sockeye for its deep red color, consistent size and full flavour. Smokers like the high fat content, which helps them to produce a moist and exceptionally palatable product.

Sockeye do not support recreational fisheries as these fish do not normally take bait or hooks.

Nutrition data:
(100 g edible weight)

Water	70.2 g
Calories	163 kcal
Protein	21.3 g
Total lipid (fat)	8.6 g
Omega-3	0.7 mg

SOCKEYE SALMON *Oncorhynchus nerka*

FISHING METHODS

 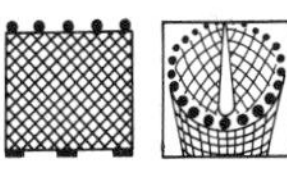 

MOST IMPORTANT FISHING NATIONS
United States, Canada

PREPARATION

USED FOR

EATING QUALITIES
Sockeye salmon has the reddest and oiliest meat of any salmon. It is moist and very tasty. It is ideal for broiling, grilling or other dry-cooking methods. Sockeye is used for cold smoked salmon, similar to traditional European style smoked salmon. It is also hot smoked, especially on the Pacific coast of the USA and Canada, making a product which, with retort pouch packaging, can be distributed without refrigeration and is finding new markets for smoked salmon.

Sockeye is Japan's favourite salmon, although Japan itself produces very little. The market has grown since imports were liberalized in the 1960s. Now, salted, smoked or frozen fillets of sockeye are an important and valued food item for the Japanese family.

IMPORTANCE
The United States produces 70 percent of all sockeye. Almost all of this comes from Alaska, where landings have steadily increased to record levels thanks to successful hatcheries and enhancement programs. Total catches average 200,000 tons, with an upward trend. Canadian production is important but much smaller. Russia has limited production, though some experts believe that Russian catches of sockeye could be substantially increased.

Canned red salmon is made from sockeye; this is the premium canned salmon product, an important export product for the USA and Canada.

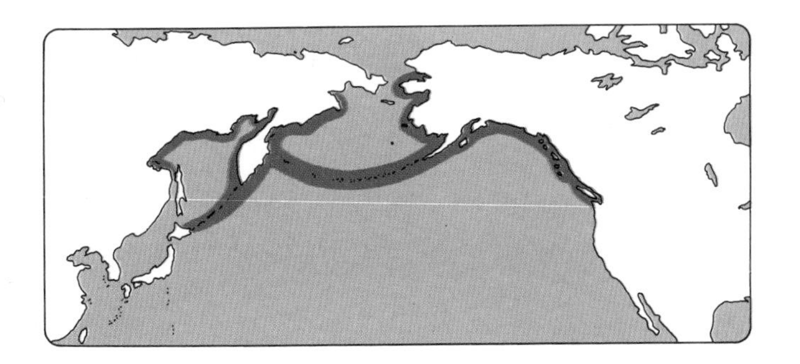

DK: Laksen er svær at opdrætte, men nye teknikker har forbedret vækstbetingelserne og dermed udbyttet. Fisken har en høj markedsværdi med dens mørkerøde farve, homogene størrelse og flavour

D: Der Lachs lässt sich nur schwer züchten, aber neue Techniken haben die Wachstumsbedingungen und damit den Ertrag verbessert. Die rote Farbe, die einheitliche Grösse und der Wohlgeschmack erhöhen den Marktwert des Fisches

E: La cría del salmón es difícil, pero nuevas técnicas han mejorado las condiciones del habitat, dando rendimientos más altos. El precio del pescado en el mercado es alto gracias a su color rojo oscuro, tamaños parejos y buen sabor

F: L'élevage du saumon est difficile, mais des techniques nouvelles ont permis d'améliorer les conditions de croissance et, donc, le rendement. Il doit sa haute valeur commerciale à son goût, sa chair rouge foncée et sa taille homogène

I: Il salmone è difficile da allevare, ma le nuove tecniche hanno migliorato le condizioni di crescita e quindi il rendimento. Ha un alto valore commerciale per il bel color rosso delle carni, la grandezza omogenea e la sua aroma

SOUTHERN BLUE WHITING

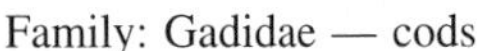

Scientific name:
Micromesistius australis

Family: Gadidae — cods
Typical size: 40 to 60 cm, 400 to 500 g, up to 90 cm

Two major fisheries for southern blue whiting, have developed in recent years, producing several hundred thousand tons. One is off New Zealand, the other and much larger off Argentina, the Falkland Islands and neighbouring Antarctic waters.

FISHING METHODS

MOST IMPORTANT FISHING NATIONS
Russia, Poland, Spain, Argentina, New Zealand

PREPARATION
Fishmeal 

USED FOR

EATING QUALITIES
The southern blue whiting has soft, white meat with delicate texture and low fat content. It can be used as fillets, which are similar to hoki (*Macruronus novaezelandiae*), but most of the catch is made into blocks, which are later manufactured into portions and other products.

COMMON NAMES
D: Südlicher, blauer Wittling
DK: Sydlig sortmund
E: Polaca austral
F: Merlan bleu austral
GR: Prosfygáki tis Australías
I: Melu australe
J: Patagonia-minamidara
NL: Zuidelijke blauwe wijting
P: Verdinho austral
RU: Yuzhnaya putassu
US: Blue whiting

LOCAL NAMES
AR: Polaca
CL: Merluza de tres aletas

Nutrition data:
(100 g edible weight)

Water	79.2 g	Total lipid	
Calories	83 kcal	(fat)	0.8 g
Protein	18.9g	Omega-3	0.2 mg

HIGHLIGHTS
Increasingly exploited by factory fleets from European countries, the southern blue whiting resources provide large quantities of inexpensive fish for processors. The fish is small and soft. It is filleted and skinned with specially designed equipment.

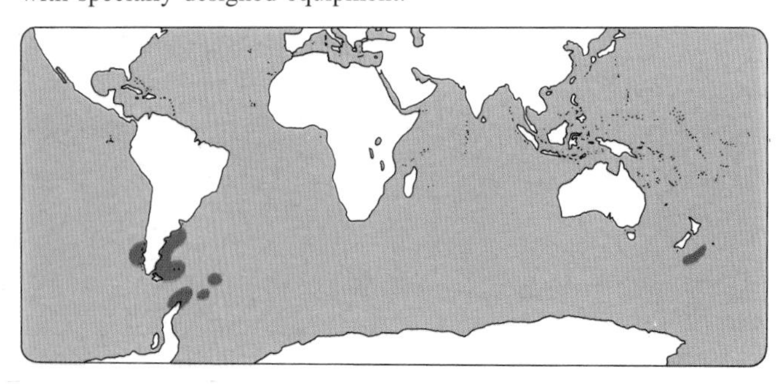

DK: Den enorme bestand udnyttes i stigende grad af europæiske fabriks-trawlere, der leverer store mængder af denne billige fisk til industrien

D: Der enorme Bestand wird zunehmend von europäischen Verarbeitungstrawlern ausgenutzt, die grosse Mengen dieses billigen Fisches an die Industrie liefern

E: Las enormes existencias son usadas cada vez más por bous-factoría de los países europeos que suministran a la industria grandes cantidades de este barato pescado

F: L'énorme population est de plus en plus exploitée par des chalutiers-usines européens approvisionnant l'industrie de grandes quantités de ce poisson bon marché

I: Le grandi risorse di questo pesce vengono sfruttate sempre di più dai trawler europei che ne consegnano grossi quantitativi a basso prezzo all'industria

SOUTHERN BLUEFIN TUNA

Scientific name:
Thunnus maccoyii

Family: Scombridae — tunas and mackerels
Typical size: 150 to 200 cm, up to 158 kg

Southern bluefin are smaller than their northern relatives; they are circumpolar, based on a single stock which spawns in the Indian Ocean and is believed to be under considerable pressure. Caught and handled individually, carefully handled fish are flown to Tokyo for sale fresh from fishing grounds off South Australia and New Zealand.

COMMON NAMES

D: Südlicher Blauflossenthun
DK: Sydlig tun
E: Atún del sur, atún rojo
F: Thon rouge du Sud
GR: Tónos
I: Tonno
IS: Sudræni túnfiskur
J: Indomaguro
NL: Zuidelijke blauwvintonijn
P: Atum do sul
RU: Avstralijskaya tunets
US: Bluefin tuna

FISHING METHODS

MOST IMPORTANT FISHING NATIONS

Japan, Australia, Taiwan, Korea, New Zealand

PREPARATION

Sashimi

USED FOR

EATING QUALITIES

The pink to red meat is highly regarded for eating raw as sashimi. It has firm texture, high fat and low moisture. The fattest (pre-spawning) fish fetch much higher prices than low-fat spent fish.

Nutrition data:
(100 g edible weight)

Water	67.0 g	Total lipid	
Calories	166 kcal	(fat)	8.0 g
Protein	23.5 g	Omega-3	1.6 mg

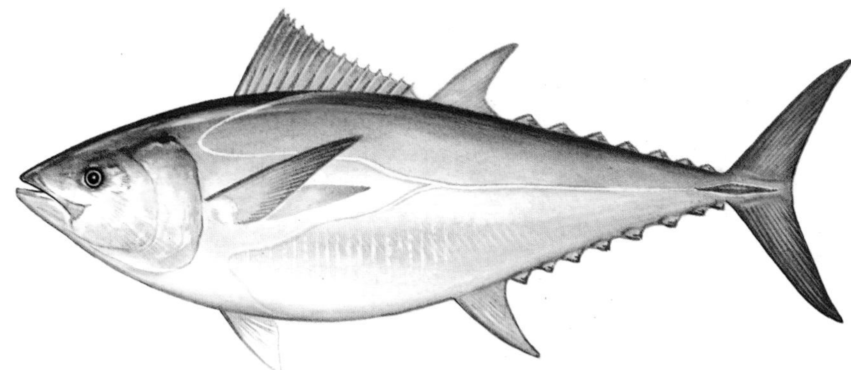

HIGHLIGHTS

Young fish are caught and grown in coastal pounds by Australian fish farmers, while scientists try to develop techniques for breeding and hatching these valuable tunas. International agreement to limit fishing has not yet restored the stock to previous healthy levels.

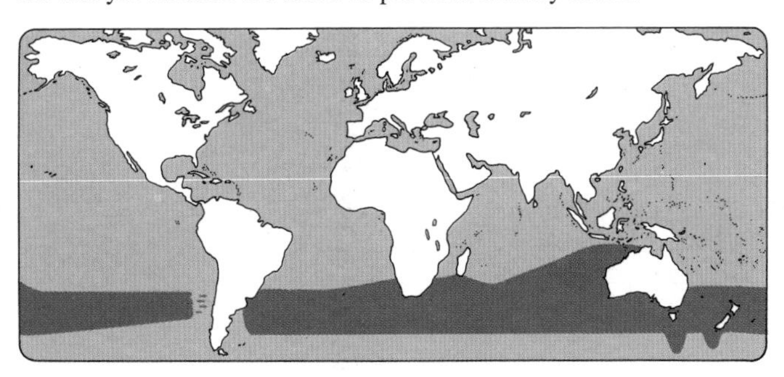

DK: Unge fisk indfanges til opdræt i australske havbrug. Samtidig forskes der i at få denne værdifulde fisk til at yngle i fangenskab

D: Jungfische werden in Australien zur Aufzucht eingefangen. Wissenschaftler versuchen, diesen wertvollen Fisch zum Laichen in Fischteichen zu bewegen

E: Los alevines son capturados para criarlos en piscifactorías australianas. Se está investigando para conseguir que este valioso pescado procree en cautividad

F: Les jeunes de ce poisson précieux sont capturés pour des fermes marines en Australie. Des recherches sont en cours pour la réproduction en captivité

I: I pesci giovani vengono catturati e allevati in vivai australiani. Si cerca di sviluppare le possibilità di riproduzione in cattività di questo pesce prezioso

SOUTHERN FLOUNDER

Scientific name:
Paralichthys lethostigma

Family: Bothidae — left-eye flounders
Typical size: 55 cm 1.5 kg

Southern flounder is a prized flatfish in the eastern United States, where traditional commercial fishing is increasingly being overcome by the burgeoning recreational fishery.

It is available close to shore throughout the year, which adds to its value to both commercial and sport fishermen.

DESCRIPTION
Southern flounder are similar to fluke (*Paralichthys dentatus*) but smaller, with most fish about one kilo in weight. It is almost identical to the Gulf flounder (*P. albigutta*), which is caught with southern flounder and seldom distinguished from it commercially, despite the taxonomic distinction.

Southern flounder is found along the southern part of the Atlantic coast of the United States, from North Carolina to the southern tip of Florida. There is a separate population in the Gulf of Mexico, from Sanibel Island in Florida to Corpus Christi, Texas. Unlike most Atlantic flounders, it tolerates low salinities: it is frequently found in brackish bays and estuaries, even on occasion in fresh water.

COMMON NAMES
E: Lenguado de Florida
F: Cardeau de Floride
RU: Yuzhnaya kambala
US: Southern flounder, fluke

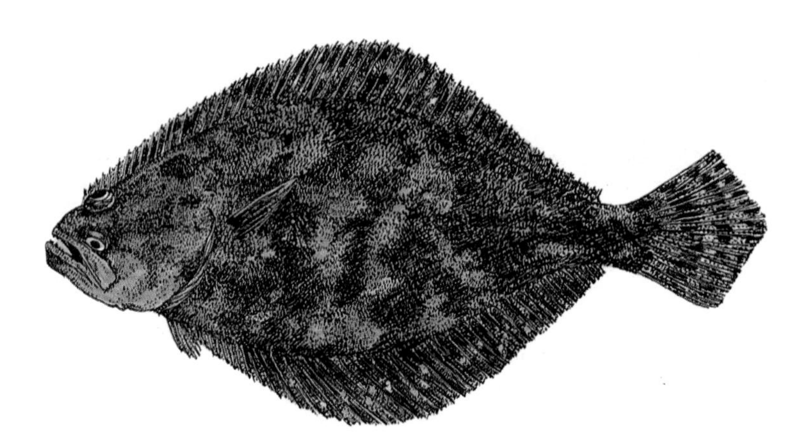

HIGHLIGHTS
Summer flounder, like other flounders and soles, has the ability to change its colour to blend into its surroundings. The blotches on its body gradually change to merge the fish into the colour of the mostly sandy bottoms where it likes to hide.

Smaller fish are dressed and trimmed and sold fresh or frozen. Larger fish are filleted and usually skinned. The underside (white) skin is sometimes left on, but the thicker and darkly coloured skin of the top side is mostly removed before the fish is cooked.

Nutrition data:
(100 g edible weight)

Water	78.0 g
Calories	84 kcal
Protein	20.0 g
Total lipid (fat)	0.5 g
Omega-3	0.1mg

FISHING METHODS

MOST IMPORTANT FISHING NATIONS

United States

PREPARATION

USED FOR

EATING QUALITIES

Summer flounder has firm, very white, meat. Like most flounders, the flake is small and dense, but the texture of the meat is light and the flavour very sweet. Fillets are excellent battered and fried. Smaller, whole fish can be pan dressed and baked or fried, when the meat slips easily from the backbone.

The backbones can also be fried crisply and eaten, a delicacy among some Asian populations.

IMPORTANCE

Southern flounder resources have been more intensively fished in recent years as other, traditionally more familiar, flounders and soles declined in abundance. There are increasing management controls on east coast fisheries in the United States, partly because of unprovable fears of declining resources, partly because of political pressure from recreational interests.

The species is important recreationally in many parts of its range, as it is readily taken close inshore from bridges, jetties and small boats. It moves to deeper water in winter, but is still easily accessible.

Because the fish is generally small, it does not fetch such a high price as the big and meaty fluke. Nevertheless, it is a valuable species and its excellent quality meat is well regarded.

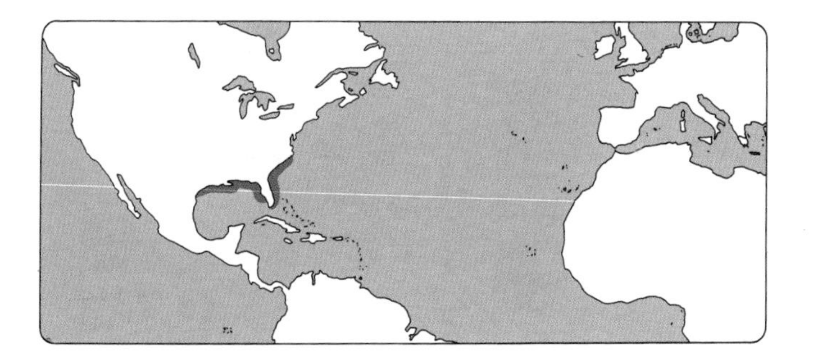

DK: Som andre flyndere og tunger, tilpasser den farven efter omgivelserne. Pletterne ændrer gradvist farve efter sandbunden, der er dens foretrukne skjule-sted

D: Wie andere Butte und Zungen passt er seine Färbung der Umgebung an. Die Flecken ändern ihre Färbung abgestuft nach der Farbe des Sandbodens, der ihr vorgezogenes Versteck ist

E: Como otras peludas y lenguados, cambia su color según el ambiente. Las manchas cambian su color gradualmente según la arena del fondo, su escondite preferido

F: Comme les autres cardeaux et soles, il adapte sa couleur en fonction du milieu environnant. Les taches changent graduellement de couleur en fonction du fond de sable qui est son lieu de cachette préféré

I: Come le altre passere e sogliole, è capace di adattare il proprio colore alle condizioni ambientali. Le macchie cambiano colore gradualmente a secondo del fondo sabbioso che è il suo nascondiglio preferito

SPINY DOGFISH

Scientific name:
Squalus acanthias

Synonyms: Blue dog, picked dogfish
Family: Sqalidae — dogfish sharks
Typical size: 75 to 105 cm, 3 to 4.5 kg

Spiny dogfish, called picked dogfish by FAO, is one of the most important sharks and one of the few regularly exploited in colder waters.

Like all sharks, dogfish must be bled and handled with great care to avoid ammonia taint of the meat. Dogfish are used not only for the meat, but also for the fins, livers and other parts.

DESCRIPTION
This small shark lives in temperate waters of the Atlantic and Pacific Oceans and is also found in Australia and New Zealand. It prefers temperatures between about 7°C and 15°C, migrating seasonally to remain in this range. Although spiny dogfish have been tagged travelling across the Pacific from Washington state, USA, to Honshu, Japan, most of the populations are distinct, with only a little mixing.

Dogfish travel in huge packs (hence the name) of fish of similar size, usually all of one sex. These packs can do immense damage to nets or to fishing operations. If targeted, they can be caught in large numbers on longlines, in trawls or with gillnets.

COMMON NAMES
D: Gemeiner Dornhai, Speerhai
DK: Pighaj
E: Mielga, galludo, pinchorro
F: Aiguillat, chien de mer
GR: Skylópsaro, stiktokentróni
I: Spinarolo
IS: Háfur
J: Abura-tsunozame, tsunozame
N: Pigghå
NL: Doornhaai
P: Galhudo malhado, melga
RU: Korotkoperaya kolyu. akula
S: Pigghaj
SF: Piikkihai
TR: Mahmuzlu camgöz
US: Spiny dogfish

LOCAL NAMES
AU: White-spotted dogfish
EG: Irsh
IL: Qozan qetan qoz
MO: Agugliat
TU: Qtat bou soka

HIGHLIGHTS
A slow moving and slow growing species, the dogfish is generally regarded as under-exploited, although slow recruitment probably limits the opportunities to increase catches substantially.

The meat is popular in Europe, where it is used for fish-and-chips and other dishes. Smoked belly flaps are an important product in Germany. In the United States, which is a major supply source, dogfish are only rarely eaten, usually in specialty restaurants looking for an "exotic" alternative fish.

Nutrition data:
(100 g edible weight)

Water	68.3 g
Calories	154 kcal
Protein	16.6 g
Total lipid (fat)	9.7 g
Omega-3	1.4 mg

SPINY DOGFISH — *Squalus acanthias*

FISHING METHODS

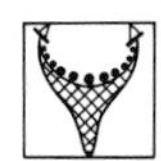 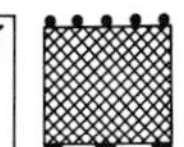 

MOST IMPORTANT FISHING NATIONS

United States, United Kingdom, France, Norway, Canada

PREPARATION

USED FOR

EATING QUALITIES

Properly handled, dogfish provide white, sweet, well flavoured, boneless meat which is ideal for fish-and-chips and other basic preparations. The large back fillets are graded by size and frozen for markets in Europe and Australia. Belly flaps are smoked to a hard consistency in Germany for a gourmet preparation called *Schillerlocken*, named because the golden pieces curl like ringlets.

IMPORTANCE

Spiny dogfish support fisheries throughout their large range; recorded catches of less than 50,000 tons annually certainly understate the economic importance of the species, which in the Pacific can grow as large as 130 cm and 9 kg. Probably, catches are reported with those of other small sharks.

Dogfish are used for food in Europe and Australia; the fins are valued for their gelatinous needles in Chinese cuisines, especially in Asia; the livers, once used for oil and later as a source of vitamin A, are still a source of squalene; large numbers of carcasses are used for dissection by biology students; and the corneas are thought to have potential uses in human eye surgery.

Dogfish cause economic damage by destroying resources of more valuable fishes, especially cod. They interfere with fishing operations by taking bait and sometimes by preying on fish that have been hooked or netted.

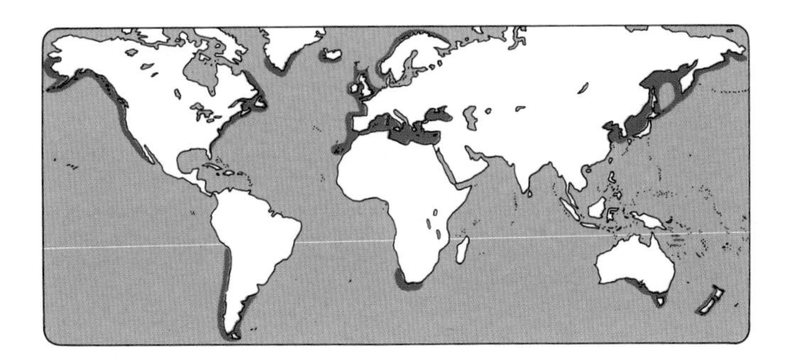

DK: Den vigtigste, lille koldtvandshaj. I Europa anvendes det søde, hvide, benløse kød til færdigretter. Stort marked i Tyskland for røgede bugstykker

D: Der wichtigste, kleine Kaltwasserhai. In Europa wird sein süsses, weisses, grätenloses Fleisch zu Fertiggerichten verarbeitet. Geräucherte Bauchstücke finden in Deutschland grossen Absatz

E: El tiburón pequeño más importante de aguas frías. En Europa, su carne dulce, blanca y sin espinas es usada para platos precocinados. En Alemania, mercado importante de trozos de tripa ahumados

F: Le plus important des petits requins des mers froides. En Europe, la chair blanche et douce sans os est utilisée pour des plats préparés. Des morceaux de chair fumée trouvent un marché important en Allemagne

I: Il più importante fra i piccoli squali d'acqua fredda. In Europa si usano le sue carni dolci, bianche e senza spine per piatti pronti. Importante mercato in Germania come prodotto affumicato

SPOTTED WEAKFISH

Synonym: Spotted sea trout
Family: Sciaenidae — drums
Typical size: 45 cm, 2 kg

Commonly known as "trout" in the southern United States, where it is a popular fish, harvesting of the species is increasingly restricted as more United States resources are reserved for recreational fisheries. Most production comes from the Gulf of Mexico.

FISHING METHODS

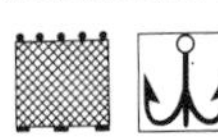

MOST IMPORTANT FISHING NATIONS

United States, Mexico

PREPARATION

USED FOR

EATING QUALITIES

The meat is lean and delicately flavoured, with a small flake like that of trout — explaining its common American name. The species does not keep well, needing careful handling and thorough icing. If the fish is bled, the meat colour is lighter and more attractive.

Scientific name:
Cynoscion nebulosus

COMMON NAMES

D: Gefleckter Umberfisch
DK: Trommefisk
E: Corvinata pintada
F: Acoupa pintade
GR: Sienída
I: Ombrina dentata
IS: Blettadodi
NL: Gevlekte ombervis
P: Corvinata pintada
RU: Pyatnisty gorbyl
SF: Pilkkuveltto
US: Spotted weakfish

Nutrition data:
(100 g edible weight)

Water	78.1 g	Total lipid	
Calories	99 kcal	(fat)	3.6 g
Protein	16.7 g	Omega-3	0.4 mg

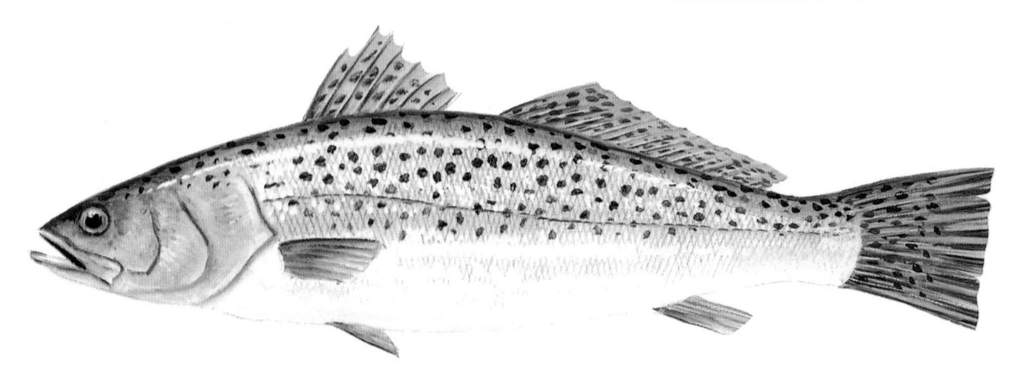

HIGHLIGHTS

Mexican production is growing to replace increasingly restricted harvesting in US waters. Mexico supplies excellent product, often size graded, which sells at a premium. Large numbers of spotted sea trout which remain inshore in winter are sometimes killed by the cold.

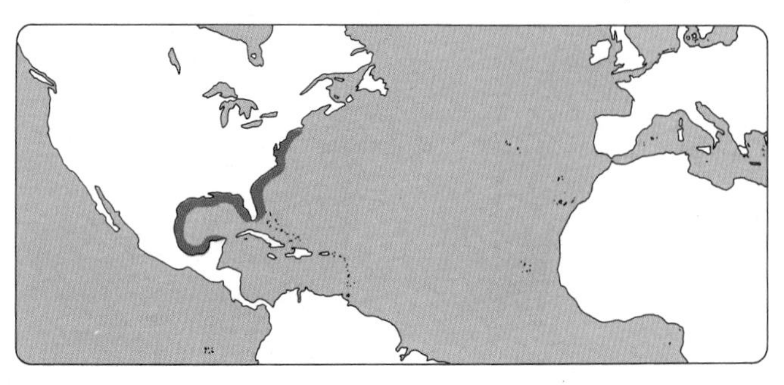

DK: Mexico har intensiveret fiskeriet i takt med USA's stærkt faldende fangster. Mexicanerne leverer et fortrinligt, håndsorteret produkt

D: Mexiko hat seine Fischerei im Zuge der zurückgehenden Fänge der USA intensiviert. Mexiko liefert ein ausgezeichnetes von Hand sortiertes Produkt

E: México ha intensificado la pesca al compás que han bajado mucho las capturas en EE.UU. Los mexicanos suministran un excelente producto, clasificado a mano

F: Au rythme de la forte chute des pêches aux E.-U., le Mexique a intensifié ses prises. Les mexicains livrent un excellent produit trié à la main

I: Il Messico ne ha intensificato la pesca di pari passo con la riduzione della cattura degli USA e fornisce un ottimo prodotto classificato a mano

SPRAT

Scientific name:
Sprattus sprattus

Synonym: Brisling
Family: Clupeidae — herrings, sardines
Typical size: 12 to 16 cm

Sprats are found in the northeast Atlantic and the Mediterranean. Total landings of about 250,000 tons are 80 percent from the Atlantic, where the take has declined. Mediterranean landings have increased in recent years. In Norway, sprats are trapped in large nets in the fiords, kept alive until needed for processing.

FISHING METHODS

MOST IMPORTANT FISHING NATIONS
Denmark, Russia, Sweden, Poland, Norway

PREPARATION

USED FOR

EATING QUALITIES
Small sprats are one of the species used in the UK as whitebait, fried whole. Canned sprats and brisling are delicately flavoured, less "fishy" than herring. Smoked sprats are a gourmet item in Europe.

COMMON NAMES
D: Sprotte, Breitling
DK: Brisling
E: Espadín, trancho
F: Sprat, esprot
GR: Papalína
I: Papalina, spratto
IS: Brislingur
N: Brisling
NL: Sprot
P: Espadilha, lavadilha
RU: Shprot
S: Skarpsill, vassbuk
SF: Kilohaili
TR: Çaça-platika

Nutrition data:
(100 g edible weight)

Water	66.3 g	Total lipid	
Calories	172 kcal	(fat)	11.0 g
Protein	18.3 g	Omega-3	1.3 mg

HIGHLIGHTS
The tiny sprat is an important fish throughout northern Europe, where there are numerous traditional preparations based on the species. Sprats tolerate salinity as low as 4 parts per thousand, so thrive in the Baltic where they are a vital fishery.

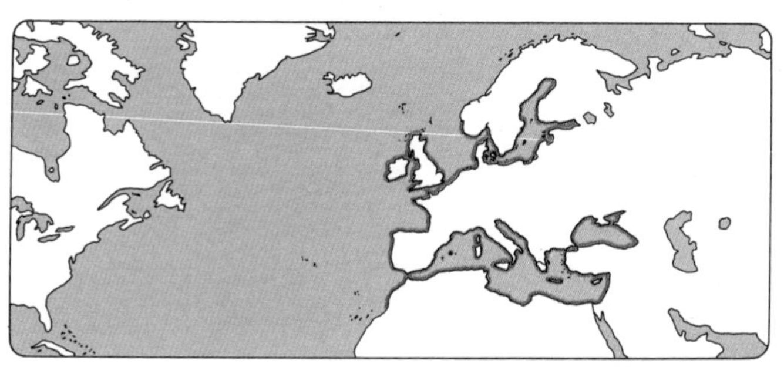

DK: Vigtig fisk overalt i Nordeuropa, især i Østersøen. I Norge holdes fiskene i live i net lige indtil de forarbejdes til brisling

D: In ganz Nordeuropa wichtiger Fisch, besonders in der Ostsee. In Norwegen wird der Fisch bis zu seiner Verarbeitung lebend in Netzen aufbewahrt

E: Pescado importante en todo el norte de Europa, especialmente en el Báltico. En Noruega se mantiene vivo en redes hasta ser transformado en sardineta

F: Poisson important en Europe du Nord, surtout dans la mer Baltique. En Norvège, les sprats sont maintenus en vie dans des filets jusqu'à être préparés en brisling

I: Pesce importante in tutta l'Europa settentrionale, in particolare nel Baltico. In Norvegia si mantengono vivi nelle reti fino alla trasformazione in spratto

SQUETEAGUE

Scientific name: *Cynoscion regalis*

Synonym: Grey weakfish
Family: Sciaenidae — drums
Typical size: 45 cm, 2 kg

The squeteague or gray trout ranges the coast of the western Atlantic from Cape Cod to Florida, but is mostly caught in the more northerly part of its range. Recreational fishing is increasingly supplanting commercial catches.

FISHING METHODS

 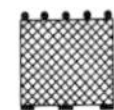 

MOST IMPORTANT FISHING NATIONS
United States

PREPARATION

USED FOR

EATING QUALITIES
The squeteague has trout-like meat with an excellent, delicate flavour and a small flake. The colour is greyish, but it lightens when cooked. The meat is also lighter if the fish is bled when caught. It does not keep well, so must be handled speedily and with plenty of ice.

COMMON NAMES
D: Königs-Corvina
DK: Trommefisk
E: Corvinata real
F: Acoupa royal
GR: Sienída
I: Ombrina dentata
IS: Dodi, aumingi
NL: Witte ombervis
RU: Korolevsky gorbyl
SF: Veltto
US: Squeteague, seatrout

Nutrition data:
(100 g edible weight)

Water	78.1 g	Total lipid	
Calories	99 kcal	(fat)	3.6 g
Protein	16.7 g	Omega-3	0.4 mg

HIGHLIGHTS
Like the spotted weakfish (*C. nebulosus*), this fish has a weak mouth which can easily lose a hook. Its delicate meat is prized in coastal communities familiar with the species. Average sizes have greatly declined in the last 50 years.

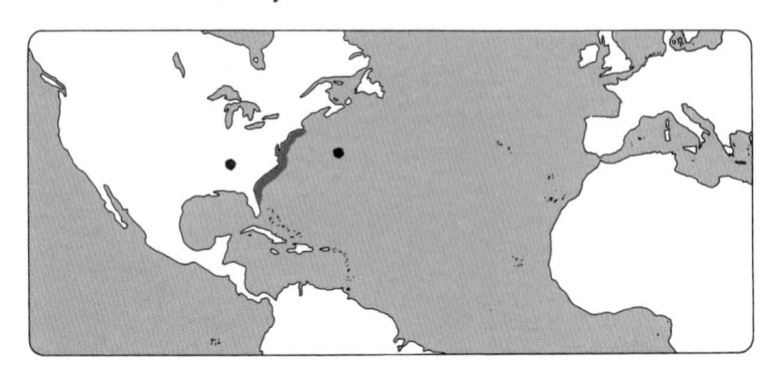

DK: Svært at fastholde krogen i fiskens skrøbelige mund. Det delikate kød er højt værdsat overalt i kystområderne, hvor den forekommer

D: Der Fisch fällt leicht vom Haken wegen seines schwachen Mauls. Das delikate Fleisch ist hoch geschätzt überall in den Küstengebieten, wo er vorkommt

E: Es difícil que el anzuelo se mantenga clavado en la débil boca de este pescado. Su carne delicada es muy apreciada en todas las zonas litorales donde vive

F: La gueule de ce poisson est si fragile qu'il est difficile d'y faire accrocher un hameçon. Chair délicate, très appréciée dans les zones côtières où il vit

I: Difficile da prendere con l'amo in quanto ha la bocca fragile. Le sue carni pregiate sono apprezzate dappertutto nelle zone costiere dove vive

STARRY FLOUNDER

Scientific name: *Platichthys stellatus*

Synonyms: Rough jacket, grindstone, long-jaw flounder
Family: Pleuronectidae — right-eye flounders
Typical size: up to 90 cm, 9 kg

The starry flounder is classified as a right-eye flounder. Despite this, almost all the Asian populations and up to 60 percent of those in North America actually have eyes on the left side. It is found along coasts throughout the North Pacific, ranging as deep as 275 m.

FISHING METHODS

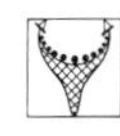

MOST IMPORTANT FISHING NATIONS
Japan, Korea, Canada, USA

PREPARATION

USED FOR

EATING QUALITIES
Starry flounder is one of the tastier flounders of the north Pacific, but is more highly regarded in Asian countries than in North America. The meat is white and tender, with the typical small flake of flounder; it lends itself well to stuffing and baking.

COMMON NAMES
D: Sternflunder
DK: Stjerneflynder
E: Platija del Pacífico
F: Plie du Pacifique
GR: Astrokalkáni
I: Passera stellata
IS: Stjörnukoli
J: Numagarei
NL: Sterreschol
P: Solha estrelado do Pacíf.
RU: Zvezdchataya kambala
SF: Tähtikampela
US: Starry flounder

Nutrition data:
(100 g edible weight)

Water	79.4 g	Total lipid	
Calories	90 kcal	(fat)	3.3 g
Protein	15.2 g	Omega-3	0.7 mg

HIGHLIGHTS
Sometimes called grindstone because of the roughness of its skin, the starry flounder is a moderately abundant flatfish, easily identified from the broad bands on the fins. It is most heavily exploited in the Asian segments of its range.

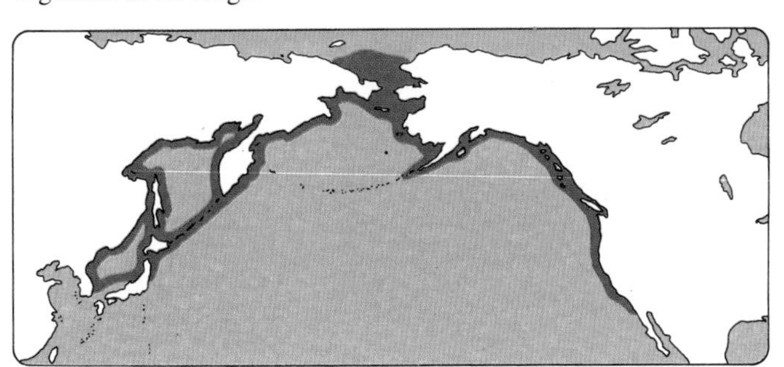

DK: Kaldes også slibesten på grund af det ru skind. En fladfisk med moderat udbredelse; genkendelig på de brede bånd på finnerne

D: Wird wegen seiner rauhen Haut auch Schleifstein genannt. Ein mässig verbreiteter Plattfisch, der leicht an dem breiten Band auf den Flossen zu erkennen ist

E: También se llama piedra de afilar, debido a su piel rugosa. Pleuronecto de abundancia moderada. Se identifica por las anchas rayas de sus aletas

F: A cause de sa peau rugueuse, ce poisson plat est aussi parfois appelé la meule. On l'identifie à l'aide des larges bandes sur les nageoires

I: Caratteristico per la sua pelle ruvida. Un pleuronettiforme di limitata diffusione, riconoscibile dai larghi nastri sulle pinne

STRIPED BASS

Scientific name:
Morone saxatilis

Synonyms: Rockfish, rock
Family: Percichthyidae — temperate basses
Typical size: 120 cm, 12 kg

There are a few landlocked populations of striped bass, but most are anadromous, swimming into major rivers in spring to spawn. They are voracious feeders and spirited fighters, making them favoured targets for recreational fishermen, who take them with hooks at the surf line.

Pollution has been blamed for the massive fall in the Atlantic population of striped bass. Many estuaries and rivers became unsuitable for spawning, or so stressed the young that few survived. Improved environment and stringent fishing regulations, including a total ban for some years, have helped the species re-establish itself.

COMMON NAMES
D: Felsenbarsch
DK: Stribet bars
E: Lubina americana
F: Bar d'Amérique
GR: Grammotó lavráki
I: Persicospigola striata
NL: Gestreepte zeebaars
P: Robalo muge
RU: Polosaty lavrak
SF: Juovabassi
US: Striped bass

DESCRIPTION
Striped bass are large, streamlined fish with prominent stripes along the sides. The hybrid form (see below) can be distinguished by the breaks in these lines.

Originally, the striper was found from the St. Lawrence to the St. John's River in Florida. Man has helped it extend its range, so it is now fished in the Gulf of Mexico from Florida to Louisiana. Introd-uced to the west coast, it is now found from Canada to California.

HIGHLIGHTS
An anadromous native of the east coast of the United States, the striped bass was successfully introduced to the Pacific coast in the 19th century. It has also been established in South Africa and Russia.

Recently, a hybrid with the freshwater white bass, *Morone chrysops*, was developed. This is being farmed successfully, showing strong growth and food conversion rates. The hybrid is sometimes called the sunshine bass, and is attracting increasing market attention.

Nutrition data:
(100 g edible weight)

Water	79.2 g
Calories	92 kcal
Protein	17.7 g
Total lipid (fat)	2.3 g
Omega-3	0.8 mg

STRIPED BASS *Morone saxatilis*

FISHING METHODS

 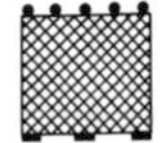 

MOST IMPORTANT FISHING NATIONS
United States

PREPARATION

USED FOR

EATING QUALITIES
Striped bass meat, as usually presented to the consumer, is greyish and soft, with a large flake. It often has an underlying sour taste, which it develops rapidly: this is a species that must be handled well and speedily if it is to be acceptable. Small fish tend to be much better than large ones.

The problems arise from the importance of the species in the United States as a game fish. Anglers sold their catches after displaying them, with the guts still inside, to their friends for several days.

The hybrid variety available from aquaculturists has fine taste and texture. It tends to have whiter meat and is altogether more attractive and palatable. This is due as much to good handling as to the intrinsic nature of the fish.

IMPORTANCE
Catches of wild striped bass are very small. Most of the now limited resources on the east coast of the USA are reserved for recreational fishermen, and there is little interest in consuming the species on the west coast.

Nevertheless, the striped bass remains an important game fish. It is renowned for its fighting abilities and its size (it may reach 25 kg). As stocks recover, more fishing is being permitted.

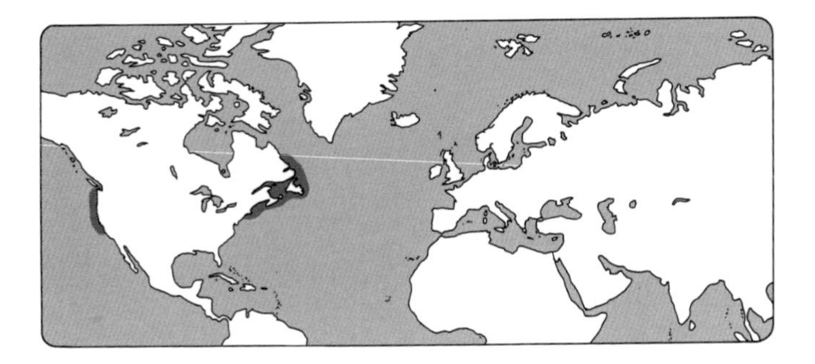

DK: En ny krydsning mellem stribet bars og ferskvandsarten hvid bars, Morone chrysops, opdrættes med et fint resultat. 'Sunshine bass', som den kaldes, vokser hurtigt og er en fremragende spisefisk

D: Eine Kreuzung des Felsenbarsches mit der Süsswasserart Weisser Sägebarsch wird mit guten Ergebnissen gezüchtet. Der sogenannte 'Sunshine bass' wächst schnell und ist ein ausgezeichneter Speisefisch

E: Nuevo cruce entre la lubina rayada y la especie de agua dulce, la lubina blanca, Morone chrysops. Se cría con buen resultado. 'Sunshine bass', como se la llama, crece rápidamente, y es un exquisito pescado comestible

F: Un hybride nouveau entre le bar d'Amérique et le bar blanc d'eau douce, le Morone chrysops, dont l'élevage est un grand succès. Le 'sunshine bass', comme il est appelé, grandit vite, et c'est un excellent poisson de table

I: Un nuovo incrocio tra la persicospigola striata e quella bianca, Morone chrysops, viene allevato con buoni risultati, in quanto cresce rapidamente ed è un ottimo pesce commestibile. L'ibrido viene chiamato 'Sunshine bass'

STURGEON

Scientific name:
Acipenser sturio

Family: Acipenseridae — sturgeons
Typical size: 3 m, 200 kg

This is the anadromous European sturgeon, also known as Baltic, common and Atlantic sturgeon. It was once found throughout Europe, from Norway's North Cape to the northern shores of the Mediterranean and Black Seas, from southern Iceland to northwest Africa. It spawned in all major rivers of this range and there were landlocked populations in many lakes.

Now, only the Danube has enough sturgeon to support a fishery and a very small amount of caviar production. Even the once-abundant population in the Gironde has almost disappeared.

English kings long reserved all sturgeon for their own table. The tradition is still honoured today: on the rare occasions a sturgeon is caught, it is offered to the monarch.

DESCRIPTION
Like all sturgeons, this is a primitive fish which has cartilage, rather than bones for part of its skeleton. The head is covered with bony plates and there are bony protuberances along the back and sides known as scutes or plates.

COMMON NAMES
D: Stör, gemeiner Stör
DK: Stør
E: Esturión
F: Esturgeon
GR: Mouroúna, stourióni
I: Storione
IS: Styrja
J: Chôzame
N: Stør
NL: Steur
P: Esturjao
RU: Osetr
S: Stör
SF: Sampi
TR: Kolan
US: Sturgeon

LOCAL NAMES
PO: Jesiotr

HIGHLIGHTS
The sturgeon, once an important species throughout the European continent, faced extinction. It is still struggling to survive as a species, but appears to be on the way back.

A huge, long-lived and slow growing fish, this species has not yet been domesticated as readily as some other sturgeon species. Other sturgeons are now farmed on an increasing scale in Asia, Europe and North America for meat and are experimentally producing farmed caviar as well.

Nutrition data:
(100 g edible weight)

Water	79.0 g
Calories	94 kcal
Protein	16.3 g
Total lipid (fat)	3.2 g
Omega-3	0.2 mg

STURGEON *Acipenser sturio*

FISHING METHODS

MOST IMPORTANT FISHING NATIONS
Rumania, France

PREPARATION
Caviar

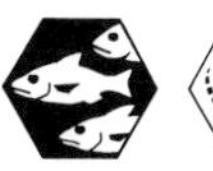 

USED FOR

EATING QUALITIES
Sturgeon meat is firm and meaty, resembling veal in texture. It is pinkish in colour, although it whitens when cooked. It is excellent smoked. The Atlantic sturgeon of North America (*A. oxyrhynchus*) was harvested in large quantities in the nineteenth century as a cheap alternative to meat; it was popularly known as "Albany beef" after the Hudson River city in New York which supplied a great deal of the fish.

Despite the high quality of the meat, sturgeons are valued more for their roe, which when salted is best known as caviar. The European sturgeon has grey eggs which are fairly large, generally between 2 and 2.5 mm.

The European sturgeon produces between 200,000 and 6 million eggs, with the numbers increasing as the fish gets older. A fish of 150 kg might have 4 million eggs. A fish of 100 kg is reported to yield 50 kg of meat and 12 kg of caviar. Note that the fish is between 8 and 14 years old at first spawning, with a length of 120 to 180 cm. The species may live as long as 100 years and grow as large as 300 kg.

IMPORTANCE
The European sturgeon is no longer economically important, although there are some signs that it is slowly returning to some rivers as pollution decreases. Fishing for all but the largest specimens is almost universally banned, allowing the stocks a chance of recovery.

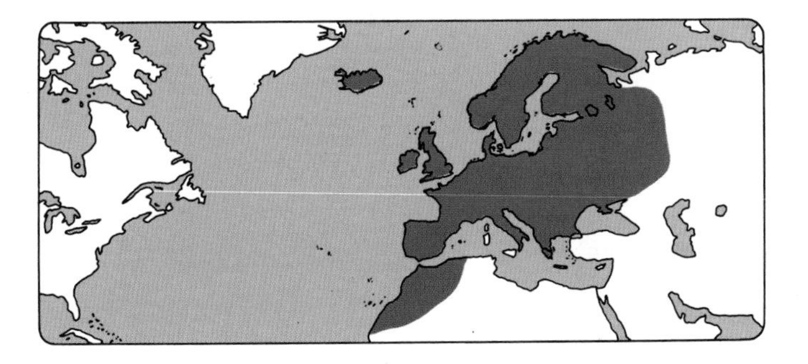

DK: Dårlige betingelser for opdræt af denne langsomtvoksende fisk. Andre størarter opdrættes til gengæld med et godt resultat, både hvad angår kød og kaviar

D: Die Zuchtbedingungen für diesen langsam wachsenden Fisch sind schlecht. Dafür werden andere Störarten mit gutem Ergebnis gezüchtet, was das Fleisch und den Kaviar betrifft

E: Son malas las condiciones de cría de este pescado, que crece lentamente. Otros esturiones, en cambio, se crían con buen resultado, tanto en cuanto a la carne como al caviar

F: Les conditions d'élevage de ce poisson de croissance lente sont assez mauvaises. Par contre, des esturgeons d'autres espèces sont élevés avec de bons résultats, tant en ce qui concerne la chair que le caviar

I: Questo pesce a crescita lenta si presta poco all'allevamento. Altri storioni invece vengono allevati con buoni risultati, sia per le carni sia per il caviale

SURMULLET/RED MULLET

Scientific name:
Mullus surmuletus

Synonym: Woodcock of the sea
Family: Mullidae — goatfishes
Typical size: 20 to 25 cm, up to 40 cm

French fishermen in Brittany and Gascony, as well as along the Mediterranean coast, target this species. The catch is all sold fresh, either whole or dressed. Supplies from more distant parts of the Mediterranean are shipped by air to French and Spanish markets.

FISHING METHODS

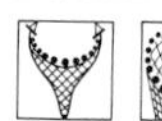 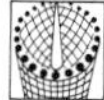

MOST IMPORTANT FISHING NATIONS
Turkey, Greece, France, Spain

PREPARATION

USED FOR

EATING QUALITIES
Red mullet is regarded as one of the highlights of Mediterranean and European fish cuisine. The flesh is delicate, moist and white. It is versatile, usable in most cooking methods. There are many classical recipes, especially in France and Spain, based on red mullet.

COMMON NAMES
D: Gestreifte Meerbarbe
DK: Gulstribet mulle
E: Salmonete de roca
F: Rouget barbet de roche
GR: Barboúni
I: Triglia di scoglio
IS: Sæskeggur
N: Mulle
NL: Mul, koning van de poon
P: Salmonete legítimo
RU: Polosataya barabulya
S: Gulstrimmig mullus
SF: Keltajuovamullo
TR: Tekir
US: Red mullet

LOCAL NAMES
TU: Mellou
EG: Barboúni
IL: Mulit happassim
MO: Rouget

Nutrition data:
(100 g edible weight)

Water	72.7 g	Total lipid	
Calories	109 kcal	(fat)	3.8 g
Protein	18.7 g	Omega-3	0.8 mg

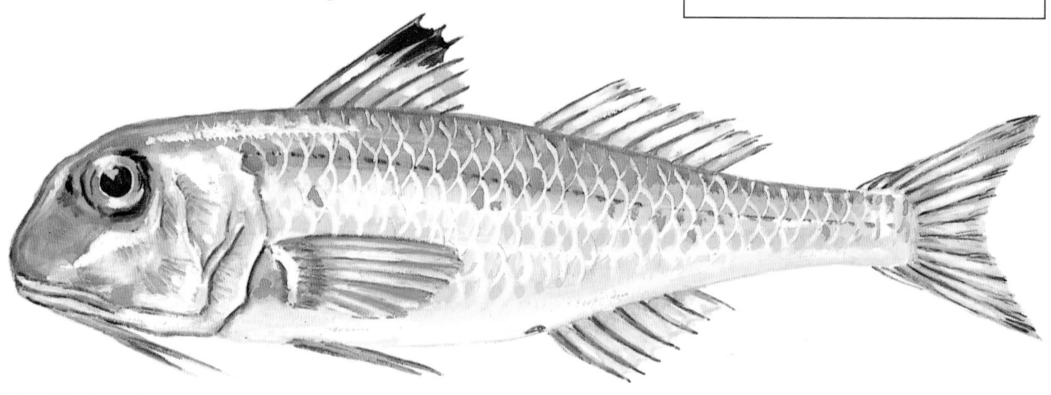

HIGHLIGHTS
Brightly coloured with pinkish flesh which cooks pearly white and contains few bones, the red mullet is highly prized in Europe. It is notable that similar species in north American waters are unexploited for lack of local demand.

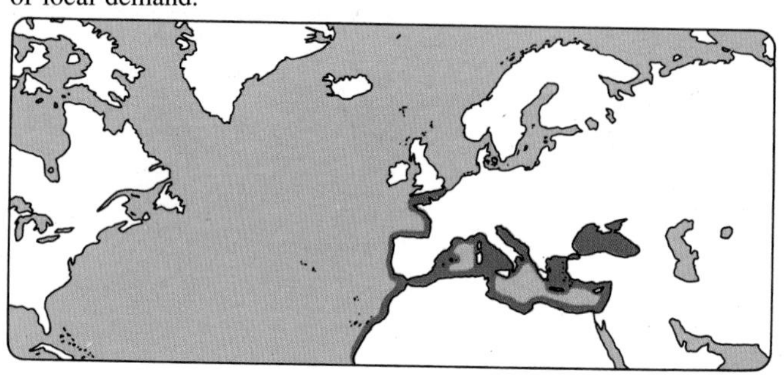

DK: Hvor gulstribet mulle er en højt estimeret spisefisk i Europa, er lignende nordamerikanske arter underfisket på grund af manglende lokal efterspørgsel

D: Während die Gestreifte Meerbarbe ein geschätzter Speisefisch in Europa ist, bleiben ähnliche nordamerikanische Arten wegen dort fehlender Nachfrage ungenutzt

E: El salmonete es muy apreciado en Europa, mientras que especies similares de Norteamérica no son aprovechadas porque no hay demanda local

F: Si ce rouget est un poisson de table très apprécié en Europe, les espèces similaires en Amérique du Nord sont sous-exploitées à défaut de demandes locales

I: Mentre la triglia di scoglio è un pesce commestibile altamente apprezzato in Europa, le simili specie dell'America Settentrionale sono poco richieste

TADPOLE MORA

Scientific name:
Salilota australis

Synonyms: Patagonian rockcod, tadpole codling
Family: Moridae — codlings or morid cods
Typical size: 50 cm

Called tadpole codling by FAO, this small greenish-brown species is trawled in Antarctic and neighbouring waters. It is mostly used for fishmeal, because of its small size.

Recorded catches are insignificant, but it is thought that the resource could support substantial fishing.

COMMON NAMES

E: Mora renacuajo
F: More têtard
J: Sarirota, akadara
RU: Solilota

LOCAL NAMES

AR: Bacalao austral
CL: Renacuajo de mar

FISHING METHODS

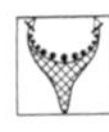

MOST IMPORTANT FISHING NATIONS

Falkland Islands

PREPARATION

Fishmeal

USED FOR

EATING QUALITIES

Small fillets machine-processed on factory trawlers are reported to be acceptable for breading and other uses which add coatings or sauces.

Nutrition data:
(100 g edible weight)

Water	81.0 g	Total lipid	
Calories	73 kcal	(fat)	0.6 g
Protein	16.9 g	Omega-3	0.3 mg

HIGHLIGHTS

The tadpole mora is a small relative of the cod, found in Antarctic regions and as far north as the Falkland Islands. It is used for fishmeal, but can also be filleted for blocks or other processing needs. Because of its small size, filleting is difficult and wasteful.

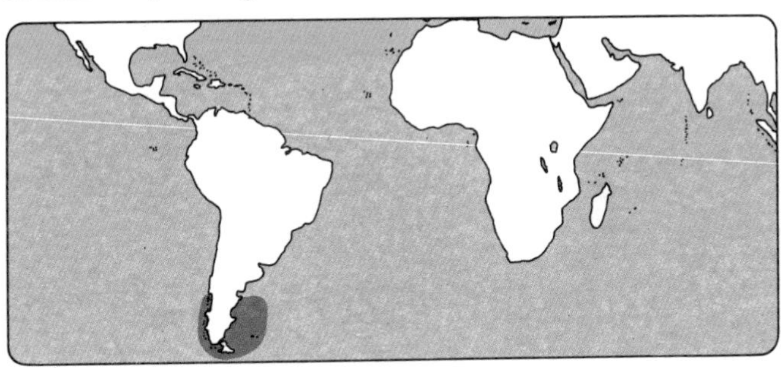

DK: En lille antarktisk torskeart, som anvendes til primært fiskemel, men også filetter. Formentlig væsentlige, uudnyttede ressourcer

D: Eine kleine antarktische Dorschart, die hauptsächlich zu Fischmehl verarbeitet wird, aber auch zu Filets. Vermutlich gibt es bedeutende ungenutzte Bestände

E: Pequeña especie antártica de bacalao. Sirve ante todo para harina de pescado, pero también para filetes. Probablemente hay grandes existencias sin aprovechar

F: Petite espèce de morue antarctique surtout utilisée pour farine mais aussi préparée en filets. Les ressources non exploitées sont probablement substantielles

I: Piccola specie antartica di merluzzo. Usato prevalentemente per farina di pesce ma anche per filetti. Probabilmente le risorse non sfruttate sono notevoli

TAIMEN

Scientific name:
Hucho taimen

Synonym: Danube salmon
Family: Salmonidae — salmonids
Typical size: 10 kg

The taimen is a freshwater species, widely distributed in rivers and lakes of northern Asia between the Volga and the Pacific. It is closely related to the Danube basin huchen (*Hucho hucho*), which is highly valued as a game fish by anglers. The taimen is locally important for food, taken in nets. Its range is somewhat inaccessible for western recreational fishermen.

COMMON NAMES
D: Huchen, sibirischer Huchen
DK: Taimen laks
E: Salmón del Danubio
F: Huchon, saumon du danube
I: Salmone del danubio
N: Taimen laks
NL: Donauzalm
P: Salmao do danúbio
RU: Taimen
S: Danube
SF: Jokinieriä
TR: Alabalik türü
US: Taimen

FISHING METHODS

MOST IMPORTANT FISHING NATIONS
Russia, China

PREPARATION

USED FOR

EATING QUALITIES
The taimen has fine quality, salmon-like meat with a pale colour. It can be smoked or salted.

Nutrition data:
(100 g edible weight)

Water	76.7 g	Total lipid	
Calories	125 kcal	(fat)	5.2 g
Protein	19.6 g	Omega-3	1.0 mg

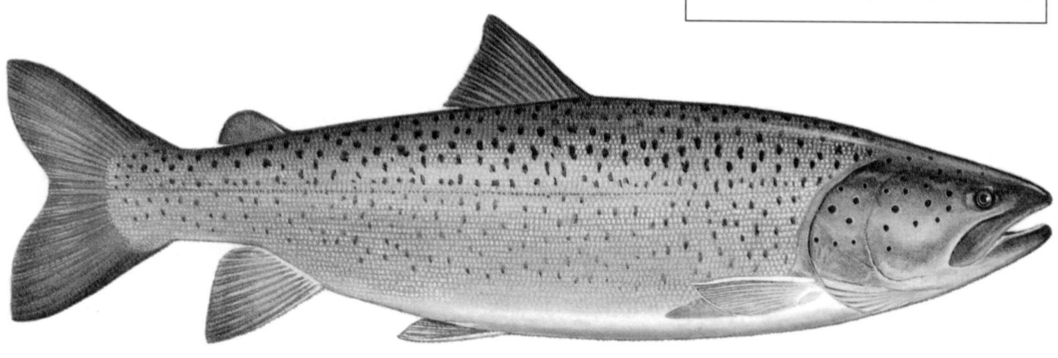

HIGHLIGHTS
The taimen is the largest salmonid, sometimes reaching 90 kg. In the Amur River on the border of Siberia and China it regularly reaches 45 kg. Similar to the lake trout of North America, it is now being farmed experimentally.

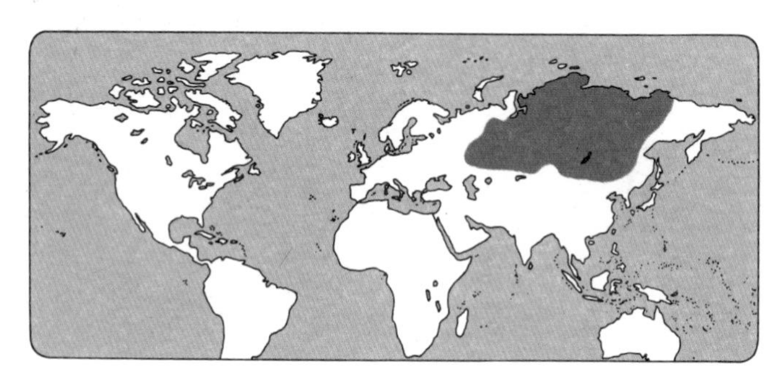

DK: Den største lakseart med eksemplarer på indtil 90 kg. Opdrættes på forsøgsbasis. En populær sportsfisk

D: Die grösste Lachsart mit Exemplaren von bis zu 90 kg. Aufzuchtversuche. Beliebt bei Sportfischern

E: Es la especie más grande de los salmones, con ejemplares de hasta 90 kg. Se cría experimentalmente. Popular pescado deportivo

F: La plus grande espèce des saumons avec des exemplaires qui pèsent jusqu'à 90 kg. Elle fait l'objet d'élevage à titre d'essai. Poisson sportif très populaire

I: La specie più grande fra i salmoni con esemplari che raggiungono i 90 kg. Allevato su base sperimentale. Un pesce sportivo popolare

TENCH

Scientific name:
Tinca tinca

Family: Cyprinidae — carps and minnows
Typical size: up to 64 cm, 5 kg

Tench are found in still lakes and slow-moving rivers throughout Eurasia; they have been introduced to North America, Australia and New Zealand. They can survive low oxygen and live out of water, protected by thick mucus on their bodies, for a long time.

FISHING METHODS

MOST IMPORTANT FISHING NATIONS
France, Czech Republic, Hungary, Spain

PREPARATION

USED FOR

EATING QUALITIES
Tench has tasty meat with an off-white colour. It is similar to carp in its uses, being good fried, baked or grilled. Some observers state that it can develop a muddy or sour taste in certain waters. It is especially popular in Germany and eastern Europe.

COMMON NAMES
D: Schlei, Schuster
DK: Suder
E: Tenca, tinca aguijón
F: Tanche, tenca, tiche
GR: Glíni
I: Tinca
IS: Grunnungur
N: Suter, sudre
NL: Zeelt, lauw, muithond
P: Tenca, godiao
RU: Lin
S: Sutare, lindare
SF: Suutari
TR: Kadife baligi, yesil sazan
US: Tench

Nutrition data:
(100 g edible weight)

Water	80.0 g	Total lipid (fat)	0.8 g
Calories	78 kcal		
Protein	17.7 g	Omega-3	0.2 mg

HIGHLIGHTS
Tench have been farmed in Europe for even longer than carp. Most of the production now is from polyculture with the more valuable carp, at the rate of one tench to ten carp to maximise use of the pond and the feed.

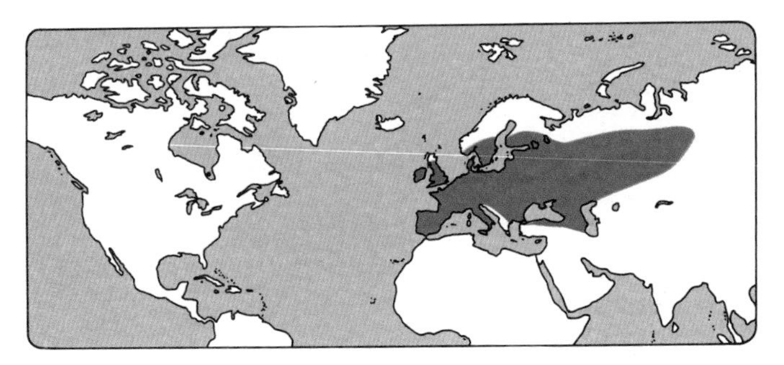

DK: Suder har været opdrættet endnu længere end den værdifulde karpe, men for at forøge udnyttelsesgraden indgår begge arter idag i et kombineret opdræt

D: Die Zucht der Schleie reicht länger zurück als die der wertvollen Karpfen, aber um den Ausnutzungsgrad zu verbessern, wird die Zucht beider Arten kombiniert

E: La tenca ha sido criada más tiempo que la valiosa carpa, pero para aumentar el rendimiento hoy, forman ambas especies parte de una cría combinada

F: La tanche a été cultivée plus longtemps que la précieuse carpe, mais afin d'augmenter le rendement, les 2 espèces sont maintenant élevées en polyculture

I: La tinca è stata allevata ancora più a lungo della preziosa carpa, ma per aumentare la produttività si fa un allevamento combinato delle due specie

TILEFISH

Scientific name:
Lopholatilus chamaeleonticeps

Synonym: Great northern tilefish
Family: Malacanthidae — tilefishes
Typical size: up to 120 cm, 30 kg

Tilefish inhabit deep water from Nova Scotia to the Gulf of Mexico. Specimens are occasionally seen off Venezuela and Surinam. They are sensitive to cold water: the species was almost wiped out off New York in 1882 when part of the Labrador Current temproarily moved south over their habitat.

FISHING METHODS

MOST IMPORTANT FISHING NATIONS
United States

PREPARATION

USED FOR

EATING QUALITIES
Tilefish has moist, white meat with a large flake. It has a mild flavour and the bones are easily removed. It is especially popular in the mid-Atlantic states. Most fish available commercially are under 10 kg.

COMMON NAMES
D: Blauer Ziegelbarsch
DK: Teglfisk
E: Blanquillo camello
F: Tile chameau
GR: Plakolepidópsaro
I: Tile gibboso
IS: Flögufiskur
J: Amadai
NL: Blauwe tegelvis
P: Peixe paleta camelo
RU: Grebnegolov, tailfish
SF: Tiilikala
US: Tilefish

LOCAL NAMES
CH: Ma tau

Nutrition data:
(100 g edible weight)

Water	78.9 g	Total lipid	
Calories	91 kcal	(fat)	2.3 g
Protein	17.5 g	Omega-3	0.4 mg

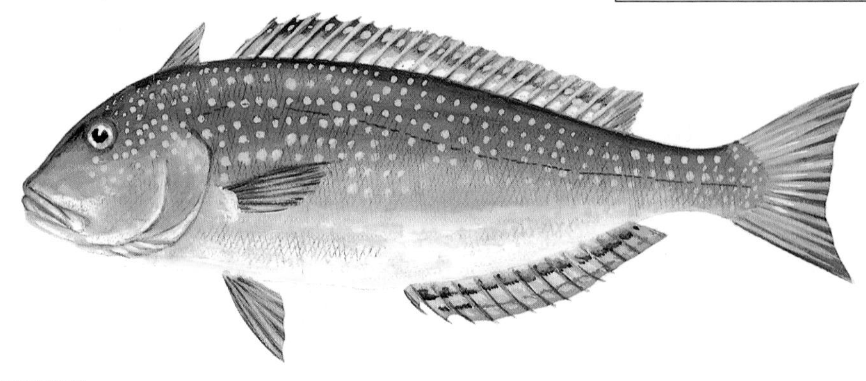

HIGHLIGHTS
Tilefish are large and colourful, attracting recreational fishermen despite the difficulties of angling the species from the deep water it prefers. The commercial fishery suffers considerable fluctuations in the resource. Demand also appears to vary erratically.

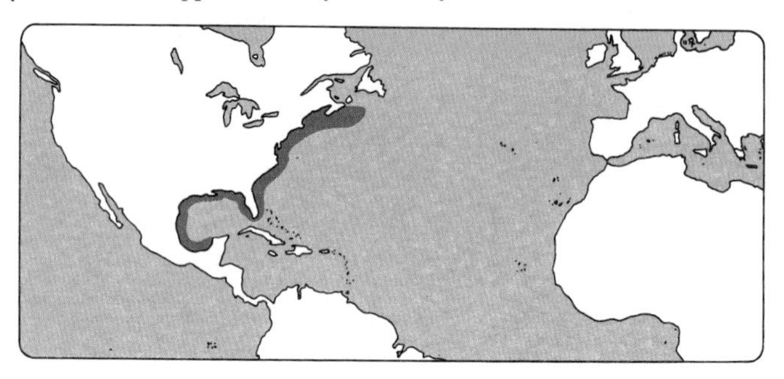

DK: Omend svært tilgængelig, er den store og farverige dybhavsfisk alligevel populær hos lystfiskere

D: Wenn auch schwer zugänglich, so ist der grosse und farbenprächtige Tiefseefisch trotzdem beliebt bei Sportfischern

E: Es un pescado grande y rico en colores, que vive en aguas profundas. Aunque es difícil de capturar, es popular entre los pescadores deportivos

F: Bien que difficilement accessible, ce grand poisson multicolore des grandes profondeurs est très populaire chez les pêcheurs sportifs

I: Questo pesce abissale, grande e colorito, è popolare tra i pescatori sportivi anche se è difficile da prendere

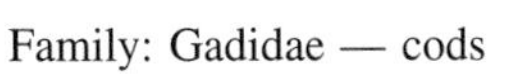

TORSK/TUSK/CUSK

Scientific name:
Brosme brosme

Family: Gadidae — cods
Typical size: 60 to 85 cm, 8 kg

This species lives in cold northern waters, between 0^0 and 10^0C. Norway produces 75 percent of the average catch and uses much of it for superior quality saltfish. The fish is slow growing; it may be 8 to 10 years old when it first spawns.

FISHING METHODS

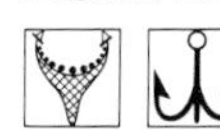

MOST IMPORTANT FISHING NATIONS
Norway, France, Iceland, Canada, United States

PREPARATION

USED FOR

EATING QUALITIES
Cusk provides an excellent firm, white meat, versatile for many uses. Line-caught fish is better than trawled; it is often used for saltfish, which requires top quality raw material without bruises or other blemishes.

COMMON NAMES
D: Lumb, Torsk, Brosme
DK: Brosme
E: Brosmio
F: Brosme, assiette, tusk
GR: Brósmios
I: Brosmio
IS: Keila
J: Atsukawadara, torusuku
N: Brosme
NL: Lom, torsk, lomp
P: Bolota
RU: Menjok
S: Lubb, brosme
SF: Keila
US: Cusk

Nutrition data:
(100 g edible weight)

Water	76.4 g	Total lipid	
Calories	82 kcal	(fat)	0.7 g
Protein	19.0 g	Omega-3	0.1 mg

HIGHLIGHTS
Torsk or cusk is a top quality cod-like fish. Catches have declined in recent years, partly because longlining has given way to trawling and these fish live on rough bottoms unsuitable for trawl nets, and partly because of increased predation by seals.

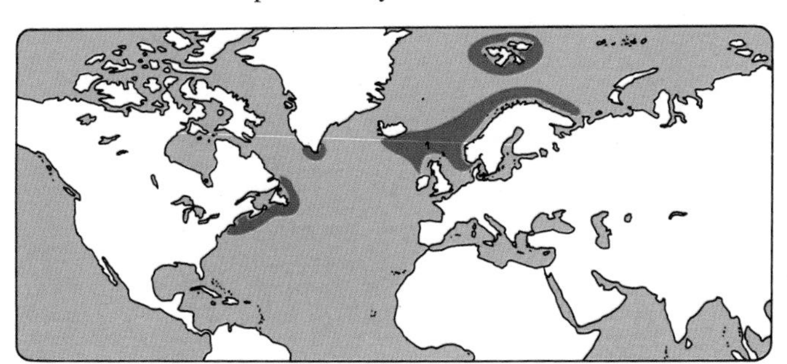

DK: En fremragende bundfisk, meget lig torsk. Sælernes rovdrift og det faktum, at den er vanskelig at fange med trawl, betyder faldende fangster

D: Ein ausgezeichneter, dorschähnlicher Bodenfisch. Auf Grund des Raubbaus durch Robben und weil er schwierig mit Schleppnetzen zu fangen ist, sinken die Fänge

E: Exquisito pescado de fondo, similar al bacalao. Las capturas han bajado porque es difícil capturarlo con red, y por las bajas causadas por las focas

F: Excellent poisson des fonds qui ressemble beaucoup au cabillaud. Dévoré par les phoques et du fait qu'il est difficile à pêcher au chalut, la pêche est en déclin

I: Un pesce di fondo squisito simile al merluzzo. Essendo preda delle foghe e essendo difficile da pescare con la rete a strascico, le catture stanno diminuendo

TURBOT

Scientific name:
Colistium nudipinnis

Family: Pleuronectidae — right-eye flounders
Typical size: 25 to 45 cm, 1 to 1.5 kg

COMMON NAMES
RU: Afrikanskij psettod
US: Flounder, brill

New Zealand turbot is not much like the better known and more valuable European turbot, (see Turbot, *Psetta maximus*). A smallish right eyed flounder which occasionally reaches 90 cm, this southern species is restricted to the west coast of New Zealand.

FISHING METHODS

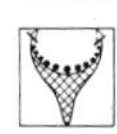

MOST IMPORTANT FISHING NATIONS
New Zealand

PREPARATION

USED FOR

EATING QUALITIES
The New Zealand turbot has lightly textured meat which is peach coloured when raw, but turns white when cooked. The flavour is light and delicate. A by-catch of the trawl fishery in shallow water, supplies are irregular and limited.

Nutrition data:
(100 g edible weight)

Water	78.7 g	Total lipid	
Calories	95 kcal	(fat)	2.7 g
Protein	17.7 g	Omega-3	0.5 mg

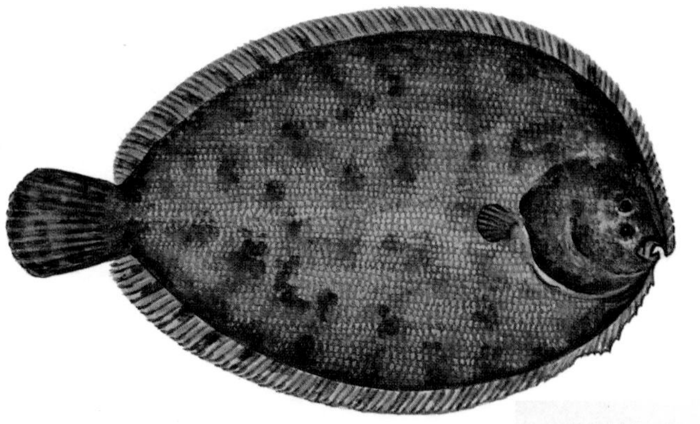

HIGHLIGHTS
Delicately flavoured but lacking the satisfying, meaty quality of European turbot, it is not clear how this uncommon species was given the name of a prime European fish. It is not exported as quantities available are small.

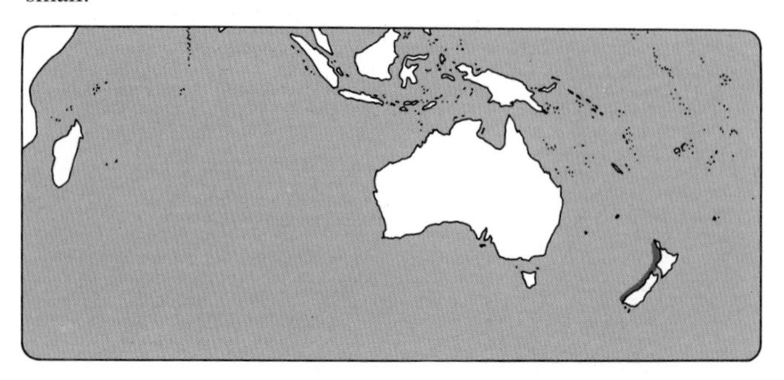

DK: Det er uvist hvorfra denne sjældne art har fået sit navn. Fisken har en fin smag, men kvaliteten er ikke så god som hos den europæiske art

D: Es ist ungewiss, woher diese seltene Art ihren Namen erhalten hat. Sie ist schmackhaft, aber von der Qualität her nicht so gut wie die europäische Art

E: No se sabe muy bien de dónde proviene el nombre de esta especie poco frecuente. Tiene sabor exquisito, pero la calidad no es tan buena como la especie europea

F: Personne ne sait exactement d'où cette espèce rare tient son nom. Elle a très bon goût mais la qualité n'égale pas celle de l'espèce européenne

I: Il nome di questa specie rara è di origine ignota. Ha un sapore squisito, ma la qualità è un po' meno buona di quella della specie europea

TURBOT

Scientific name:
Psetta maxima

Synonyms: Brett, butt, britt
Family: Bothidae — left-eye flounders
Typical size: 40 to 50 cm, up to 1 m

Turbot is a large, inshore flatfish found throughout the Mediterranean north to England and Ireland. Small populations are found further north, in Norway and Iceland.

Fast growing in warm water, turbot may reach 25 kg.

FISHING METHODS

MOST IMPORTANT FISHING NATIONS
Turkey, Netherlands, Denmark

PREPARATION

USED FOR

EATING QUALITIES
Turbot has very firm, gleaming white flesh with excellent flavour. It provides thick, meaty portions. It is one of the best of all flatfish and is considered a gourmet's fish by many experts.

COMMON NAMES
D: Steinbutt, Dornbutt
DK: Pighvarre
E: Rodaballo, parracho, rémol
F: Turbot
GR: Kalkáni, siáki
I: Rombo chiodato
IS: Sandhverfa
N: Piggvar
NL: Tarbot
P: Pregado
RU: Psettod
S: Piggvar
SF: Piikkikampela
TR: Kalkan baligi
US: Turbot

LOCAL NAMES
TU: Syaks

Nutrition data:
(100 g edible weight)

Water	78.7 g	Total lipid	
Calories	95 kcal	(fat)	2.7 g
Protein	17.7 g	Omega-3	0.5 mg

HIGHLIGHTS
Turbot are highly prized and expensive, esteemed for their fine flavour and texture. Catches averaging 10,000 tons a year do not keep the market fully supplied. Farmers are now trying to fill the gap, producing turbot in many different areas.

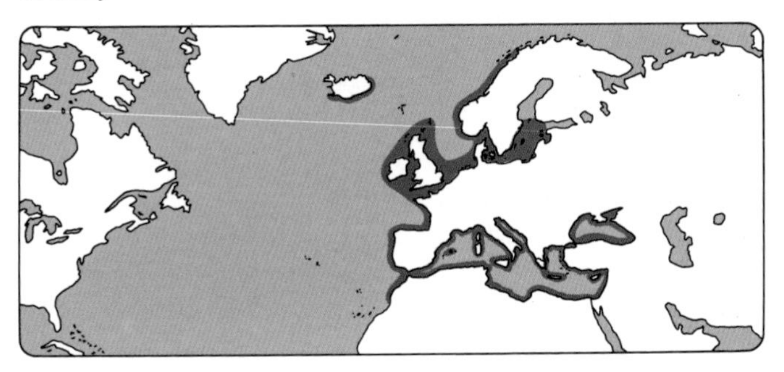

DK: En dyr og højt værdsat fisk med fin smag og konsistens. For at dække efterspørgslen opdrættes arten i stigende grad

D: Ein teurer, sehr geschätzter, schmackhafter Fisch mit guter Konsistenz. Um der Nachfrage gerecht zu werden, wird die Art zunehmend gezüchtet

E: Pescado caro y muy apreciado, de exquisito sabor y consistencia. Para satisfacer la demanda, se está intensificando mucho la cría de esta especie

F: Le turbot est un poisson cher et très apprécié pour sa chair délicate. Pour satisfaire à la demande, on commence a en faire un élevage intensif

I: Un pesce caro e altamente apprezzato di consistenza e sapore squisiti. Per coprire la richiesta si ricorre sempre di più all'allevamento

WALLEYE

Scientific name:
Stizostedion vitreum

Family: Percidae — perches
Typical size: 90 cm; up to 11 kg

The walleye is closely related to the European zander and very similar to it in eating characteristics. Populations in different lakes show wide variations in colour. It is also very similar to its smaller relative, the sauger (*S. canadense*). Walleye may be infested with tapeworm, which is harmful to man, and to black-spot and yellow grub parasites under the skin which are not harmful, but unsightly. It should be properly cooked to ensure any parasites are killed before the fish is eaten.

DESCRIPTION

The walleye prefers large, shallow lakes with high turbidity; in clear water, they restrict feeding to night-time, when their sensitive eyes are not disturbed by too much light.

Walleye is now found from Quebec to Alabama, through most of the eastern United States and Canada to the Mackenzie River and James Bay. The original range of the species has been greatly extended by man since it was successfully introduced into New York as early as 1830. Although not widely farmed for commercial sale as a food fish, large numbers are hatched and raised for stocking lakes for anglers.

COMMON NAMES

D: Amerikanischer Zander
DK: Hvidøjet sandart
E: Lucioperca americana
F: Doré jaune
GR: Potamolávrako tis Amerikís
I: Sandra americana
IS: Kanavidnir
NL: Amerikaanse snoekbaars
P: Picao verde
SF: Valkosilmäkuha
US: Walleye, yellow pike

HIGHLIGHTS

The walleye derives its name from its glassy eyes, which may mislead consumers into thinking that the fish is stale because the eyes do not look fresh. These eyes have developed to enable the fish to hunt at night in the gloom of lake bottoms.

Almost exclusively a recreational species in the United States, Canadian fishermen in the far north are still able to supply markets with substantial quantities of processed yellow pike, which is especially popular in the mid-sections of both Canada and the USA.

Nutrition data:
(100 g edible weight)

Water	79.3 g
Calories	87 kcal
Protein	19.1 g
Total lipid (fat)	1.2 g
Omega-3	0.3 mg

WALLEYE

Stizostedion vitreum

FISHING METHODS

  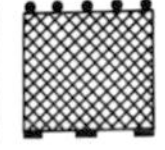 

MOST IMPORTANT FISHING NATIONS

Canada, United States

PREPARATION

USED FOR

EATING QUALITIES

Like its close relative, the zander, *S. lucioperca* (see that entry), the walleye is one of the finest of all fish. It has clean, white meat with a small flake and excellent, sweet flavour. It is adaptable to any cooking method and can be used for any of the classic recipes for sole, although its flavour is different. It is regarded as at least the equal of trout among freshwater fish.

Best known in the United Sates as yellow pike, although it is not a pike, walleye is also used in gefilte fish and for quenelles.

Walleye is available commercially in numerous forms: dressed, pan-ready, as fillets with or without skin and bones, minced blocks and ready breaded. The fillets have few bones, adding to the popularity of the fish, which has excellent keeping qualities, both fresh and frozen. It has a comparatively small visceral cavity, which means that the yield of meat from the carcass is quite high.

IMPORTANCE

Walleye is probably the most valuable freshwater fish in Canada. It is caught, mainly in winter, with gillnets set under the ice of frozen lakes. Even more important, in both Canada and the USA, is its recreational value. The pressure of game fishermen and their desire to catch the species has pushed commercial fisheriess into the far north and the largest lakes.

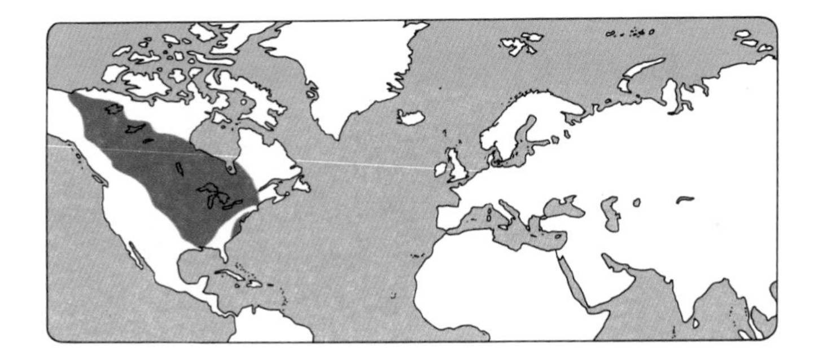

DK: Fisken har fået navn efter de glasagtige øjne, som fejlagtigt giver forbrugeren det indtryk, at fisken ikke er frisk. Øjnene gør det muligt for fisken at jage om natten og i uklart vand

D: Der Fisch ist nach den glasigen Augen benannt worden, die dem Verbraucher den falschen Eindruck vermitteln, der Fisch sei nicht frisch. Die Augen ermöglichen es ihm, nachts und in trübem Wasser zu jagen

E: El pescado debe su nombre a sus ojos vítreos que pueden dar al consumidor la impresión errónea de que no sea fresco. Gracias a estos ojos puede ir de caza de noche y en agua turbia

F: Le poisson a été dénommé d'après ses yeux vitreux qui donnent l'impression au consommateur que le poisson n'est pas frais. Le yeux permettent au poisson de chasser sa proie la nuit et dans des eaux troubles

I: Il nome di questo pesce deriva dagli occhi vitrei che danno ai consumatori l'impressione sbagliata che il pesce non sia fresco. Gli occhi gli permettono di cercare il cibo di notte e in acqua torbida

WHITEBAIT

Scientific name:
Galaxias maculatus

Family: Galaxiidae — galaxiids
Typical size: 5 cm, 0.45 gm

Different from European whitebait, which are young herring and sprats, galaxiids are found in the southern hemisphere. This species is catadromous, although there are also landlocked populations. Most species grow to less than 20 cm.

FISHING METHODS

MOST IMPORTANT FISHING NATIONS
New Zealand

PREPARATION

USED FOR

EATING QUALITIES
These tiny fish are eaten whole. The flesh is delicate and lightly flavoured. Normally, they are mixed with seasoned flour and fried quickly to crisp the outside. Most of the catch is recreational, though there is a limited commercial fishery in New Zealand.

COMMON NAMES
I: Bianchetti
J: Shirasu
P: Puf
RU: Snetok
SF: Täplämeltti

LOCAL NAMES
AU: Jollytail
NZ: Inanga

Nutrition data:
(100 g edible weight)

Water	78.8 g	Total lipid	
Calories	92 kcal	(fat)	2.4 g
Protein	17.6 g	Omega-3	0.7 mg

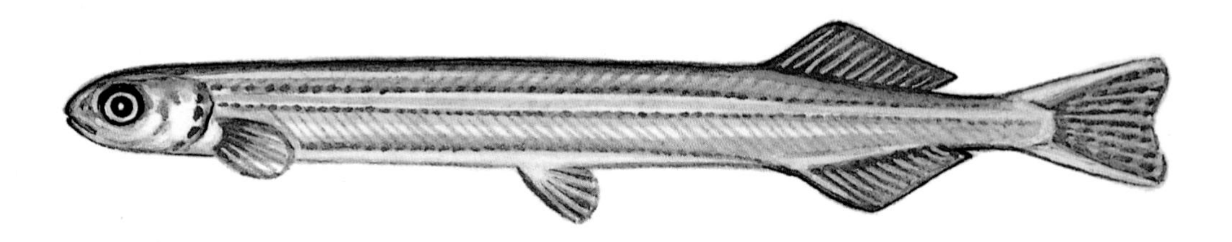

HIGHLIGHTS
Whitebait are the young of a tiny fish (which reaches only 16 cm), caught and eaten as they migrate into rivers from the sea when about six months old. *G maculatus* is one of a number of similar species which support popular sport fisheries in New Zealand and Tasmania.

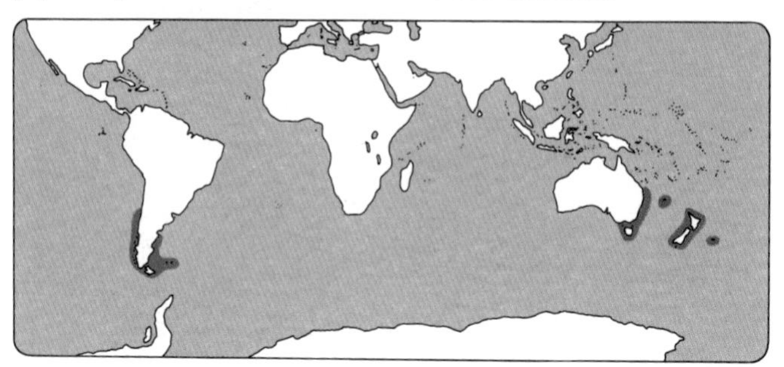

DK: Unger af en lillebitte fisk, som kun bliver 16 cm lang. De fiskes når de, ca. 6 mdr. gamle, vandrer fra havet op ad floderne

D: Die Jungen eines sehr kleinen Fisches, der nur 16 cm lang wird. Sie werden gefangen, wenn sie, ca. 6 Monate alt, vom Meer aus die Flüsse hinaufsteigen

E: Alevines de un pescado pequeñísimo, que sólo llega a tener 16 cm de largo. Al tener unos 6 meses, son capturados cuando migran del mar a los ríos

F: Les jeunes d'un tout petit poisson qui devient seulement 16 cm de long. Ils sont pêchés lorsque, vers 6 mois, ils quittent la mer pour monter dans les rivières

I: I giovani di questo piccolo pesce, che diventa lungo soltanto 16 cm, sono pescati quando all'età di 6 mesi, lasciano il mare per risalire i fiumi

WHITE HAKE

Synonym: Mud hake
Family: Gadidae — cods
Typical size: 70 cm, up to 120 cm

White hake is caught in the Gulf of Maine and off Newfoundland, often when it is moving inshore to spawn. It is not highly regarded in North America, but has more value exported to Europe, where this and other hakes are better appreciated.

FISHING METHODS

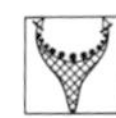  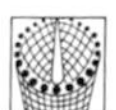 

MOST IMPORTANT FISHING NATIONS

Canada, United States

PREPARATION

USED FOR

EATING QUALITIES

White hake has delicate, white, sweet meat but soft texture and a tendency to fall apart. Provided the skin is kept on, it works well for a wide variety of recipes and dressed fish can be smoked. It does not keep well and must be handled quickly.

Scientific name:
Urophycis tenuis

COMMON NAMES

D: Weisser Gabeldorsch
DK: Hvid skægbrosme
E: Locha blanca
F: Merluche blanche
GR: Lefkós bakaliáros
I: Musdea americana
IS: Stóra brosma
NL: Witte heek
P: Abrótea branca
RU: Belyj nalim
US: White hake

Nutrition data:
(100 g edible weight)

Water	82.1 g	Total lipid	
Calories	68 kcal	(fat)	0.4 g
Protein	16.1 g	Omega-3	0.1 mg

HIGHLIGHTS

Landings average around 20,000 tons, mostly from Canada. Larger fish are filleted, smaller ones sold headless dressed and the smallest fish are used for animal feed. Fresh white hake is now being shipped by air to Europe from the United States.

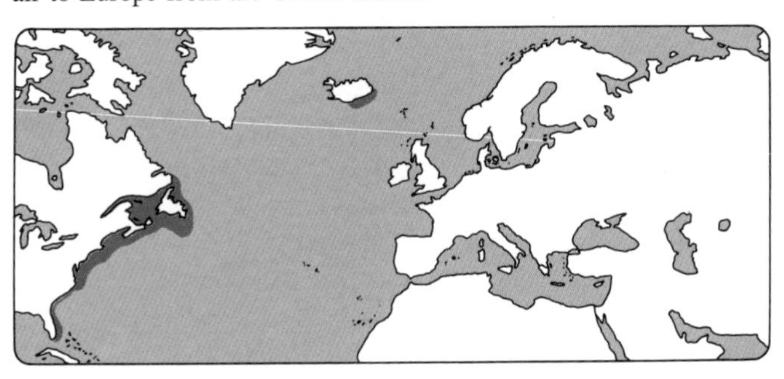

DK: Hver størrelse sit formål: enten som filet, hele renset eller til fremstilling af fiskemel. Den ferske fisk sendes med fly til Europa

D: Jede Grösse dient ihrem Zweck: als Filet, im ganzen Stück gesäubert oder zur Herstellung von Fischmehl. Beförderung von Frischfisch per Flugzeug nach Europa

E: Según su tamaño, este pescado es consumido en filetes, entero, o usado para la producción de harina de pescado. El pescado fresco es enviado por avión a Europa

F: En fonction de la taille, il est utilisé pour filets, entier, ou pour la production de farine de poisson. Le poisson frais est expédié par avion vers l'Europe

I: A seconda della grandezza, questo pesce è usato per filetti, intero o per la produzione di farina di pesce. Il pesce fresco è spedito in aereo in Europa

WHITE STURGEON

Scientific name:
Acipenser transmontanus

Family: Acipenseridae — sturgeons
Typical size: Up to 6 m, 680 kg

Rare in the wild, the white sturgeon is now readily available to consumers thanks to successful farming of the species in North America. Farmed white sturgeon are generally harvested smaller than 10 kg. The bony scutes, also called plates or buttons, are removed and the carcasses sold headless and dressed, or prepared as steaks, portions or slabs. The harvest is still small and is sold fresh, often distributed by air, or smoked, which is a particularly palatable and tasty product.

COMMON NAMES
D: Weisser Stör
DK: Hvid stør
E: Esturión blanco
F: Esturgeon blanc
GR: Asprostoúriono
I: Storione bianco
IS: Hvítstyrja
NL: Witte steur
P: Esturjao branco
RU: Bely osetr
SF: Valkosampi
US: White sturgeon

DESCRIPTION

The white sturgeon is an anadromous fish which now spawns in rivers from the Sacramento north to the Skeena. It is also found in Gray's Harbor and Willapa Bay, Washington. Many individuals of this species spend all their lives in fresh water; they may migrate up and down the river seasonally, but do not enter the ocean. White sturgeon have 28 to 44 bony scutes or plates along each side.

This is the largest fish species found in inland waters of North America. Sturgeon live to great ages: some experts estimate that the largest white sturgeon may be 80 or more years old.

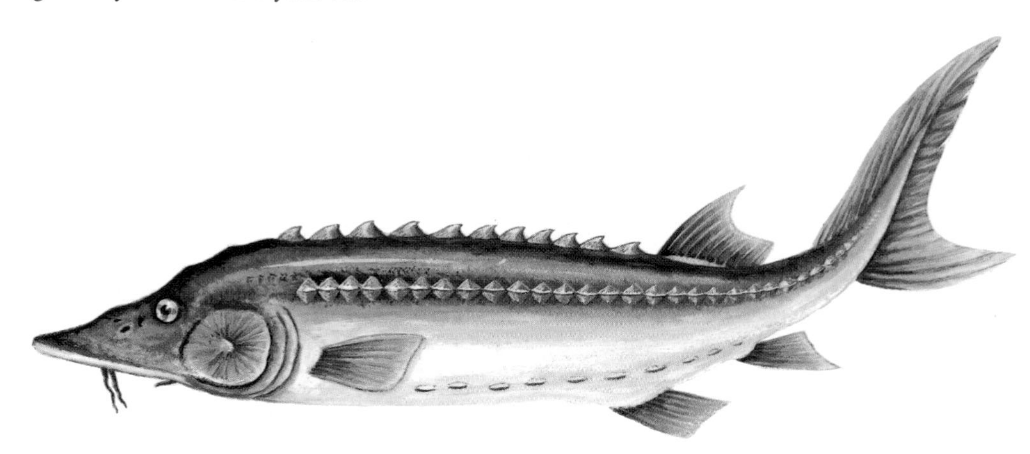

HIGHLIGHTS

The small wild populations of white sturgeon remaining on the Pacific coast of North America are now the most important sturgeons of the continent. Fisheries are greatly reduced and heavily restricted.

White sturgeon is being farmed for food and farmers hope to be able to produce caviar from the species also. However, females mature very late and intervals between spawning may be as long as 11 years; other sturgeons, especially the hybrid bester, offer more attractive potential for caviar farming.

Nutrition data:
(100 g edible weight)

Water	78.4 g
Calories	100 kcal
Protein	16.1 g
Total lipid (fat)	4.0 g
Omega-3	0.3 mg

WHITE STURGEON *Acipenser transmontanus*

FISHING METHODS

 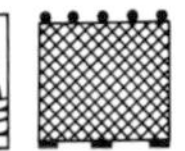 

MOST IMPORTANT FISHING NATIONS

United States, Canada

PREPARATION

Caviar   

USED FOR

EATING QUALITIES

White sturgeon meat is firm and very tasty. It is greyish white in colour, with a meaty rather than a fishy texture. It is excellent smoked and also makes good soup, stews and other dishes that require lengthy cooking; the meat stands up to such treatment.

Historically, the white sturgeon was not much used for caviar because the logistics of the fishery meant that the eggs were not fresh enough to be processed when they reached suitable facilities. The swim bladder was used for isinglass, used to clarify wines.

IMPORTANCE

The United States allows landings of about 200 tons a year, mostly from the Columbia River. Canadian catches, mostly from the Fraser River, are often less than 10 tons. Fish large enough to breed are protected, as are small fish. There are few, if any, white sturgeons left in an anadromous range that once extended from California to the Gulf of Alaska.

In the last century, there was a brief period of excess in the fishery. In 1892, four years after the commercial sturgeon fishery started on the Columbia, 2,500 tons of meat was shipped to markets in the eastern United States. A decade later the catch had fallen to 45 tons. On the Fraser, landings of over 500 tons in 1897 fell to 9 tons in 1905, though the Fraser fishery has always been wildly erratic.

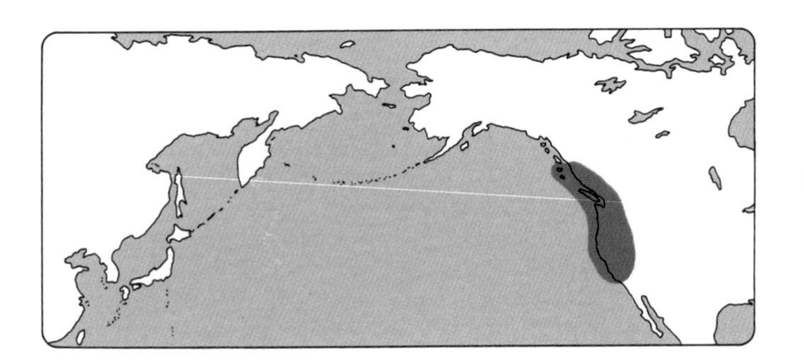

DK: Fisken opdrættes som konsumfisk, men man håber senere at kunne udnytte kaviaren, når fisken 11 år gammel er kønsmoden. Andre arter foretrækkes nu, idet de bliver langt hurtigere kønsmodne

D: Der Fisch wird als Speisefisch gezüchtet, aber man hofft, den Kaviar künftig nutzen zu können, wenn der Fisch 11 Jahre alt und geschlechtsreif ist. Andere Arten werden jetzt bevorzugt, da sie viel früher geschlechtsreif werden

E: Se cría como pescado de consumo, pudiéndose aprovechar el caviar, cuando tiene 11 años y es capaz de reproducirse. Ahora son más atractivas otras especies que maduran más rápidamente

F: Le poisson est cultivé pour la consommation, mais à l'avenir, il est envisagé d'exploiter le caviar à la maturité du poisson vers onze ans. Actuellement, on préfère d'autres espèces qui atteignent la maturité plus rapidement

I: La specie viene allevata come pesce commestibile, ma si spera di poter usare anche il caviale, quando all'età di 11 anni diventa sessualmente maturo. Comunque ora si preferiscono altre specie che maturano più rapidamente

WHITING

Scientific name:
Merlangius merlangus

Family: Gadidae — cods
Typical size: 25 cm, sometimes up to 70 cm

Whiting is a small relative of the cod found in limited coastal areas of the northeast Atlantic, as well as in the eastern Mediterranean and Black Seas. Although it is generally a low-priced species, the delicate meat is valued in some countries.

FISHING METHODS

 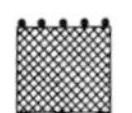 

MOST IMPORTANT FISHING NATIONS
United Kingdom, France, Turkey

PREPARATION

USED FOR

EATING QUALITIES
Rather soft but very sweet, white meat with a small flake charac-terises the whiting's flavour and texture. The meat falls apart unless it is treated with care, so is usually cooked with the skin on to hold the flesh in place. Whiting are occasionally dry-salted.

COMMON NAMES
D: Wittling, Merlan
DK: Hvilling
E: Plegonero, bacaladilla
F: Merlan, valet
GR: Daoúki tou Atlantikoú
I: Merlano, nasello atlantico
IS: Lysa
N: Hvitting
NL: Wijting
P: Badejo
RU: Merlang
S: Vitling
SF: Valkoturska
TR: Bakalyaro
US: Whiting

LOCAL NAMES
IL: Merlan
MO: Peskadil'ia
PO: Witlinek
TU: Nazalli

Nutrition data:
(100 g edible weight)

Water	80.7 g	Total lipid	
Calories	81 kcal	(fat)	0.7 g
Protein	18.7 g	Omega-3	0.1 mg

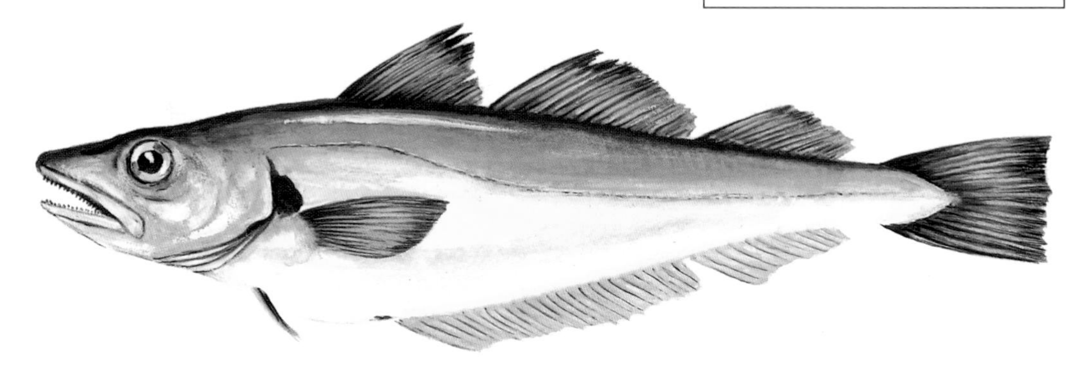

HIGHLIGHTS
Catches of well over 100,000 tons a year are taken mainly by the United Kingdom and France. The delicate meat is appreciated especially in these countries. Turkey's catches in the Black Sea are often used for feeding more valuable farmed species.

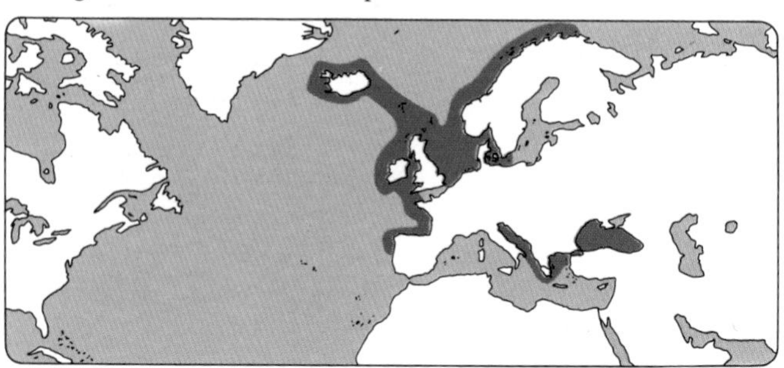

DK: Populær fisk i specielt Storbritannien og Frankrig, som også står for størsteparten af de årlige fangster på langt over 100.000 tons

D: Ein besonders in Grossbritannien und Frankreich beliebter Fisch. Diese Länder fangen den überwiegenden Teil der jährlichen Fänge von weit über 100.000 Tonnen

E: Pescado popular, especialmente en Gran Bretaña y Francia, países que también realizan las mayores capturas anuales de más de 100.000 toneladas

F: Très coté surtout en Grande-Bretagne et en France, pays qui s'inscrivent aussi pour la majeure partie des pêches qui dépassent de loin 100.000 t par an

I: Pesce popolare specialmente in Gran Bretagna e Francia, paesi che registrano le maggiori catture annuali che superano di gran lunga le 100.000 tonnellate

WIDOW ROCKFISH

Scientific name:
Sebastes entomelas

Family: Scorpaenidae — scorpionfishes
Typical size: up to 59 cm

Pacific rockfish are classified in the marketplace between those with red skin (which tend to have pinkish coloured flesh and are preferred) and those with brown or greenish skin, which have darker meat and are considered less desirable. Widow rock is a brown-skinned species.

FISHING METHODS

MOST IMPORTANT FISHING NATIONS
United States, Canada

PREPARATION

USED FOR

EATING QUALITIES
Widow rockfish has rather dark meat which whitens when cooked. It has almost no taste, but the lightly textured meat absorbs flavours from sauces and spices when cooking, making it easy to adapt to popular recipes.

COMMON NAMES
D: Witwen-Drachenkopf
DK: Enkerødfisk
E: Gallineta rocote
F: Sébaste veuf
GR: Skoúro kokkinópsaro
I: Sebaste bruno
IS: Brúnkarfi
NL: Weduweroodbaars
P: Cantarilho viúvo
US: Widow rockfish

Nutrition data:
(100 g edible weight)

Water	79.3 g	Total lipid	
Calories	90 kcal	(fat)	1.6 g
Protein	18.6 g	Omega-3	0.3 mg

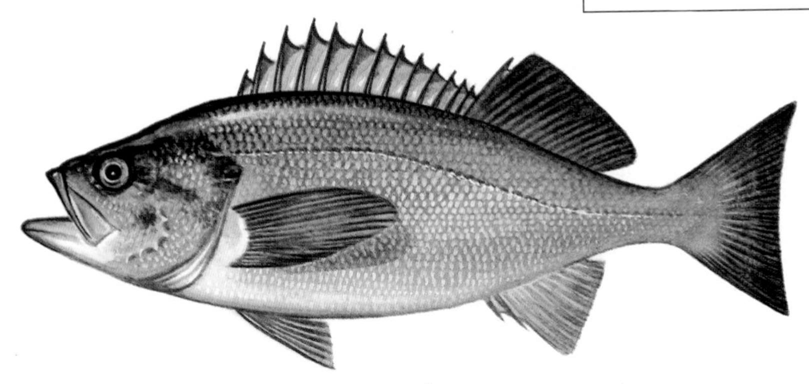

HIGHLIGHTS
Found from Baja California to Southeast Alaska, the widow rockfish is an important commercial species, targeted by the mid-water trawl fisheries of California, Oregon and Washington. Alaska produces small quantities of line-caught fish, which sells for a premium.

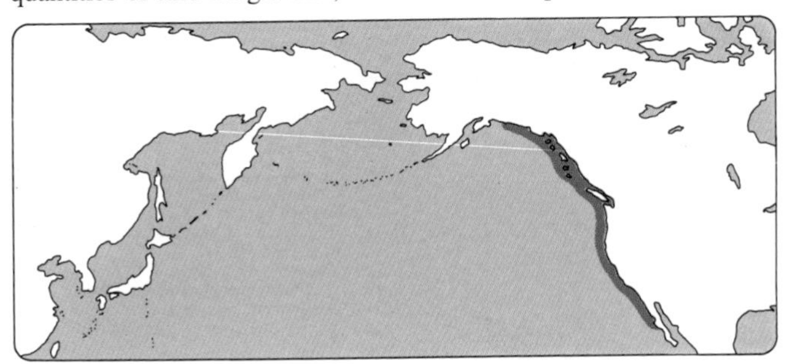

DK: Stillehavsarten klassificeres på fisketorvet efter skindets farve: 1) rød og 2) brun eller grøn

D: Die pazifische Art wird auf dem Fischmarkt nach der Farbe des Fleisches eingestuft: 1) rot (wird bevorzugt) und 2) braun oder grünlich

E: La especie del Océano Pacífico se clasifica en el mercado de pescado según el color de la piel: 1) rojo y 2) marrón o verde

F: Aux marchés aux poissons, cette espèce, qui vit dans l'océan Pacifique, est classée d'après la couleur de sa peau: 1) rouge et 2) brune ou verte

I: La specie del Pacifico viene classificata sul mercato secondo il colore della pelle in: 1) rosso e 2) marrone o verdognolo

WITCH

Scientific name:
Glyptocephalus cynoglossus

Synonyms: Witch flounder, craig fluke, pale dab
Family: Pleuronectidae — right-eye flounders
Typical size: 60 cm, 2 kg

Witch flounder, called gray sole in the United States, is now much less abundant than it was in the 1960s and 70s, when the fishery was at its height. Despite limited supplies, it remains an important species because of its high price.

FISHING METHODS

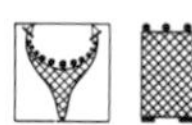 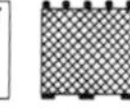

MOST IMPORTANT FISHING NATIONS
Canada, Iceland, Spain, Portugal, United Kingdom

PREPARATION

USED FOR

EATING QUALITIES
Gray sole is regarded in the USA as the best eating of all the Atlantic flounders and soles. The fillets are long and rather thin, but have a smooth texture and a sweet taste. The meat when cooked is extremely white, an important factor in its appeal to American consumers.

COMMON NAMES
D: Zungenbutt, Rotzunge
DK: Skærising
E: Mendo, falso lenguado
F: Plie cynoglosse
GR: Kalkáni
I: Passera lingua di cane
IS: Langlúra
J: Akashitabirame
N: Mareflyndre, smørflyndre
NL: Witje, hondstong
P: Solhao
RU: Dlinnaya kambala
S: Rödtunga
SF: Mustaeväkampela
US: Gray sole

LOCAL NAMES
PO: Ptastugi

Nutrition data:
(100 g edible weight)

Water	76.8 g	Total lipid	
Calories	107 kcal	(fat)	4.0 g
Protein	17.7 g	Omega-3	0.8 mg

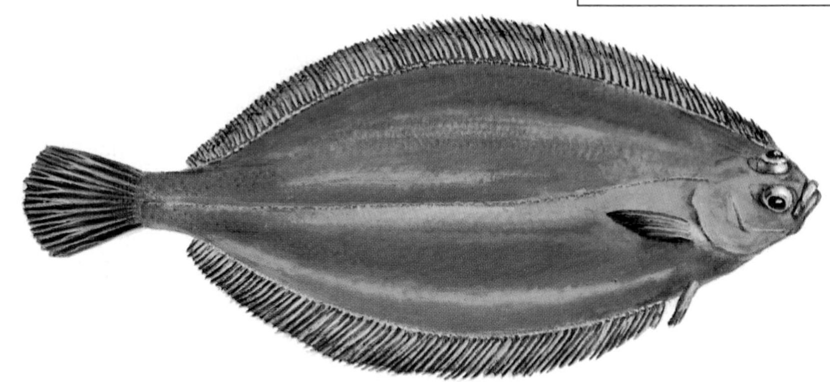

HIGHLIGHTS
A valued species from both sides of the North Atlantic, the witch provides about 25,000 tons of high value flatfish to world markets. It is mostly sold as fillets, either skinless or skin on. Smaller fish are sometimes sold dressed. Fresh fillets often cost more than frozen.

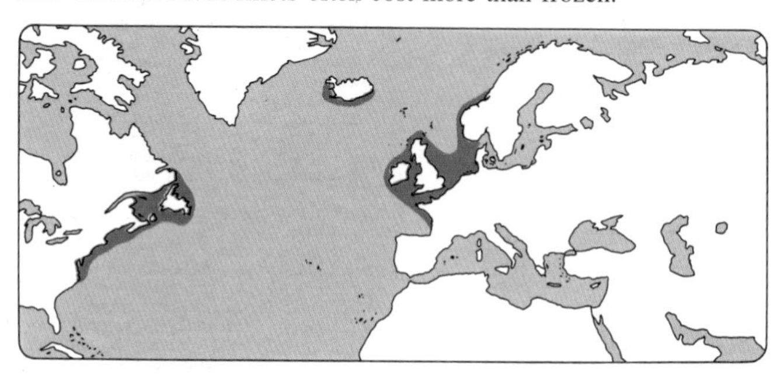

DK: Fiskeri på begge sider af Nordatlanten efter denne yndede fladfisk - årlig fangst på 25.000 tons. God afsætning af filetter til høje priser

D: Dieser beliebte Plattfisch wird beiderseits des Nordatlantiks gefangen. Der jährliche Fang beträgt 25.000 Tonnen. Guter Absatz der Filets zu hohen Preisen

E: Pleuronecto muy apreciado a ambos lados del Atlántico Norte. Se capturan hasta 25 mil toneladas al año. Existe un buen mercado de filetes a precios elevados

F: Poisson plat très apprécié, pêché des deux côtés de l'Atlantique Nord. La pêche atteint 25.000 t/an. Les filets sont vendus à bon prix sur les marchés mondiaux

I: Ne vengono pescate fino a 25.000 tonnellate all'anno sui due lati del Nordatlantico di questa specie dei pleuronettiformi. Venduto come filetti a prezzi alti

WRECKFISH

Scientific name:
Polyprion americanus

Synonyms: Stone bass, wreck bass
Family: Serranidae — sea basses
Typical size: up to 2 m, 45 kg

Wreckfish are found in small numbers on both sides of the Atlantic. Populations off the United States were rapidly reduced by enthusiastic sport fishermen targeting these large fish. Current controls prevent fishing; this situation is likely to continue for some time.

FISHING METHODS

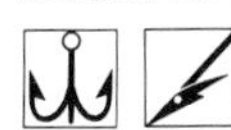

MOST IMPORTANT FISHING NATIONS
Portugal, Greece

PREPARATION

USED FOR

EATING QUALITIES
Wreckfish is similar to grouper in texture, flavour and other edibility characteristics. The meat is firm and a little grey, but it has excellent flavour and is versatile in almost any cooking method. It is prized, when available, in Mediterranean countries.

COMMON NAMES
D: Wrackbarsch
DK: Vragfisk
E: Cherna, mero de roca
F: Cernier commun
GR: Vláchos
I: Cernia di fondale
IS: Blákarpi
J: Aruzentin-Ohata
N: Vrakfisk
NL: Wrakbaars, wrakvis
P: Cherne
RU: Bury kamenny okun
S: Vrakfisk, vrakabborre
SF: Hylkyahven
TR: Iskorpit hanisi
US: Wreckfish

LOCAL NAMES
SO: Bafaro
TU: Shringi

Nutrition data:
(100 g edible weight)

Water	69.9 g	Total lipid	
Calories	166 kcal	(fat)	9.9 g
Protein	19.1 g	Omega-3	2.0 mg

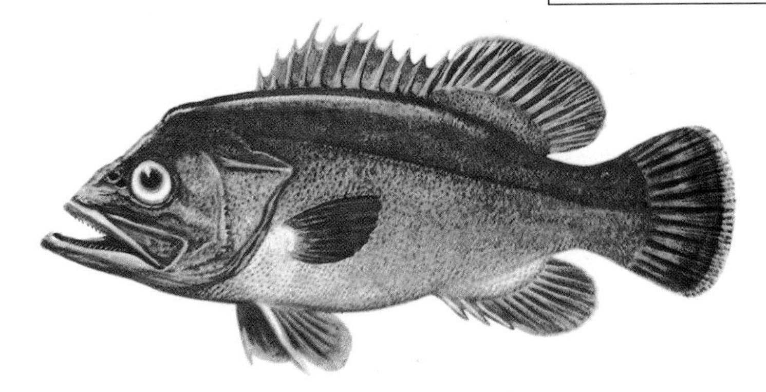

HIGHLIGHTS
Wreckfish are large, solitary fish often found sheltering in wrecks, reefs and other obstructions. Some small fish are said to retreat into shelters from which they cannot escape as they grow rapidly, trapping themselves for the benefit of passing divers armed with spears.

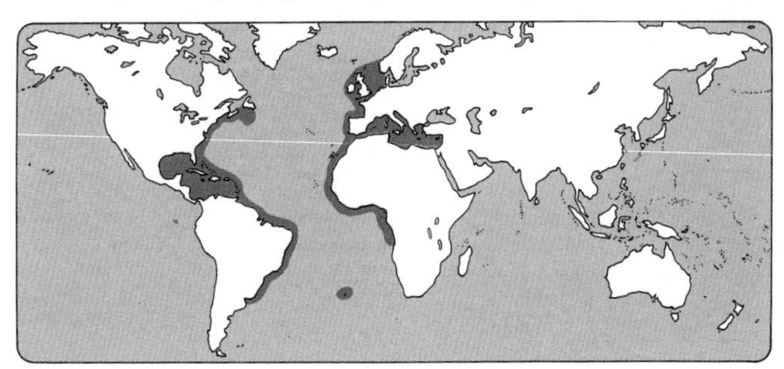

DK: Et let bytte for dykkere: det siges at de små fisk søger ly i vrag og rev, hvor de så sidder fastklemt på grund af deres hurtige vækst

D: Eine leichte Beute für Taucher: es heisst dass die kleinen Fische Schutz in Wracks und Riffen suche wo sie dann eingeklemmt werden weil sie so schnell wachsen

E: Fácil presa para los buceadores. Se dice que los pescados pequeños buscan abrigo en arrecifes etc. de donde no pueden escapar porque crecen rápidamente

F: Une proie facile pour les plongeurs. Petits, ils sont réputés chercher refuge dans les épaves et récifs où ils restent coincés à cause de leur croissance rapide

I: Una preda facile per i pescatori subacquei. Si dice che i pesci piccoli cercano riparo nelle scogliere dove rimangono incastrati, essendo di rapida crescita

YELLOWFIN SOLE

Scientific name:
Pleuronectes asper

Family: Pleuronectidae — right-eye flounders
Typical size: 26 cm, 200 g
Also known as *Limanda aspera*

Yellowfin sole is the second most abundant flatfish in the North Pacific. The resource, which supplies catches of about 150,000 tons a year, is heavily concentrated in the eastern Bering Sea, where it is exploited by factory trawlers and made into fillets and blocks.

FISHING METHODS

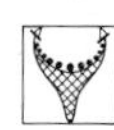

MOST IMPORTANT FISHING NATIONS
United States

PREPARATION

USED FOR

EATING QUALITIES
A light, delicate flatfish with white meat, the yellowfin sole's small fillets are rather thin, so the fish is mostly used for portions and coated products. It has a bland flavour which does not dominate accompaniments. It is widely used in North American food service.

COMMON NAMES
D: Rauhe Kliesche
DK: Japansk ising
E: Limanda japonesa
F: Limande du Japon
I: Limanda giapponese
NL: Japanse schar
P: Solha áspera
RU: Zheltopjoraya kambala
US: Yellowfin sole

Nutrition data:
(100 g edible weight)

Water	82.5 g	Total lipid	
Calories	70 kcal	(fat)	1.1 g
Protein	14.9 g	Omega-3	0.2 mg

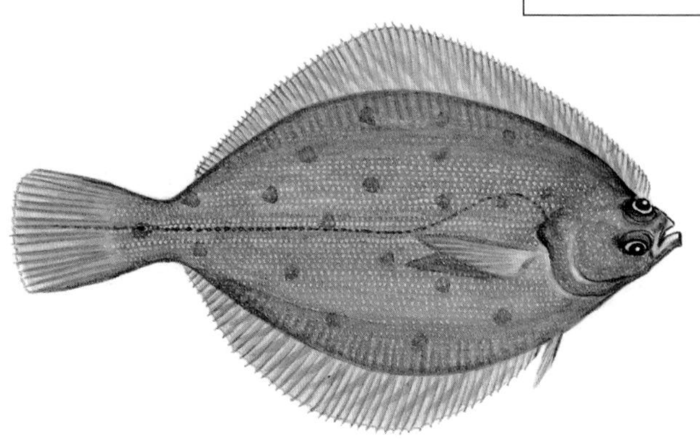

HIGHLIGHTS
Yellowfin is one of the most important commercial flatfish species, and a significant part of Alaska's huge fishing industry, although the fish rarely appears on menus under its own name. It is used mostly for consumer-ready products and is identified only as sole or flounder.

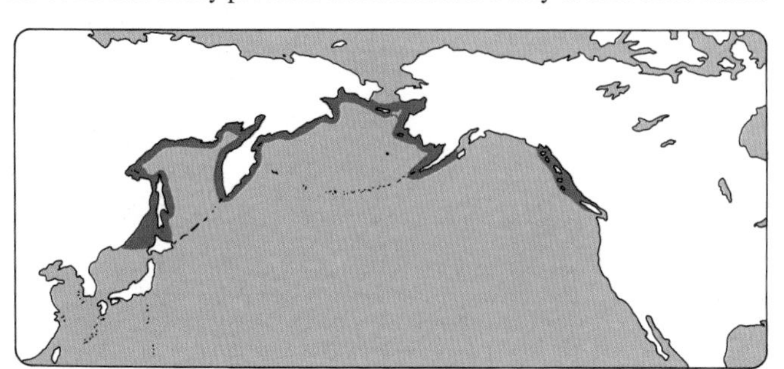

DK: I økonomisk henseende indgår arten som en af de vigtigste fladfisk i Alaska's enorme fiskeindustri

D: In wirtschaftlicher Hinsicht ist diese Art von grosser Bedeutung für die riesige Fischindustrie Alaskas

E: Desde el punto de vista económico esta especie es uno de los pleuronectos más importantes de la notable industria pesquera de Alaska

F: Sur le plan économique, cette espèce fait partie des poissons plats les plus importants de l'énorme industrie de pêche de l'Alaska

I: Sotto l'aspetto economico, la specie è una delle più importanti fra i pleuronettiformi per l'enorme industria ittica dell'Alasca

YELLOW GURNARD

Scientific name:
Trigla lucerna

Synonyms: Tub gurnard, tubfish, latchet
Family: Triglidae — searobins
Typical size: 35 cm

The yellow or tub gurnard is the most common European searobin and is sought by both commercial and recreational fishermen because of its comparatively large size. It is found from Norway to West Africa, including the Mediterranean and Black Seas.

FISHING METHODS

 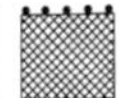 

MOST IMPORTANT FISHING NATIONS
Not available

PREPARATION

USED FOR

EATING QUALITIES
Larger than other European searobins, the yellow gurnard provides fillets of good size. The meat is a little dry, but firm and mildly flavoured. It is good as fish and chips; dressed fish may be stuffed and baked.

COMMON NAMES
D: Roter Knurrhahn
DK: Rød knurhane
E: Lucerna, golondrina, bejel
F: Grondin perlon
GR: Chelidonás, kapóni
I: Capone gallinella
IS: Knurri
N: Rødknurr
NL: Rode poon, grote poon
P: Cabra cabaço, bacamarte
RU: Morskoj petukh-zhe. trigla
S: Fenknot
SF: Isokurnusimppu
TR: Kirlangic baligi
US: Searobin

LOCAL NAMES
TU: Djaj
EG: Farkha

Nutrition data:
(100 g edible weight)

Water	72.5 g	Total lipid	
Calories	129 kcal	(fat)	5.0 g
Protein	21.0 g	Omega-3	1.0 mg

HIGHLIGHTS
The yellow gurnard has three isolated rays on the pectoral fin which serve as "legs" on which the fish rests. They also help it locate food on the soft bottom which it prefers. Specimens of the species grow as large as 60 cm with a weight of 5 kg or more.

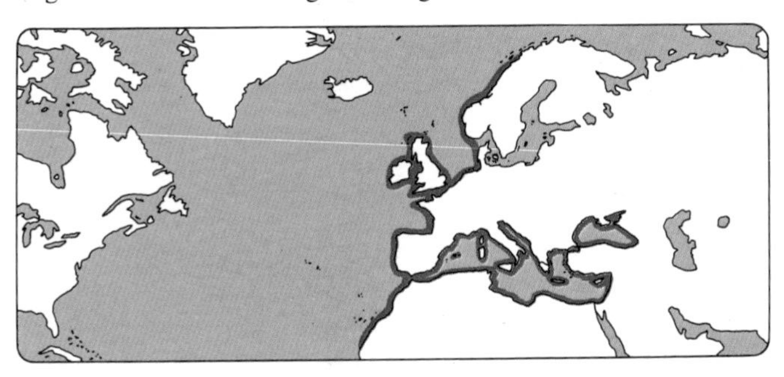

DK: Den røde knurhane har 3 isolerede stråler på bugfinnen, som virker som 'hvileben'. Anvendes endvidere til lokalisering af føde

D: Der rote Knurrhahn hat 3 Stachelstrahlen an der Bugflosse, die als 'Ruhebeine' und ausserdem zur Lokalisierung von Nahrung dienen

E: La lucerna tiene 3 espinas aisladas en la aleta ventral, que sirven como 'piernas' en que descansa el pescado. También le ayudan a localizar alimento

F: Le grondin perlan est doté de trois rayons isolés sur la ventrale qui lui servent de 'pattes' pour se reposer, et aussi pour localiser la nourriture

I: Il capone gallinella rosso ha 3 raggi isolati sulla pinna addominale che servono come 'gambe di appoggio', nonché per la localizzazione di cibo

YELLOW PERCH

Scientific name:
Perca flavescens

Family: Percidae — perches
Typical size: up to 50 cm

Yellow perch is one of the most important freshwater fishes of North America, found in lakes from Nova Scotia to South Carolina, to the western side of the Appalachians, to the upper Missouri and Montana to upper Slave Lake, James Bay and Quebec.

The species has been introduced to a more extensive region including Utah, New Mexico, Texas, Washington, Oregon, and coastal regions of British Columbia.

DESCRIPTION
The typical distinctive skin colours of the yellow perch, with its green and yellow skin and dark bars, are not universal on the species. Some populations lack certain pigments and are strikingly different. Size also varies greatly in different lakes. In Ontario, most of the fish caught are under 280 g, while in Saskatchewan fish as large as 2.5 kg have been reported.

The species adapts readily to a wide range of conditions, including large lakes, small ponds and even slow rivers. It is also found in brackish waters on the Atlantic coast and in salt lakes on the Prairies.

COMMON NAMES
D: Amerikanischer Flussbarsch
DK: Gul aborre
E: Perca canadiense
F: Perche canadienne
GR: Kitrinóperka
I: Persico dorato
IS: Kanaborri
NL: Amerikaanse gelebaars
P: Perca americana
S: Amerikansk abborre
SF: Kelta-ahven
US: Yellow perch

HIGHLIGHTS
A distinctively coloured fish, the yellow perch is identical in appearance to the European perch, *Perca fluviatilis* (see entry under Perch). Some scientists consider the two to be a single species, but the majority of expert opinion now appears to reject this thesis; the two perches, although interchangeable commercially, are two separate species taxonomically.

Together, the two species have an almost complete range around the pole in northern latitudes.

Nutrition data:
(100 g edible weight)

Water	79.1 g
Calories	86 kcal
Protein	19.4 g
Total lipid (fat)	0.9 g
Omega-3	0.3 mg

YELLOW PERCH *Perca flavescens*

FISHING METHODS

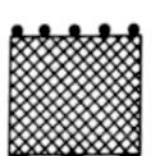 

MOST IMPORTANT FISHING NATIONS
Canada, United States

PREPARATION

USED FOR

EATING QUALITIES
Yellow perch is usually too small to fillet, so is headed, dressed and pan-fried. However, fillets are the most popular item; these, too, are usually fried, sometimes in coatings.

The meat is white and mild tasting, with little fishiness, a reason for the popularity of the species in North America.

Fillets are normally sold with the distinctive skin on, partly to aid identification. Despite this, Atlantic ocean perch and Pacific rockfish fillets (both of which are *Sebastes spp.*) are sometimes substituted for yellow perch when consumers are unwary. This is particularly easy to do if the fillets are first breaded.

IMPORTANCE
Yellow perch, a widely fluctuating resource, provides on average about 7,500 tons yearly, of which Canada produces about 6,000 tons. Most of the commercial catches come from the Great Lakes, although yellow perch is important locally in other regions.

It is valuable throughout its extensive range as a recreational species, not least because the fish feed and take bait throughout the year, so can be fished both summer and winter. In spring, yellow perch tend to shoal near to the shore, making them especially easy to locate and catch.

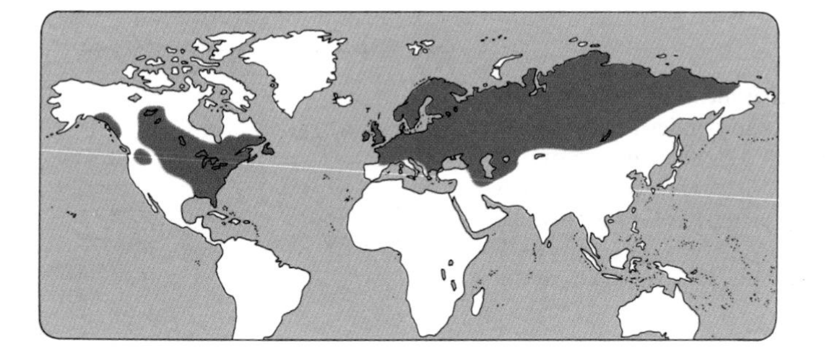

DK: Af hensyn til identifikationen sælges filetterne normalt med det karakteristiske skind på, idet de ellers nemt forveksles med langt billigere arter

D: Aus Identifizierungsgründen werden die Filets meistens mit der charakteristischen Haut verkauft, um Verwechslungen mit weit billigeren Arten vorzubeugen

E: Normalmente, los filetes se venden con su piel característica para evitar que esta especie sea confundida con otras mucho más baratas

F: Dans un but d'identification, les filets de la perche canadienne sont normalement vendus avec la peau caractéristique pour éviter toute confusion avec d'autres espèces de moindre valeur

I: Per facilitare l'identificazione, i filetti sono normalmente venduti senza che sia tolta la caratteristica pelle; altrimenti si confondono facilmente con altre specie meno costose

YELLOWTAIL

Synonym: Japanese amberjack
Family: Carangidae — jacks
Typical size: up to 150 cm, 40 kg

Scientific name:
Seriola quinqueradiata

COMMON NAMES
D: Japanische Seriola
DK: Japansk ravfisk
E: Pez límon del Japón
F: Sériole du Japon
GR: Magiátiko tis Iaponías
I: Ricciola giapponese
J: Buri
NL: Japanse seriols
P: Charuteiro do Japao
RU: Zheltokhvostaya lakerda
US: Yellowtail, amberjack

Japanese yellowtail is probably the most valuable member of the jack family. Superficially similar to the Australian yellowtail kingfish (California yellowtail) (*Seriola lalandei*), it is considered far superior as a table fish. Nevertheless, it is reported that the California species is occasionally offered as a substitute for Japanese yellowtail in United States markets.

Yellowtail caught in Australia are also sometimes used for sashimi; however, the species has not yet developed acceptance for this purpose in Japan.

DESCRIPTION
A large, silver fish with a shoaling habit, almost all production is now from farms, with limited catches of wild fish made mainly off the coasts of Central Japan.

Amberjacks, of which this is one, are frequently subject to substantial parasite infestations, including long worms in the major muscles of the sides. Farmers have learned how to prevent such problems, which would destroy the major market for the species.

HIGHLIGHTS
Yellowtail is one of the world's major farmed species; it has been grown successfully in Japanese marine waters for many years, supplying eager restaurant and department store markets for fresh, top quality sashimi.

A large, fast-swimming jack, its successful domestication is a tribute to the skill of Japanese fish farmers, who have been able to tame such a species and make it respond and grow in captivity, where it has little room to swim.

Nutrition data:
(100 g edible weight)

Water	76.3 g
Calories	102 kcal
Protein	21.4 g
Total lipid (fat)	1.2 g
Omega-3	0.3 mg

YELLOWTAIL *Seriola quinqueradiata*

FISHING METHODS

MOST IMPORTANT FISHING NATIONS

Japan

PREPARATION

Sashimi

USED FOR

EATING QUALITIES

Yellowtail has delicate, oily meat with a small flake. It is mostly used raw, sliced thinly, for sashimi in Japan. It is regarded as one of the best of all fishes for eating raw and fetches a correspondingly high price in Japanese markets.

Yellowtail can also be salted and grilled, but this treatment is usually applied only to fish that is less than perfect, perhaps after it has been out of the water for more than a few days and has lost the absolute freshness required if it is to be eaten and enjoyed raw. Cooked, the meat is white and delicious, with a tender texture and sweet taste which is brought out by the use of salt and high heat.

IMPORTANCE

Japanese fish farmers produce over 150,000 tons a year of high priced yellowtail, making it one of the most valuable of all the world's farmed species. The fish are raised in cages, mainly in the Inland Sea, carefully harvested and sold fresh for sashimi.

Special handling is used to ensure that the fish are kept thoroughly chilled and do not get bruised, since bruising substantially reduces the price that buyers are willing to pay for the fish.

Small quantities of yellowtail are airfreighted to specialist restaurants in California and other parts of the USA.

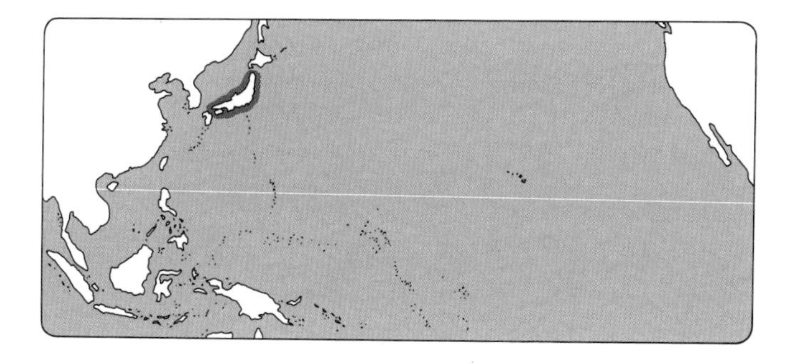

DK: Overalt en vigtig akvakulturfisk. Mangeårigt og succesrigt havopdræt af Japansk ravfisk i Japan forsyner restauranter og supermarkeder med frisk førsteklasses kød til 'sashimi'

D: Japanische Seriola ist überall ein wichtiger Aquakulturfisch. In Japan langjährige und erfolgreiche Zucht zur Belieferung von Restaurants und Supermärkten mit frischem erstklassigem Fleisch für 'Sashimi'

E: En acuicultura, uno de los pescados más importantes del mundo. La exitosa cría marina durante muchos años de este pescado en Japón suministra carne fresca de primera clase a restaurantes y supermercados para 'sashimi'

F: La sériole est partout un poisson d'aquiculture important. L'élevage de mer, qui depuis de nombreuses années connaît un grand succès, approvisionne les restaurants et supermarchés japonais en chair fraîche pour préparer le 'sashimi'

I: Dappartutto una specie di allevamento importante. La piscicoltura marina, praticata nel Giappone da tanti anni con grande successo, fornisce ai ristoranti e ai supermercati un prodotto di alta qualità per 'sashimi'

YELLOWTAIL FLOUNDER

Scientific name:
Limanda ferruginea

Synonyms: Rusty dab, sandy dab, yellowtail
Family: Pleuronectidae — right-eye flounders
Typical size: 50 cm, 1.5 kg

Catches of yellowtail flounder have declined substantially as the resource was increasingly pressured by heavy fishing. The fishery peaks in spring and autumn, but product, mainly fillets, is available throughout the year in frozen form.

FISHING METHODS

 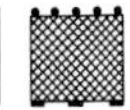

MOST IMPORTANT FISHING NATIONS
Canada, United States

PREPARATION

USED FOR

EATING QUALITIES
The meat of yellowtail flounder is sweet and firm and is regarded as the standard against which other flounders are measured. The thin skin from the white side of the fish is often left on, but the darker, thicker skin from the top side is normally removed.

COMMON NAMES
D: Gelbschwanzflunder
DK: Gulhalet ising
E: Limanda nórdica
F: Limande à queue jaune
GR: Chomatída me kítrini ourá
I: Limanda
IS: Gulspordur
J: Karei
N: Sandflyndre
NL: Zandschar
P: Solha dos mares do norte
RU: Zheltokhvostaya kambala
SF: Ruostekampela
US: Yellowtail flounder

LOCAL NAMES
PO: Zimnica, zolcica

Nutrition data:
(100 g edible weight)

Water	77.3 g	Total lipid	
Calories	89 kc	(fat)	0.8 g
Protein	20.4 g	Omega-3	0.2 mg

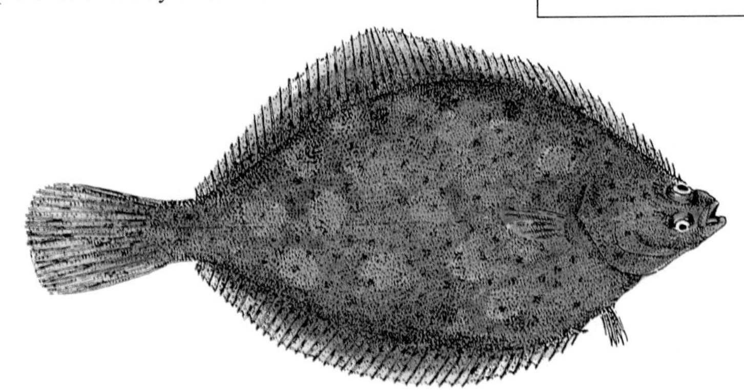

HIGHLIGHTS
Yellowtail flounder is an offshore species of the northwest Atlantic, preferring shallow water on the great fishing banks. Economically, it is still the most important flounder of the region. Many other flounders are processed and sold as yellowtail.

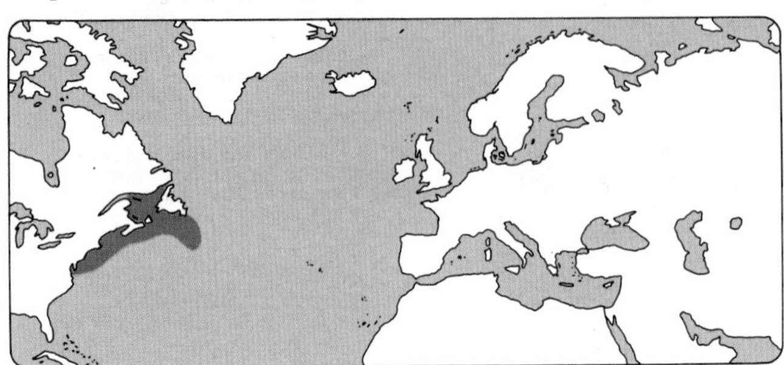

DK: Fangsterne er faldet væsentligt som følge af overfiskeri. Fisken fås året rundt i frossen form

D: Die Fänge sind wegen Überfischens stark zurückgegangen. Der Fisch ist ganzjährig in gefrorenem Zustand erhältlich

E: Las capturas han bajado notablemente por la pesca excesiva. Este pescado se vende congelado todo el año

F: Les pêches sont en déclin, essentiellement en raison de la surpêche. On trouve la limande à queue jaune toute l'année à l'état congelé

I: Le catture sono diminuite sensibilmente in seguito allo sfruttamento esagerato delle risorse. Il pesce è disponibile tutto l'anno come prodotto congelato

ZANDER/PIKE-PERCH

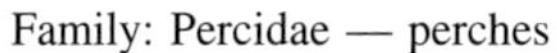

Scientific name:
Stizostedion lucioperca

Family: Percidae — perches
Typical size: 60 cm; up to 130 cm, 12 kg

The zander, originally native to the rivers of the Baltic, Black and Caspian Seas, has been introduced throughout most of Europe and has become a pest in eastern England, where it is not much valued for food, although it is targeted by recreational anglers.

FISHING METHODS

MOST IMPORTANT FISHING NATIONS
Turkey, Russia, Rumania, Denmark, Germany, Poland

PREPARATION

USED FOR

EATING QUALITIES
Zander has a small flake and delicate texture with a mild, sweet taste. It can be used in recipes designed for sole. It is appreciated most in the areas of its original range and is farmed in many parts of central and eastern Europe.

COMMON NAMES
D: Zander, Hechtbarsch
DK: Sandart
E: Lucioperca
F: Sandre, perche brochet
GR: Potamolávrako
I: Luccioperca, sandra
IS: Gedduborri, vatnavidnir
N: Gjørs
NL: Snoekbaars, zander
P: Lucioperca
RU: Sudak
S: Gös
SF: Kuha
TR: Levrek, sudak

LOCAL NAMES
PO: Sandacz

Nutrition data:
(100 g edible weight)

Water	78.6 g	Total lipid	
Calories	83 kcal	(fat)	0.7 g
Protein	19.2 g	Omega-3	0.1 mg

HIGHLIGHTS
An aggressive predator, the zander has depleted stocks of native fish in some of the areas to which angling interests have introduced it. Boneless fillets of farmed zander are frozen and sold to U.S. markets where they are sold under the name of yellow pike.

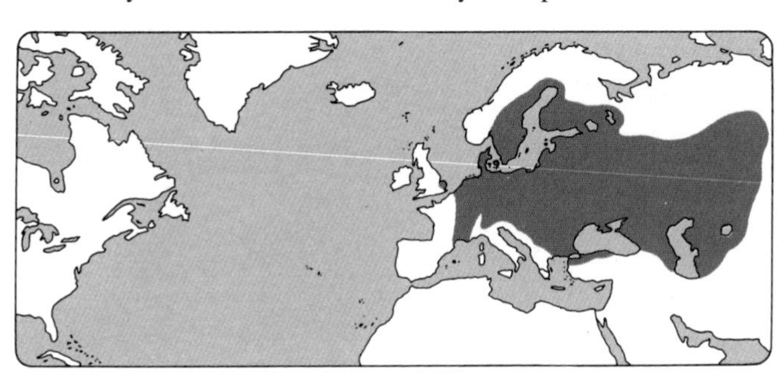

DK: I visse områder, hvor denne aggressive rovfisk er udsat til sportsfiskeri, har den næsten fortrængt de naturlige forekomster

D: In bestimmten Gebieten, wo dieser aggressive Raubfisch zur Sportfischerei ausgesetzt worden ist, hat er die natürlichen Bestände fast verdrängt

E: La cría con fines deportivos de este voraz y agresivo pez ha hecho casi desaparecer las especies naturales de las zonas en que vive

F: Dans les eaux où cet agressif poisson carnassier a été introduit dans l'intérêt de la pêche sportive, il a pratiquement éliminé les populations naturelles

I: Un pesce predatore aggressivo, introdotto per la pesca sportiva in certe zone dove ha quasi eliminato le popolazioni originarie

This Bibliography mentions only the more important of the many hundreds of documents consulted, which included scientific articles, magazines, conference proceedings and other material. The Authors, contacted through the Publishers, will be pleased to suggest further sources on specific subjects.

1981. *Guide Book to New Zealand Commercial Fish Species*. Wellington, New Zealand: New Zealand Fishing Industry Board.

1988. *Fish List, The.* FDA Guide to Acceptable Market Names for Food Fish Sold in Interstate Commerce. Washington, DC: Superintendent of Documents, U.S. Government Printing Office.

1988. *Recommended Marketing Names for Fish*. Canberra, Australia: Australian Government Publishing Service.

1993. *Seafood List, The.* FDA's Guide to Acceptable Market Names for Seafood Sold in Interstate Commerce. Washington, DC: Superintendent of Documents, U.S. Government Printing Office.

Beveridge, Malcolm. 1987. *Cage Aquaculture.* England: Fishing News Books Ltd.

Bigelow, Henry B. and William C. Schroeder. 1964. *Fishes of the Gulf of Maine*. Cambridge, Massachusetts: Museum of Comparative Zoology, Harvard University.

Browning, Robert J. 1980. *Fisheries of the North Pacific*. Anchorage, Alaska: Alaska Northwest Publishing Company.

Bykov, V.P., Ed. 1972 *Marine Fishes*. Moscow: Pishchevaya Promyshlennost' Publishers.

Castro, José 1983. *The Sharks of North American Waters*. College Station, Texas: Texas A & M University Press.

Commission of the European Communities. 1993. *Multilingual Illustrated Dictionary of Aquatic Plants and Animals*. Oxford, U.K. Fishing News Books Ltd.

Cobo, Mario and Sheyla Massay. 1989. *Lista de los Peces Marinos del Ecuador*. Guayaquil, Ecuador: Instituto Nacional de Pesca del Ecuador.

Davidson, Alan. 1977. *Seafood of South-East Asia*. Singapore: Federal Publications (S) Pte Ltd.

Davidson, Alan. 1980. *North Atlantic Seafood*. New York: The Viking Press.

Davidson, Alan. 1981. *Mediterranean Seafood*. Baton Rouge, Louisiana: Louisiana State University Press.

Davidson, Alan. 1989. *Seafood: A Connoisseur's Guide and Cookbook*. New York: Simon and Schuster.

Doré, Ian. 1984. *Fresh Seafood — The Commercial Buyer's Guide*. New York: Van Nostrand Reinhold/Osprey Books.

Doré, Ian. 1989. *The New Frozen Seafood Handbook, A Complete Reference for the Seafood Business*. New York: Van Nostrand Reinhold/Osprey Books.

Doré, Ian. 1990. *Making the Most of Your Catch — An Angler's Guide*. New York: Van Nostrand Reinhold/Osprey Books.

Doré, Ian. 1990. *Salmon — The Illustrated Handbook for Commercial Users*. New York: Van Nostrand Reinhold/Osprey Books.

Doré, Ian. 1991. *The New Fresh Seafood Buyer's Guide — A manual for distributors, restaurants and retailers*. New York: Van Nostrand Reinhold/Osprey Books.

Exler, Jacob. 1987. *Composition of Foods: Finfish and Shellfish Products*. Washington, D.C: United States Department of Agriculture.

Food and Agriculture Organization of the United Nations. *FAO Species Catalogue*. Volumes 2 to 12, 14 and 15. Rome, Italy: F.A.O.

Food and Agriculture Organization of the United Nations. 1991 (annual). *Fishery Statistics. Catches and Landings*. Rome, Italy: F.A.O.

Frimodt, Claus. 1955-1992. *Multilingual Posters of Fish and Shellfish*. (Series). Hedehusene, Denmark: Scandinavian Fishing Year Book.

Gousset, J. and G. Tixerant. *Les Produits de la Pêche: Poissons — Crustaces — Mollusques*. Paris, France: Ministère de l'Agriculture.

Grant, E.M. 1985. *Guide to Fishes*. Brisbane, Australia: Department of Harbours and Marine.

Gulland, J.A. 1970. *The Fish Resources of the Ocean*. Rome, Italy: Food and Agriculture Organisation of the United Nations.

Hart, J.L. 1988. *Pacific Fishes of Canada*. Ottawa, Ontario: Department of Fisheries and Oceans.

Hoese, H. Dickson and Richard H. Moore. 1977. *Fishes of the Gulf of Mexico*. College Station, Texas: Texas A & M University Press.

Huet, Marcel. 1986. *Textbook of Fish Culture*. Farnham, Surrey, England: Fishing News Books Ltd.

Innes, William T. 1966. *Exotic Aquarium Fishes, 19th ed.* Maywood, NJ: Metaframe Corp.

Joseph, James, Witold Klawe and Pat Murphy. 1988. *Tuna and Billfish*. La Jolla, California: Inter - American Tropical Tuna Commission.

Kailola, P.J. et al. 1993. *Australian Fisheries Resources*. Canberra, Australia: Department of Primary Industries.

Kirk, R. 1987. *A History of Marine Fish Culture in Europe and North America*. England: Fishing News Books Ltd.

Krane, Willibald. 1986. *Fish: Five-language Dictionary of Fish, Crustaceans and Molluscs*. New York: Van Nostrand Reinhold.

Lamb, Andy and Phil Edgell. 1986. *Coastal Fishes of the Pacific Northwest*. Madiera Park, British Columbia: Harbour Publishing.

Manooch, Charles S. 1988. *Fisherman's Guide — Fishes of the Southeastern United States*. Raleigh, North Carolina: North Carolina State Museum of Natural History.

Masuda, H., et al. 1984. *The Fishes of the Japanese Archipelago*. Tokyo: Tokai University Press.

McClane, A. J. 1978. *Field Guide to Freshwater Fishes*. New York: Holt, Rinehart and Winston.

McClane, A. J. 1978. *Field Guide to Freshwater Fishes*. New York: Holt, Rinehart and Winston.

McClane, A. J. 1974. *New Standard Fishing Encyclopedia*. New York: Holt, Rinehart and Winston.

McClane, A. J. 1977. *The Encyclopedia of Fish Cookery*. New York: Holt, Rinehart and Winston.

Migdalski, Edward C. and George S. Fichter. 1983. *The Fresh and Salt Water Fishes of the World*. New York: Greenwich House.

Myers, Robert F. 1991. *Micronesian Reef Fishes*. Guam, USA: Coral Graphics.

Negedly, Robert, Ed. 1990. *Elsevier's Dictionary of Fishery. Processing Fish and Shellfish Names of the World*. Amsterdam: Elsevier Science Publishers.

Nelson, Joseph S. 1984. *Fishes of the World, 2nd Edition*. New York: John Wiley & Sons.

Organisation for Economic Co-operation and Development. 1990. *Multilingual Dictionary of Fish and Fish Products*. UK: Fishing News Books.

Paust, Brian and Ronald Smith. 1986. *Salmon Shark Manual*. Fairbanks, Alaska: Alaska Sea Grant College Program, University of Alaska.

Pennington, Jean A. T. and Helen Nichols Church. 1985. *Food Values of Portions Commonly Used*. New York: Harper and Row.

Poissons et Fruits de Mer de France. Paris, France: F.I.O.M.

Quero, Jean-Claude. 1984. *Les Poissons de Mer*. France: Jacques Grancher.

Rau, Norbert and Anke Rau. 1980. *Commercial Fishes of the Philippines*. Eschborn, Germany: Deutsche Gesselschaft für Technische Zusammenarbeit (GTZ) GmbH.

Robins, C. Richard et al. 1991. *A List of Common and Scientific Names of Fishes from the United States and Canada*. Fifth edition. Bethesda, Maryland: American Fisheries Society.

Sainsbury, Keith J., Patricia J. Kailola and Guy G. Leyland. 1984. *Continental Shelf Fishes of Northern and North-Western Australia*. Australia: CSIRO Division of Fisheries Research.

Scandinavian Fishing Year Book. 1955-1992. *Multilingual Posters of Fish and Shellfish*. (Series). Hedehusene, Denmark: Scandinavian Fishing Year Book.

Scott, J.S. 1959. *An Introduction to the Sea Fishes of Malaya*. Kuala Lumpur, Malaya: Ministry of Agriculture, Federation of Malaya.

Scott, W.B. and E.J. Crossman. 1985. *Freshwater Fishes of Canada*. Ottawa, Ontario: Fisheries Research Board of Canada.

Scott, W.B. and M.G. Scott. 1988. *Atlantic Fishes of Canada*. Toronto, Ontario: University of Toronto Press.

Suvatti, Chote. 1981. *Fishes of Thailand*. Bangkok: Royal Institute Thailand.

Teubner Edition. 1987. *Das grosse Buch vom Fisch*. Füssen, Germany. Teubner Edition.

Van der Elst, Rudy. 1990. *A Guide to the Common Sea Fishes of South Africa*. Cape Town, South Africa: Struik Publishers.

Walford, Lionel A. 1974. *Marine Game Fishes of the Pacific coast from Alaska to the Equator*. Washington, DC: Smithsonian Institution Press.

Wheeler, Alwyne. 1975. *Fishes of the World, An Illustrated Dictionary*. New York: Macmillan Publishing Co., Inc.

Wheeler, Alwyne. 1978. *Key to the Fishes of Northern Europe*. London, Frederick Warne Ltd.

Wheeler, Alwyne. 1992. *Saltwater Fishes of Britain and Europe*. Limpsfield, U.K: Dragon's World Ltd.

Wheeler, Alwyne. 1992. *Freshwater Fishes of Britain and Europe*. Limpsfield, U.K: Dragon's World Ltd.

Whitehead, P.J.P. et al. (eds.) 1986. *Fishes of the Northeastern Atlantic and the Mediterranean Volume III*. Paris, France: UNESCO.

Whitehead, P.J.P. et al. (eds.) 1986. *Fishes of the Northeastern Atlantic and the Mediterranean Volume II*. Paris, France: UNESCO.

Whitehead, P.J.P. et al. (eds.) 1984. *Fishes of the Northeastern Atlantic and the Mediterranean Volume I*. Paris, France: UNESCO.

Yudkin, John. 1986. *The Penguin Encyclopeadia of Nutrition*. UK: Penguin Books Ltd.